**WITHDRAWN**

LIVERPOOL
JOHN MOORES UNIVERSITY
AVRIL ROBARTS LRC
TITHEBARN STREET
LIVERPOOL L2 2ER
TEL. 0151 231 4022

# Road Vehicle Suspensions

# Road Vehicle Suspensions

by

**Wolfgang Matschinsky**

TRANSLATION EDITED BY ALAN BAKER

Professional Engineering Publishing Limited
London and Bury St Edmunds, UK

First Published in English 2000

This publication in English is copyright under the Berne Convention and the International Copyright Convention. All rights reserved. Apart from any fair dealing for the purpose of private study, research, criticism or review, as permitted under the Copyright, Designs and Patents Act, 1988, no part may be reproduced, stored in a retrieval system, or transmitted in any form or by any means, electronic, electrical, chemical, mechanical, photocopying, recording or otherwise, without the prior permission of the copyright owners. Unlicensed multiple copying of the contents of this publication is illegal. Inquiries should be addressed to: The Publishing Editor, Professional Engineering Publishing Limited, Northgate Avenue, Bury St Edmunds, Suffolk, IP32 6BW, UK.

ISBN 1 86058 202 8

Translation from the German language edition:
*Radführungen der Straßenfahrzeuge* by Wolfgang Matschinsky
Copyright © Springer-Verlag Berlin Heidelberg 1998
All Rights Reserved

A CIP catalogue record for this book is available from the British Library.

The Publishers are not responsible for any statement made in this publication. Data, discussion, and conclusions developed by the Author are for information only and are not intended for use without independent substantiating investigation on the part of potential users. Opinions expressed are those of the Author and are not necessarily those of the Institution of Mechanical Engineers or its Publishers.

Printed and bound in Great Britain by St Edmundsbury Press Limited, Suffolk, UK

# Contents

| | |
|---|---|
| **Preface** | ix |
| **Related titles of interest** | x |
| **Notation** | xi |
| **1 Introduction** | 1 |
| **2 Basic characteristics of wheel suspensions** | 5 |
| 2.1 Degrees of freedom of wheel suspensions | 5 |
| 2.2 Elements of wheel suspensions | 6 |
|     2.2.1 The wheel carrier | 6 |
|     2.2.2 Joints for wheel suspensions | 6 |
|     2.2.3 Suspension links | 8 |
|     2.2.4 The kinematic chain | 9 |
| 2.3 Basic types of wheel suspensions | 10 |
|     2.3.1 Evolution from the statically determinate suspension of a spatial body | 10 |
|     2.3.2 Independent suspension layouts | 12 |
|     2.3.3 Rigid-axle suspensions | 14 |
|     2.3.4 Tandem wheel carrier | 17 |
|     2.3.5 Compound suspensions | 17 |
| **3 Kinematic analysis of wheel suspensions** | 19 |
| 3.1 Basic rules of planar kinematics | 19 |
| 3.2 Basic rules of vector calculation | 25 |
| 3.3 Considerations of wheel-suspension systematics | 28 |
| 3.4 Motion analysis of the wheel carrier | 32 |
| 3.5 External and internal forces | 38 |
| 3.6 Influence of driveshafts and reduction gears | 42 |
| 3.7 Attitude and motion of the wheel | 50 |
| **4 The tyre** | 53 |

## 5 Springs and dampers — 59

- 5.1 Purpose of springing and damping — 59
- 5.2 Vehicle vibrations — 60
    - 5.2.1 Basic system — 60
    - 5.2.2 System of two masses — 66
    - 5.2.3 Pitching and rolling — 67
- 5.3 Interaction of springs — 73
- 5.4 Interaction of springs and suspension — 78
- 5.5 Vehicle springs — 84
    - 5.5.1 General remarks — 84
    - 5.5.2 Leaf springs — 84
    - 5.5.3 Torsion bars — 89
    - 5.5.4 Coil springs — 92
    - 5.5.5 Rubber springs — 95
    - 5.5.6 Pneumatic springs — 98
- 5.6 Hydraulic dampers — 101
- 5.7 Controlled suspension systems — 103
- 5.8 The wheel-travel angle — 106

## 6 Traction and braking — 109

- 6.1 Steady-state accelerating and braking — 109
- 6.2 The support angle — 111
    - 6.2.1 General remarks — 111
    - 6.2.2 'Torque support' at the wheel carrier — 114
    - 6.2.3 'Torque support' at the vehicle chassis — 118
    - 6.2.4 Special cases — 122
    - 6.2.5 Effective support angles — 125
- 6.3 Traction and braking pitching — 126
    - 6.3.1 Static and dynamic pitching under traction and braking — 126
    - 6.3.2 Single-axle traction and braking — 128
    - 6.3.3 Torque transmission via driveshafts — 130
    - 6.3.4 Hub reduction gear — 133
    - 6.3.5 Effect of longitudinal forces on the spring rate — 134
    - 6.3.6 Non-symmetrical vehicle attitude — 136
    - 6.3.7 Influence of unsprung and rotating masses — 137
    - 6.3.8 Influence of flexible mountings — 138
- 6.4 Tandem axles — 140

## 7 Cornering — 143

  7.1 Camber and steering angle with wheel displacement — 143
  7.2 Forces and moments under lateral acceleration — 145
  7.3 The roll centre — 152
    7.3.1 The vehicle at extremely low lateral accelerations — 152
    7.3.2 High lateral accelerations — 157
    7.3.3 Simple method for suspension-geometry approximation in cornering — 164
  7.4 Vehicle attitude in steady-state cornering — 167
  7.5 Kinematic self-steering — 176
  7.6 Driving stability of two-track vehicles — 185
  7.7 Cornering of motorcycles — 191

## 8 The steering — 195

  8.1 Basic systems — 195
  8.2 Steering boxes — 196
  8.3 Characteristics of steering geometry — 199
    8.3.1 Conventional definitions and physical interpretations — 199
    8.3.2 Generalized definitions relating to spatial geometry — 204
  8.4 The steering linkage — 220
    8.4.1 Basic types — 220
    8.4.2 Steering geometry — 224
    8.4.3 Steering returnability — 231
    8.4.4 Steering-system vibrations — 238
  8.5 Self-aligning steering systems — 239

## 9 Elasto-kinematics — 243

  9.1 Principal considerations — 243
  9.2 Elasto-kinematics of independent wheel suspensions — 251
    9.2.1 Elastic behaviour of the suspension mechanism — 251
    9.2.2 Compliantly mounted subframes — 261
  9.3 Statically over-constrained systems — 266

## 10 Synthesis and design — 269

  10.1 General remarks — 269
  10.2 Planar wheel suspensions — 270
  10.3 Spatial wheel suspensions — 273
  10.4 Design considerations — 278

## 11 Motorcycle suspensions 293

## 12 Independent wheel suspensions 303
    12.1 General remarks 303
    12.2 Front suspensions 303
    12.3 Rear suspensions 318

## 13 Rigid-axle suspensions 335
    13.1 General remarks 335
    13.2 Kinematically exact systems 338
    13.3 Over-constrained systems 343

## 14 Compound suspensions 345

**Final remarks** 353

**References** 355

**Index** 357

# Preface

During more than 100 years of motor car production, wheel suspensions have passed through a remarkable evolution. Soon the simple springing systems adopted from horse-drawn carriages no longer met the requirements of increasing speed; exact and reproducible wheel motion began to be achieved by the introduction of kinematic suspension mechanisms. With growing knowledge of the vibrational behaviour of the vehicle, and of vehicle-dynamic laws, refinement of suspension geometry came to the fore, so today a great variety of suspension types is known, with differing targets and properties. Efficient measuring methods and computers have made possible the deliberate elasto-kinematic harmonization of suspensions, to achieve optimum driving stability along with optimum ride comfort.

In this book I have endeavoured to compile and to arrange the tasks and methods that appeared to me essential and useful from my own experience as a suspension designer. The purpose of the book is not only to inform the reader of the principles involved but to provide him also with the knowledge about design features and methods that he needs in order to design a suspension system.

In view of the software available today for solving the problems of spatial geometry, I have deliberately limited the equations to those that are essential to explain the suspension problems and to solve them, and have renounced many of the graphical methods that dominated the 1987 first (German) edition.

I am grateful to all those manufacturers who made technical documents and illustrations available to me, and to the Executive Board of the BMW company for permission to publish the book.

Special thanks I owe to Sheril Leich for systematic and helpful promotion of the project, and to Alan Baker who struggled through my own poor attempt at a translation and converted it into idiomatic English.

Munich, autumn 1999　　　　　　　　　　　　　　　　　　Wolfgang Matschinsky

# RELATED TITLES OF INTEREST

| Title | Editor/Author | ISBN |
|---|---|---|
| Advances in Vehicle Design | John Fenton | 1 86058 181 1 |
| Design Techniques for Engine Manifolds – Wave Action Methods for IC Engines | D E Winterbone and R J Pearson | 1 86058 179 X |
| Automotive Braking: Recent Developments and Future Trends | D Barton and M Haigh (Eds) | 1 86058 131 5 |
| Handbook of Automotive Body Construction and Design Analysis | John Fenton | 1 86058 073 4 |
| Handbook of Automotive Body and Systems Design | John Fenton | 1 86058 067 X |
| Brakes and Friction Materials – the History and Development of the Technologies | Graham A Harper | 1 86058 127 7 |
| IMechE Engineers' Data Book | Clifford Matthews | 1 86058 175 7 |
| Journal of Automobile Engineering | Proceedings of the IMechE, Part D | ISSN 0307/6490 |

For the full range of titles published by Professional Engineering Publishing contact:

    Sales Department
    Professional Engineering Publishing Limited
    Northgate Avenue
    Bury St Edmunds
    Suffolk
    IP32 6BW
    UK
    Tel: +44 (0) 1284 724384
    Fax: +44 (0) 1284 718692

# Notation

Vectors are characterized by **bold** letters:
**v** = velocity vector; v = scalar amount of the velocity
**a · b** = c 'interior' or 'scalar' product of the vectors **a** and **b**; the result is a scalar c.
**a × b** = **c** 'exterior' or 'vector' product of the vectors **a** and **b**; the result is a vector **c** which is perpendicular to **a** and **b**.

*Special terms* that are frequently used in the book are explained at the end of this list.

| | | |
|---|---|---|
| A | mm$^2$ | area; surface |
| A | | tyre contact point |
| a | mm/s$^2$ | acceleration |
| $a_q$ | mm/s$^2$ | lateral acceleration |
| b | mm | track width |
| $b_F$ | mm | spring track width |
| c | N/mm | spring rate |
| $c_F$ | N/mm | rate of the spring element |
| $c_{FA}$ | N/mm | resultant spring rate 'reduced' to the tyre contact point |
| $c_S$ | N/mm | stabilizer spring rate per wheel |
| $c_A$ | N/mm | compound spring rate per wheel |
| $c_\varphi$ | Nmm/rad | angular spring rate |
| D | | degree of damping |
| D | | intersection point of the kingpin axis and the road surface |
| D | | centre of rotation |
| d | | kingpin axis; axis of rotation |
| E | N/mm$^2$ | elasticity modulus |
| e | | unit vector (vector of amount '1') |
| e | mm | edge distance of a profile section |
| F | | degree of freedom of a mechanism |
| F | N | force |
| $F_F$ | N | spring force |
| f | | degree of freedom of a joint |
| f | mm | spring deflection, spring travel |
| $f_R$ | | rolling-resistance factor |
| G | N | vehicle weight |
| G | N/mm$^2$ | shear modulus |
| G | | moving polode |

| | | |
|---|---|---|
| $g$ | mm/s² | gravitational acceleration |
| $g$ | | number of joints of a mechanism |
| $H$ | | auxiliary point on the wheel axis |
| $h$ | mm | centre-of-gravity height |
| $h$ | mm | pitch of the instantaneous screw |
| $h_{RZ}$ | mm | roll-centre height above ground |
| $I$ | mm⁴ | geometrical moment of inertia (second moment of area) |
| $i$ | | transmission ratio; reduction ratio |
| $i_D$ | | damper transmission ratio |
| $i_F$ | | spring transmission ratio |
| $i_H$ | | steering-gear transmission ratio |
| $i_L$ | | steering-linkage transmission ratio |
| $i_S$ | | overall steering ratio |
| $i$ | mm | radius of gyration |
| $K$ | | wheel carrier |
| $k$ | | number of wheel carriers of a wheel suspension |
| $k_D$ | Ns/mm | damper rate |
| $L$ | | longitudinal pole (pole in the vehicle's side-view) |
| $L$ | | steering gearbox |
| $L$ | mm²kg/s | angular moment; moment of momentum |
| $l$ | | number of links of a wheel suspension |
| $l$ | mm | length |
| $l$ | mm | wheelbase |
| $M$ | | wheel-centre point |
| $M$ | Nmm | moment, torque |
| $M_B$ | Nmm | bending moment |
| $M_D$ | Nmm | torque |
| $M_H$ | Nmm | steering-wheel torque |
| $M_{RB}$ | Nmm | scrub torque of the tyre |
| $M_{RS}$ | Nmm | restoring torque of the tyre resulting from sideslip angle |
| $m$ | | instantaneous axis |
| $m_p$ | | instantaneous axis for parallel wheel travel |
| $m_w$ | | instantaneous axis for antimetric wheel travel |
| $m$ | kg | mass, inertia |
| $n$ | | normal vector, perpendicular unit vector |
| $n$ | | polytropic exponent |
| $n$ | mm | castor offset |
| $n_R$ | mm | pneumatic trail |
| $n_\tau$ | mm | castor offset at wheel centre |
| $P$ | | pole, instantaneous pole |
| $p$ | Pa | gas pressure |

Notation

| | | |
|---|---|---|
| p | mm | polar distance |
| p | mm | wheel-load lever arm |
| Q | | transverse pole (pole in the vehicle's cross-section) |
| R | | fixed polode |
| R | mm | tyre radius, wheel radius |
| RZ | | roll centre |
| r | | roll axis of vehicle |
| r | | number of individual rotations of links in a mechanism |
| r | mm | radius, lever arm |
| $r_c$ | mm | wheel-centre offset (kingpin offset at wheel centre) |
| $r_S$ | mm | scrub radius (kingpin offset) |
| $r_T$ | mm | traction-force radius |
| S, SP | | centre of gravity |
| $S_F$ | | elastic centre |
| s | | instantaneous screw axis |
| s | mm | wheel travel |
| T | | pole of inertia |
| T | | shear centre |
| T | | percussion point |
| T | s | oscillation period |
| t | mm/s | pitch velocity (instantaneous screw) |
| t | s | time |
| U | J | energy |
| u | mm/s | peripheral velocity, circumferential velocity |
| V | mm$^3$ | volume |
| v | mm/s | velocity |
| $v_M$ | mm/s | velocity of the wheel centre |
| $v_A^*$ | mm/s | virtual velocity of the tyre contact point assuming a 'locked' torque support at the wheel carrier |
| $v_A^{**}$ | mm/s | virtual velocity of the tyre contact point assuming a 'locked' torque support at the vehicle chassis and torque transmission via driveshafts |
| $W_b$ | mm$^3$ | axial section modulus |
| $W_d$ | mm$^3$ | polar section modulus |
| x, y, z | | axes of the coordinate system |
| $\alpha$ | rad | angular orientation of a rubber bush |
| $\alpha$ | rad | deflection angle of a driveshaft joint |
| $\alpha$ | rad | tyre sideslip angle |
| $\beta$ | rad | attitude angle of the vehicle |
| $\gamma$ | rad | camber angle |

| | | |
|---|---|---|
| $\delta$ | rad | steering angle |
| $\delta_V$ | rad | toe-in angle |
| $\varepsilon$ | rad | wheel-travel angle |
| $\varepsilon_A$ | rad | traction-force support angle |
| $\varepsilon_B$ | rad | braking-force support angle |
| $\varepsilon_{MB}$ | rad | engine-braking-force support angle |
| $\varepsilon^*$ | rad | support angle for a torque support at the wheel carrier |
| $\varepsilon^{**}$ | rad | support angle for a torque support at the vehicle chassis and torque transmission via driveshafts |
| $\eta$ | | frequency ratio |
| $\eta$ | | efficiency |
| $\Theta$ | mm²kg | moment of inertia |
| $\vartheta$ | rad | pitch angle |
| $\kappa$ | | adiabatic exponent |
| $\kappa$ | rad | angle between the roll axis of the vehicle and the instantaneous axis of antimetric travel of a suspension |
| $\lambda$ | | slip, circumferential slip |
| $\lambda$ | rad | deflection angle of the kingpin axis from the vertical |
| $\mu$ | | friction coefficient |
| $\mu$ | rad | transmission angle in a linkage |
| $\Pi$ | | plane |
| $\Pi'$ | | side view |
| $\Pi''$ | | cross-section |
| $\Pi'''$ | | plan view |
| $\rho$ | mm | radius of curvature |
| $\sigma$ | N/mm² | normal stress |
| $\sigma$ | rad | kingpin inclination |
| $\tau$ | rad | castor angle |
| $\tau$ | N/mm² | shear stress |
| $\varphi$ | rad | roll angle |
| $\varphi_d$ | rad | angle of rotation |
| $\varphi_k$ | rad | coning angle |
| $\chi$ | | share of acceleration or braking force of the front axle |
| $\psi$ | rad | yaw angle |
| $\omega$ | rad/s | angular velocity |
| $\omega_0$ | rad/s | natural frequency |
| $\omega_K$ | rad/s | angular velocity of the wheel carrier |
| $\omega_R$ | rad/s | angular velocity of the wheel |
| $\omega_\gamma$ | rad/s | camber velocity |
| $\omega_\delta$ | rad/s | steering velocity |

## Subscripts

| | |
|---|---|
| a | outer wheel |
| a | wheel at rebound |
| e | wheel at bump |
| h | rear wheel |
| i | inner wheel |
| n | normal position |
| o | above; upper |
| u | below; lower |
| v | front wheel |

## Special terms frequently used in the book

*Antimetric motion*

Motion of two points (or parts) without any 'symmetrical' component referring to a line (or plane) of symmetry. This is more precise than the familiar terms 'non-symmetrical' and 'asymmetrical' which merely exclude pure symmetry.

*Coning angle*

Angle of deflection between the axes of the inner and outer bushes of a joint (preferably a flexible one). See Chapter 2, Section 2.2.2 and Chapter 5, Section 5.5.5.

*Degressive*

A degressive function drops from a linear function by an increasing gradient; it is the opposite of a progressive function. For example, disc springs ('saucer springs') normally show a degressive spring characteristic – a characteristic of decreasing spring rate with spring deflection.

*Geometrical moment of inertia*

The geometrical moment of inertia (or 'second moment of area') is the integral over the products of the infinitesimal elements of an area (for example, the cross-section of a beam) and the square of their distances from an axis or a pole.

*Support angle*

Angle between the road surface and the line of action of a force at the tyre contact point which is reacted by the suspension mechanism alone without loading the spring. The support angle is explained in Chapter 3, Section 3.5, and defined in detail in Chapter 6, Section 6.2.

*Torque support*

The constructional part which in a special case of external force (traction or braking) generates, reacts or transforms a torque; examples are the engine, gearbox, clutch and brake (see Chapter 6, Section 6.2).

*Traction-force radius*

The effective lever arm of the traction force with respect to the steering linkage of a steered wheel. While the 'wheel-centre offset' (or 'kingpin offset at wheel centre') approximates to the lever arm of the traction force only in a position of the suspension where the driveshafts are not deflected (and if no hub reduction gear is applied), the traction-force radius defined in Chapter 8, Section 8.3, always facilitates precise investigation.

*Wheel-load lever arm*

The effective lever arm of the vertical wheel-load force with respect to the steering linkage of a steered wheel (see Chapter 8, Section 8.3). This 'steering characteristic' is, for example, responsible for the self-restoring torque due to the vehicle's sprung weight on the steered axle.

*Wheel-travel angle*

The angle between the vertical and the path (or the tangent of the path) of the wheel centre with wheel travel in the side view of the vehicle; it depends on the design of the suspension - see Chapter 5, Section 5.8. This angle is especially obvious on a motorcycle telescopic fork.

# Chapter 1

# Introduction

The suspension system connecting a vehicle body to the wheel and its tyre allows the wheel to move in an essentially vertical direction in response to road surface irregularities; a spring element temporarily stores and releases energy, thus insulating the vehicle body from acceleration peaks. A shock-absorber or 'damper' ensures that oscillations induced by road unevenness or aerodynamic forces (or by accelerating, braking or lateral forces), which would impair ride comfort and roadholding, die away quickly.

Road vehicles are usually steered by the front wheels. To this end, one link of the wheel suspension mechanism, the 'track rod', can be displaced by means of a steering box operated by the driver via the steering wheel.

The transmission of the wheel load and the traction, braking and cornering forces from the road to the vehicle body by the wheel suspension offers the possibility of diminishing unwanted side-effects (such as 'pitching' and 'rolling' movements) on the body through the appropriate design of the suspension geometry and the springing system. Examples of such measures are 'anti-dive' or 'anti-squat' arrangements and anti-roll torsion bars. Because all geometry, springing and damping measures affect the wheel or tyre attitude with respect to the road, the available knowledge of vehicle dynamics must be fully utilized during suspension design and development, and it may force compromises between the respective demands of optimum handling and optimum ride comfort.

The tyre as the connecting link between road and vehicle is of outstanding importance to handling and comfort, and at the same time the most difficult suspension factor to control. This is not only because it is made of highly deformable and mainly organic material, but also because it is a 'wearing part'; its properties change with wear and its working reliability depends essentially on proper maintenance by the vehicle owner, being thus outside the influence of the manufacturer.

Being a springing element itself, the tyre on a moving vehicle generates high-frequency oscillations. For this reason the wheel suspension - at least on passenger cars - is usually insulated from the vehicle body by resilient mountings. Consequently the wheel suspension system becomes compliant to some extent, and the tuning of these 'elasto-kinematic' elements must be done with care, especially on fast vehicles.

Not every mechanism meeting the kinematic demands of wheel suspension geometry also has suitable elasto-kinematic properties, so the number of acceptable suspension layouts for fast and comfortable road vehicles is noticeably smaller than for racing cars, for example.

**Fig. 1.1** shows schematically an independent wheel suspension system. Component 1 is the wheel with the tyre, 2 is the 'wheel carrier', which accommodates the wheel bearings and determines the attitude of the wheel in relation to the vehicle body. The wheel carrier also usually incorporates the static part of the braking system - e.g. the brake caliper - and sometimes a reduction gear.

The wheel carrier 2 is the 'coupler' of the three-dimensional wheel-suspension linkage. Component 3 is a 'wishbone link' or 'A-arm' and 4 is a transverse link, both connected to the vehicle body and to the wheel carrier by flexible joints. A tension link 5 triangulates the transverse link 4 to the vehicle body by a joint that is very compliant for good noise isolation. Another link of the suspension is the 'track rod' 6, the inboard end of which is moved by a steering gearbox 7. A spring 8 and a damper 9 complete the system. The source of driving torque (e.g. the final-drive unit, not shown in the sketch) is assumed to be mounted on the vehicle body, and a drive shaft 10 with universal joints transmits the torque to the wheel.

This book deals mainly with the mechanism of wheel suspension and the kinematic laws and design methods for such suspensions, the steering geometry, the vehicle's springing system and its interaction with the suspension, and elasto-kinematics. Comprehensive information about tyres and the theory of vehicle motions as well as vehicle dynamics can be found in the standard works of automotive literature (**16**).

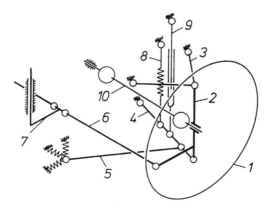

**Fig. 1.1** Schematic of a typical independent wheel suspension

# Introduction

In what follows, the kinematic laws of the mechanisms of wheel suspensions will be covered first, and subsequently methods for the kinematic analysis of the suspension geometry will be shown.

On this basis, the interaction of suspension and springing systems, the behaviour of the suspension under traction, braking and lateral (cornering) forces, the steering geometry, the influence of external forces and the influence of compliance on the functioning of the suspension will be discussed. It will be shown that for the theoretical treatment of these problems a basic knowledge of mechanics, descriptive geometry and vector analysis is fully sufficient, as the mechanisms of wheel suspensions generally prove to be 'statically determinate' systems.

After some comments on the synthesis of suspension geometry, examples of suspension systems for motorcycles and of independent suspensions, rigid-axle suspensions and compound layouts for two-track vehicles such as passenger cars and trucks will be given. Since the objective in selecting the examples was to indicate the wide variety of possible solutions, not simply to catalogue the actual suspension designs on the market, 'historical' systems are included.

In the theoretical chapters, the motion of a wheel suspension for kinematic analysis (Chapter 3) and the calculation of the kinematic characteristics (Chapters 5 to 8) are described by 'velocity vectors' **v** of certain points of the wheel carrier and its angular velocity $\omega$. While this corresponds to the usual methods of mechanism analysis, it should be appreciated that, for most of the investigations made in this book, 'velocities' are not meant to describe violent motion but rather 'virtual'. For time t, a velocity $v_z$ can also be described by the derivative $dz/dt$, and in the same manner an angular velocity $\omega_\gamma$ by $d\gamma/dt$. The quotient $\omega_\gamma/v_z$, after reduction by the time element dt, will then become the derivative $d\gamma/dz$.

In the design and development of suspension systems, as already indicated, a comprehensive knowledge of vehicle dynamics and associated aspects is of fundamental importance. Everyday design work, however, generally involves quite different problems, such as the haggling about every millimetre of space or clearance in the vehicle package, or assembly practicalities or strength analysis. Every suspension system is therefore designed on the assumption that the vehicle body is the 'fixed system' or, so to speak, the 'inertial system'. This method offers the advantage, especially for the independent suspensions predominant in passenger cars, of a definite and reproducible form of wheel travel with respect to the body of the vehicle.

The suspension calculations and descriptions in the book are founded on a vehicle-fixed coordinate system having its x-axis directed forward in the median plane of the vehicle, its y-axis directed to the left and its z-axis pointing upwards, as shown in **Fig. 1.2**. The pitch angle $\vartheta$, the roll angle $\varphi$ and the yaw angle $\psi$ are defined in the same illustration. On the left front wheel, the steering angle $\delta$ is defined as greater than zero for a left-hand corner.

Since the book is concerned primarily with the wheel suspension mechanism and its kinematic properties, suspension systems in what follows are mostly represented (especially in the schematic drawings predominant in the theoretical chapters) without showing the springs. This is because normally any type of spring can be used in any design of suspension.

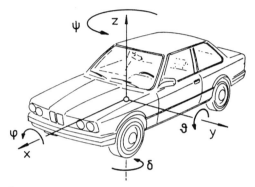

**Fig. 1.2** The coordinate system

# Chapter 2

# Basic Characteristics of Wheel Suspensions

## 2.1 Degrees of freedom of wheel suspensions

A fast road vehicle should have a mainly vertically directed motion at each wheel, one 'degree of freedom', in order to respond to road surface irregularities. A degree of freedom is the displacement of a spatial component or body following a defined and reproducible function. **Fig. 2.1** shows that this degree of freedom not only can be realized by an exactly vertical travel alone (a), but also can consist of a combination of vertical and lateral displacement and a rotation (camber change, b) or may appear as a general non-linear 'coupler movement' (c), where all the parameters of movement are at any moment strictly interdependent (so-called 'constrained motion').

If two wheels are mounted together on one wheel carrier, as on a rigid beam axle (d), this carrier needs two degrees of freedom to give each wheel one degree of freedom, permitting the axle both parallel travel and rolling motion in relation to the vehicle body. A rigid-axle suspension is therefore a mechanism with two degrees of freedom.

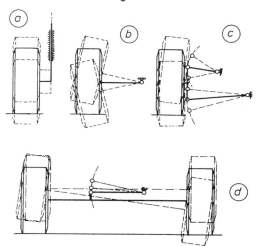

**Fig. 2.1** Wheel suspensions with one (a,b,c) and two degrees of freedom

If all wheels are always to be able to contact the road surface, not more than three of them may be fixed to one carrier, while not more than two wheel planes and not more than two wheel axes may be aligned.

## 2.2 Elements of wheel suspensions

### 2.2.1 The wheel carrier

Every wheel is attached to the suspension by a wheel bearing, normally of the rolling-element (ball or roller) type. The part of the suspension on which this bearing is mounted is called the 'wheel carrier'. In the independent suspension of Fig. 2.1b, of the swing-axle type, the swing axle itself represents the wheel carrier, which is immediately connected to the vehicle body by a single joint. The suspension of Fig. 2.1c is, on the other hand, a kinematic mechanism – a four-joint system, consisting of the wheel carrier as its 'coupler' and two links which guide the wheel carrier in non-linear motion. The rigid axle of Fig. 2.1d is simply a wheel carrier with two wheels.

As already indicated by Fig. 2.1c, wheel or axle suspensions in general are mechanisms consisting of wheel carriers, links and joints, so arranged as to provide the necessary degrees of freedom and the desired dynamic properties.

### 2.2.2 Joints for wheel suspensions

The smallest construction element of a mechanism is a joint. Joints serve either for an immediate connection of the coupler of the mechanism (here the wheel carrier) to the 'fixed part' of the mechanism (here the vehicle body) or to connect the two indirectly by means of links.

In space there are six degrees of freedom, namely three translations along three spatial curves and three rotations around three axes. One joint may provide five degrees of freedom, but one with six degrees of freedom would be nonsense since both sides of it would be freely movable independently of each other. **Fig. 2.2** shows a selection of the joint types used in wheel suspensions.

The **ball joint**, Fig. 2.2a, allows the free relative motion of both joint halves (the ball and the socket) about three independent axes of rotation and thus offers three (rotational) degrees of freedom. The degree of freedom of a joint will be characterized in the following by the lower-case 'f', the ball joint showing $f = 3$. With many such ball joints, only one rotational axis is mainly utilized, the angular displacements about the two other axes

Basic Characteristics of Wheel Suspensions 7

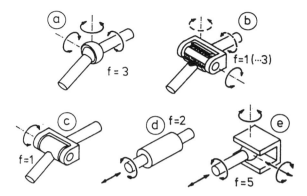

**Fig. 2.2** Joint types

being relatively small. In that case a **rubber joint** as shown in Fig. 2.2 b can be employed with the advantages of good resistance against transient overload, freedom from maintenance, better noise isolation and lower cost. In contrast with the true ball joint, both the main rotation around the axis of the bush and the secondary crossing displacement of the bush axes (the so-called 'coning angle') induce restoring moments by rubber deformation which are to be taken up by the connected suspension parts.

The true **turning joint**, Fig. 2.2c, caters for pure rotation (f = 1), while the **turning-and-sliding joint**, Fig. 2.2 d, allows both rotation about an axis and an independent translation along the same axis (f = 2). Turning joints are frequently achieved by the combination of two rubber bushes, **Fig. 2.3** (left), while the most common application of the turning-and-sliding joint is the piston in the telescopic damper of a suspension strut (Fig. 2.3 right).

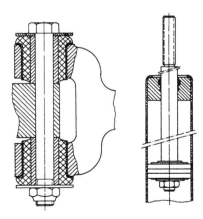

**Fig. 2.3** Examples for the turning joint and the turning-and-sliding joint

Very rarely found in wheel suspensions is the **ball-and-surface joint**, Fig. 2.2e, the surface of which may be spatially curved. As the ball is inhibited from translation only in the rectangular or normal line of the surface, this design provides five degrees of freedom: f = 5.

### 2.2.3 Suspension links

These links provide the indirect connection of a wheel carrier to the vehicle body, the most important types being illustrated in **Fig. 2.4**.

The simplest example is the **rod link**, Fig. 2.4a, with a ball joint at each end (or with equivalent rubber joints, as in Fig. 2.2b). Each ball joint contributes three degrees of freedom, so the sum of the degrees of freedom is f = 6. This would make little sense for a suspension link but, in the case of the rod, one of those six degrees of freedom is the free individual rotation (r) of the rod itself about its longitudinal axis; this may cause clearance problems in a mechanism if the rod is curved or cranked but does not affect the overall motion of the mechanism. In effect, then, a rod link in a mechanism provides five degrees of freedom. In calculating the overall degree of freedom of a mechanism, however, every individual rotation of a link has to be heeded.

The combination of a turning joint and a ball joint leads to the **triangular link** or 'A-arm', Fig. 2.4b. This, with its turning joint (f = 1) and its ball joint (f = 3), offers four degrees of freedom, thus reducing the overall degree of freedom of the mechanism by two. This link can - at least from the kinematic standpoint - also be regarded as a combination of two rod links, where two ball joints coincide; this design gives the same result, namely a reduction of the overall degree of freedom by f = 1 per rod.

Two turning joints, possibly with skew axes, form a **trapezoidal link**, Fig. 2.4c. Each joint has one degree of freedom, so the trapezoidal link reduces the degree of freedom of a mechanism by four.

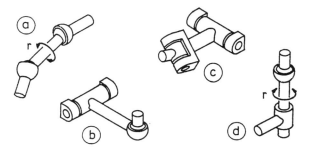

**Fig. 2.4** Types of wheel suspension links

Basic Characteristics of Wheel Suspensions

A link with a ball joint (f = 3) and a turning-and-sliding joint (f = 2) would provide five degrees of freedom. As already stated, the most frequent application of this type of link is the telescopic damper in strut suspensions - a special case of the **turning-and-sliding link**, since the ball joint lies on the central axis of the damper and thus allows the piston rod an individual rotation (r) about its axis without affecting the suspension mechanism, Fig. 2.4d. With the degrees of freedom of the ball joint and of the turning-and-sliding joint, and taking account of the individual rotations, the degree of freedom of the link results in $f = 3 + 2 - 1 = 4$, and this means that it reduces the overall degree of freedom of the mechanism by two.

### 2.2.4 The kinematic chain

Except for the case where a wheel suspension system is based on an immediate connection of the wheel carrier and the vehicle body by a joint, the suspension forms a kinematic chain, consisting of one or several wheel carriers, links and one 'fixed part'. As already specified in Chapter 1, the vehicle body will be regarded here as the fixed part. **Fig. 2.5** shows schematically such a kinematic chain, where the most important types of joints and links are utilized.

The (only) wheel carrier K is the 'coupler' of the spatial suspension mechanism, and the vehicle body S is the 'fixed part'.

The suspension consists of three links with six joints in total - i.e. one rod link 'a' with two ball joints 1 and 2, one triangular link 'b' with a ball joint 3 and a turning joint 5, and a turning-and-sliding link 'c' with a ball joint 4 and a turning-and-sliding joint 6 (e.g. the strut mounting and the piston rod of the damper in a strut-type suspension).

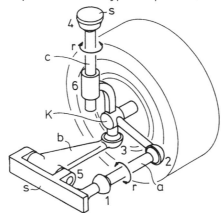

**Fig. 2.5** Kinematic chain with one degree of freedom (independent wheel suspension)

Being spatial bodies, each wheel carrier and each link possesses six degrees of freedom in space. If k is the number of wheel carriers and l the number of links, their total degrees of freedom are $6(k+l)$. A joint i with its individual degree of freedom $f_i$ reduces the overall degree of freedom of a mechanism by $(6-f_i)$. Each individual rotation r of a link (see Fig. 2.4) does not contribute to the degree of freedom of the mechanism. With g joints and r rotations of links, the balance of degrees of freedom of a mechanism **(4)** – which may here be characterized by the upper-case 'F' – results in $F = 6(k+l) - r - \sum_{1}^{g}(6-f_i)$ or

$$F = 6(k + l - g) - r + \sum_{1}^{g} f_i \qquad (2.1)$$

where k = number of wheel carriers, l = number of links, g = number of joints, r = number of individual rotations of links, $f_i$ = degree of freedom of the joint i.

The suspension shows the values k = 1, l = 3, g = 6. Four ball joints, the turning joint and the turning-and-sliding joint bring in $\Sigma f_i = 4 \times 3 + 1 + 2 = 15$. The rod link and the turning-and-sliding link can rotate individually (r = 2). Thus the degree of freedom of the suspension becomes $F = 6(1 + 3 - 6) - 2 + (4 \times 3 + 1 + 2) = 1$, as necessary for an independent wheel suspension.

## 2.3 Basic types of wheel suspensions

### 2.3.1 Evolution from the statically determinate suspension of a spatial body

As is well known from the basic laws of statics, a spatial body with its six degrees of freedom can be 'fixed' by cancelling them all with suitable elements – e. g. with six rod links, each of which locks one degree of freedom, namely the displacement of all points along their longitudinal axes. However, these six rods must not intersect a common line, because this would create an unstable system.

Such a statically determinate suspension is shown in **Fig. 2.6**a. Its degree of freedom is zero, and this corresponds to a wheel carrier which is rigidly fixed to the vehicle body, without any suspension, Fig. 2.6e.

By removing one of the rods, Fig. 2.6b, the spatial body gets one degree of freedom (F = 1) and forms the basic layout for an independent suspension with five rod links, Fig. 2.6f.

If controlled by only four rod links, Fig. 2.6c, the mechanism with two degrees of freedom is either the archetypal rigid-axle suspension, Fig. 2.6g, or the basic arrangement for a wheel carrier with two wheels in tandem (h).

Basic Characteristics of Wheel Suspensions

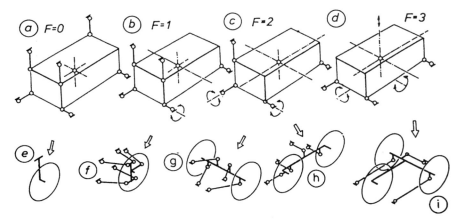

**Fig. 2.6** Evolution of wheel suspension mechanisms from a statically determinate model

Removal of one rod more results in a spatial mechanism which, with three degrees of freedom, can carry three wheels as a 'three-wheel set', Figs. 2.6 d and i.

By building up a triangular link from two rod links or a 'three-leg' component from three of them, most of the known variations of independent and rigid-axle suspensions can be derived from the five-rod and four-rod mechanisms shown in Fig. 2.6, f and g.

The number of variants can be almost infinitely extended by the following adaptations: replacing rods by equivalent joint and link connections, introducing trapezoidal links (which are not derivable from Fig. 2.6!), the addition of new links or joints and compensating for possible new degrees of freedom by further restrictive means, and the insertion of 'intermediate couplers' as (for example) rod links which interconnect other links. Splitting of the beam of a rigid axle into two parts and flexibly interconnecting the halves, or connecting the wheel carriers of two independent suspensions of an 'axle' by links, of course while heeding (and, if required, correcting) the overall degree of freedom $F = 2$ for the arrangement of two wheels together, lead to the so-called 'compound suspensions' which can be classed neither with the independent nor with the rigid-axle layouts, since their two wheel carriers are neither independent of each other nor fixed together.

For this reason, a different method of presentation will be used in the following, by tracing the evolution of wheel suspensions from simple to complicated mechanisms along a historical path.

### 2.3.2 Independent suspension layouts

The simplest way to give a wheel carrier one degree of freedom with respect to the vehicle body is to connect the two by a joint. However, the only joint with one degree of freedom is the turning joint, see Fig. 2.2c. Depending on the attitude of such a joint's axis of rotation in the vehicle, the result is a trailing-arm suspension, **Fig. 2.7**a, semi-trailing-arm suspension (b) or swing-axle suspension (c). The axes of rotation of the wheel and arm need not be parallel nor lie in a common plane; such configurations may be advantageous to the suspension properties, as will be explained later.

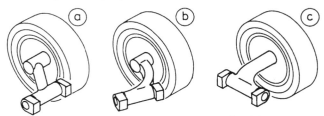

**Fig. 2.7** Independent suspension layouts based on a turning joint

A turning-and-sliding joint provides two degrees of freedom and is therefore not on its own sufficient to form an independent suspension. The surplus degree of freedom can be cancelled by a rod link, **Fig. 2.8**, and the layout hence becomes an example of the rare but still encountered vertical telescopic suspension for front wheels (a) or a semi-trailing-arm design embellished from 'planar' to 'spatial' geometry by overlaid rotation and axial shift (b); in Chapter 3, it will be shown that a 'spatial' mechanism is characterized by an instantaneous screw geometry.

If the wheel carrier and the vehicle body are immediately connected by a ball joint, which introduces three degrees of freedom, the suspension needs – for example – two additional rod links to reduce the overall degree of freedom to $F = 1$, **Fig. 2.9**. The 'double-wishbone' version, Fig. 2.9a, has quite

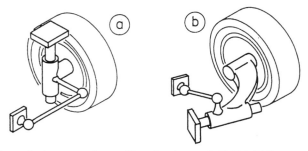

**Fig. 2.8** Independent suspensions with a turning-and-sliding joint

# Basic Characteristics of Wheel Suspensions

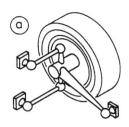
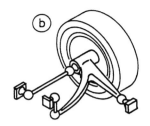

**Fig. 2.9**
'Spherical' suspension mechanisms

often been employed, while there has been only one instance of the 'semi-trailing-link' version (b).

The suspension layouts shown in Fig. 2.9 clearly represent a transition stage from a direct to an indirect connection of wheel carrier and vehicle body. As all points of the wheel carrier travel on surfaces of spheres, centred on the ball joints at the vehicle body, these suspensions can be described as 'spherical' mechanisms.

If the wheel carrier is no longer connected directly to the vehicle body, it becomes the 'coupler' of a kinematic chain.

The simplest chain mechanism results from the combination of a trapezoidal link and a rod link, **Fig. 2.10**. The axes of the two turning joints of the trapezoidal link need not be parallel nor lie in a common plane.

**Fig. 2.10** Trapezoidal-link suspension

A triangular link can be regarded as a combination of two rods, as already mentioned. By building up two triangular links from four rods, the suspension of Fig. 2.6f changes into the well-known 'double-wishbone' type with its numerous variations, **Fig. 2.11**. In the case of version a, if this is a

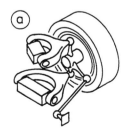
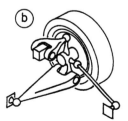

**Fig. 2.11**
Double-wishbone suspensions

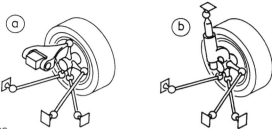

**Fig. 2.12** Four-link suspensions

front suspension, the remaining (fifth) rod serves as the 'track rod'. The version shown in b has been used for the rear suspension of racing cars.

Three rod links, **Fig. 2.12**, may be completed by a triangular link (a) when we get the basic 'four-link suspension' as used in racing cars (and sometimes in production cars). Alternatively they can be used with a turning-and-sliding link (b) to form the basic 'strut suspension'.

The most sophisticated and versatile independent suspension type consists of five rods, **Fig. 2.13**. Version 'a' is equivalent to the basic independent-suspension mechanism illustrated already in Fig. 2.6f. Version 'b', however, shows an 'intermediate coupler', namely an approximately vertical rod which connects the longitudinal rod with the upper transverse one. This version is therefore not an evolutionary derivative of the basic design of Fig. 2.6f, any more than is the trapezoidal-link suspension.

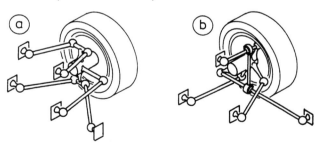

**Fig. 2.13** Five-link suspensions

### 2.3.3 Rigid-axle suspensions

The suspension of a rigid axle needs two degrees of freedom, one of them characterized by an essentially vertical motion and the other by a rotation about a horizontal axis in the centre plane of a vehicle. However, the only joint that offers these two degrees of freedom and combines a translation with a rotation - namely the turning-and-sliding joint (Fig. 2.2d) - is not

Basic Characteristics of Wheel Suspensions

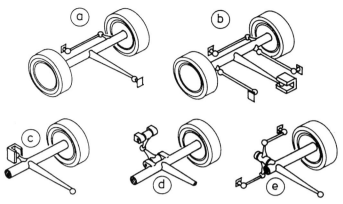

**Fig. 2.14** Rigid-axle suspensions with a thrust ball

qualified for this task on its own, since the axis of translation and the axis of rotation coincide, while a perpendicular orientation of these two axes would be necessary.

Hence, the only joint types that can be considered for the direct connection of axle beam and vehicle body are the ball joint and the ball-and-surface joint, **Fig. 2.14**.

Fig. 2.14a shows the simplest version, with a ball joint between axle and vehicle: the superfluous third degree of freedom offered by the ball joint is cancelled here by a rod which locates the axle laterally. This design is the 'thrust-ball' or 'A-bracket' axle with a 'Panhard rod', and it was and is common on commercial vehicles while still being sometimes utilized on passenger cars.

If the ball joint is replaced by a ball-and-surface joint ($f = 5$, Fig. 2.14b), the suspension needs three rods to reduce the degree of freedom to $F = 2$. The ball now serves only for reacting vertical forces and the braking and, if necessary, the traction torque, while the two longitudinal rods react the fore-and-aft forces.

Figs. 2.14c, d and e show alternative solutions to the Panhard rod to avoid the lateral movement of the axle relative to the vehicle (and therefore steering effects) due to the rod's arcuate motion with wheel travel. Fig. 2.14c shows a ball-and-surface joint, 2.14d a 'scissors' mechanism formed by two triangular links (rare in cars but familiar in aircraft applications) and 2.14e a 'Watt linkage'. In the last two cases the Panhard rod has been replaced by a complete sub-mechanism.

For an indirect connection of the axle to the vehicle, the most-used systems are the combination of a triangular link with two rods and a four-rod suspension, **Fig. 2.15**a and b.

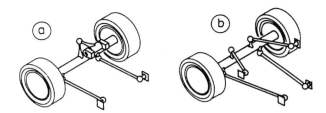

**Fig. 2.15**
Rigid-axle suspension by a linkage

In contrast with independent suspensions, the rigid-axle variety sometimes fails to provide the correct degree of freedom, two examples being given in **Fig. 2.16**.

A suspension system with five rods - namely four longitudinal rods and a Panhard rod (Fig. 2.16a) - possesses only one degree of freedom and is therefore 'over-constrained'. Problems will not normally arise in the case of parallel vertical wheel travel, since the suspension is symmetrical about the central plane of the vehicle, and the corresponding longitudinal rods on both sides move congruently. With vehicle roll or antimetrical wheel travel, however, both ends of the axle beam will try to rotate about a transverse line and in the opposite sense, unless the four longitudinal rods are parallel in the static load position and are of the same length. That tendency to contra-rotation is, of course, resisted by the axle beam's torsional stiffness and leads to a distortion of the suspension joints, which must here of course be made elastically compliant. The constraining forces can be lowered by skilful arrangement of the longitudinal rods (see Chapter 9), but this suspension does limit the choice of kinematic and elasto-kinematic properties of the system.

The suspension of Fig. 2.16b, on the other hand, has only three rods and consequently three degrees of freedom, so its mechanism is 'under-constrained'. It can be easily seen that the suspension cannot resist torques about the transverse axis - e.g. braking or traction torques, which here are reacted by the leaf springs, elastic wind-up of the axle being accepted. Such suspensions are occasionally used on very heavy trucks in order to minimize the overall loading of the joints.

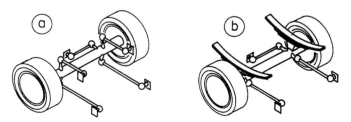

**Fig. 2.16** Imperfectly constrained rigid-axle suspensions (a = over-, b = under-constrained)

# Basic Characteristics of Wheel Suspensions

## 2.3.4 Tandem wheel carrier

Besides the usual combination of two wheels on opposite sides of the vehicle, on one wheel carrier in the form of a rigid axle, there is also the possibility of coupling two wheels in line on each side, **Fig. 2.17**. This suspension needs two degrees of freedom also, and can be realized for example by four rods as shown. The main forms of motion are a vertical parallel wheel travel and rotation about a transverse axis.

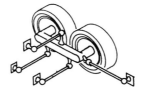

**Fig. 2.17** Tandem wheel carrier

## 2.3.5 Compound suspensions

On an 'axle' consisting of two independent wheel suspensions, each wheel can move with one degree of freedom without being influenced by the motion of the other wheel in any way. With rigid-axle suspension, in contrast, both wheels together possess two degrees of freedom but cannot change their relative positions on the axle beam.

Both these principles are special cases of a superior suspension type, where two wheels of an 'axle' together possess two degrees of freedom and where, unlike the rigid axle, mutual displacements of the wheels are possible; in contrast with independent suspensions, too, the motion of one wheel is influenced by the actual attitude of the other.

Such designs are known as 'compound suspensions' and they are intended to combine the advantages of independent and rigid-axle types and especially to provide optimal kinematic conditions in both straight-ahead driving and cornering.

Independent suspension, with '0% of compound' represents a special case of the compound axle and is, so to speak, the 'zero point' of the compound 'scale'. Rigid-beam-axle suspension has '100% of compound' but does not limit the upward 'scale' because the compound properties can be varied quite freely. Several characteristics can be envisaged to quantify the degree of compound - for instance the coupling of the inertias of the two wheels with antimetric wheel travel - but they will not be considered here.

Compound suspensions may appear to be closely derived from independent suspensions, and from rigid axles, too, although unlike the latter a compound suspension has a separate carrier for each wheel.

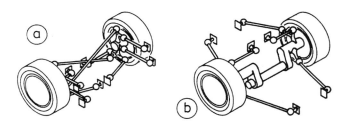

**Fig. 2.18** Examples of compound suspensions

Two rather different examples of compound suspensions are shown in **Fig. 2.18**. The suspension of Fig. 2.18a is comparable with a five-rod layout (see Fig. 2.13a); here, though, the upper transverse rods are connected not to the vehicle body but to the respective opposite wheel carriers. The mechanism has $k = 2$ wheel carriers, $l = 10$ links with overall $r = 10$ individual rotations of links, $g = 20$ ball joints with overall $\Sigma F_i = 3 \times 20 = 60$ degrees of freedom. In conformity with equation (2.1) the overall degree of freedom of the mechanism is $F = 2$, as necessary for the correct suspension of two wheels.

Unlike the version in Fig. 2.18a, the compound suspension of Fig. 2.18b shows a significant relationship to a rigid axle. The axle beam has, however, been split into two wheel carriers which are connected by a turning-and-sliding joint. As this joint introduces two new degrees of freedom to the system, two additional constraints - e.g. two rods - are necessary to offset them. This suspension therefore has six rods, compared with the four of a normal rigid axle - see Fig. 2.15b.

If the turning-and-sliding joint in Fig. 2.18b was to be replaced by a turning joint, one of those six rods could be saved. With skilful design, such a turning joint can, on the other hand, also be represented by a transverse beam which is resistent to bending but compliant to torsional moments (i.e. a bar with an 'open' profile section). This variant forms the basis of a worldwide family of compound rear suspensions for front drive cars, known as 'torsion-beam suspensions'.

The properties of compound suspensions may be aligned more either to those of independent or to those of rigid-axle suspensions. Either way, though, they have one feature in common with the latter, namely their physical extension across the full width of the vehicle.

# Chapter 3

# Kinematic Analysis of Wheel Suspensions

## 3.1 Basic rules of planar kinematics

Analysis of the three-dimensional motions of most wheel suspensions involves at least two projection views if made graphically, or a spatial co-ordinate system if made by calculation. Because a 'planar' motion – i.e. a motion that can be described in a single projection view – can be assessed more easily, however, and because the rules of planar kinematics can in many cases be analogously transferred to spatial kinematics, some essential rules of such kinematics should first be stated.

Fig. 3.1a shows a 'planar' body with a point A which has the instantaneous velocity $v_A$. The body is assumed to be rigid, so the distance of a second point B from A remains constant, and point B must have the same velocity component $v_{AB}$ in the direction A-B as point A. The absolute velocity $v_B$ follows from $v_A$ or $v_{AB}$ if the line of action t of $v_B$ is given. This situation is the basis of a well-known graphical method of determining the velocity vector diagram of a rigid body: the velocity vector $v_A$ is swivelled by 90° (vector $v_A'$), as are the line of action t of $v_B$ and all velocity vectors at the body. Hence, the unknown vector $v_{AB}$ swivels into a rectangular position with respect to the line AB and, since $v_{AB}$ is valid for both points A and B, the rectangular vector $v_B'$ results from the intersection of the parallel line to AB through the vector tip of $v_A'$ and of the rectangular line to t in B (method of 'rectangular velocity vectors'). To get the true vector $v_B$, the vector $v_B'$ has to be swivelled back by 90°.

All points on the rectangular or 'normal' line to $v_A$ in A cannot have any velocity component in the direction of A, and the same is valid for $v_B$ and B. The point P of the body which coincides with the intersection point of the normal lines to $v_A$ and $v_B$ in A and B has no velocity component in the direction to A nor that to B and consequently cannot move at all. P is therefore the 'instantaneous pole' about which the body pivots instantaneously. The velocity vectors of all points of the body are therefore directed rectangularly to their connecting lines to P, the 'polar rays', and their velocities are proportional to their distances from the pole.

20 Road Vehicle Suspensions

The rectangular velocity vector $v_C'$ of any point C may be determined via the method of rectangular velocity vectors either by using the 'polar ray' PC and one of the given velocities $v_A$ or $v_B$, or – without reference to the pole P – immediately by using the rectangular vectors $v_A'$ and $v_B'$ and intersecting the parallel lines to AC and BC through the vector tips.

A and B may be the joints of two links which are directed rectangularly to the velocity vectors $v_A$ and $v_B$ and connected to the 'fixed part' by two joints $A_0$ and $B_0$. These two links will secure the body's given velocity state. There cannot be more links, because a planar body possesses three degrees of freedom within its plane – namely two translations and one rotation – and each of the two links $AA_0$ and $BB_0$ cancels one of these degrees of

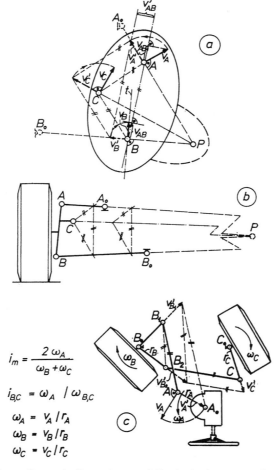

$$i_m = \frac{2\omega_A}{\omega_B + \omega_C}$$

$i_{BC} = \omega_A / \omega_{B,C}$

$\omega_A = v_A / r_A$

$\omega_B = v_B / r_B$

$\omega_C = v_C / r_C$

**Fig. 3.1** Planar kinematics: velocity vectors and the instantaneous pole

freedom. $A_0ABB_0$ form a planar four-joint chain with one degree of freedom ($F=1$), and the rigid body AB is the 'coupler' of this chain.

For graphic design work it will sometimes be necessary to determine the 'polar ray' of a point of the coupler while the pole itself is far outside the field of the drawing. The determination of $v_C$ from $v_A$ and $v_B$ by rectangular velocity vectors, or the 'theorem of proportional segments', points a way to find the polar ray PC by means of similar triangles, see Fig. 3.1b.

Fig. 3.1c shows an application of the method of rectangular velocity vectors to a (planar) steering linkage (consisting of a drag link $AB_1$ and a track rod $B_2C$) to determine the steering ratios $i_B$ and $i_C$ between the steering gearbox and the two wheels, and the overall steering ratio $i_m$.

The 'velocity vectors' $v_A$ etc. used here may also be regarded as 'infinitesimal displacement vectors' when multiplied by a time differential dt. As kinematics methods are generally related to 'velocities', this practice will be retained in what follows, though 'velocities' will sometimes be interpreted also as virtual displacements.

In **Fig. 3.2**a, the instantaneous pole P moves with the polar rays and in relation to the polar ray AP has the velocity component $v_{PA}$, which follows from $v_A$ by the theorem of proportional segments, and in relation to BP the component $v_{PB}$. The resulting 'pole velocity' $v_P$ can be found by intersection of the lines normal to the vector tips of $v_{PA}$ and $v_{PB}$.

$A_0$ and $B_0$ are the (here invariable) 'centres of curvature' of the paths of A and B; the velocity of the polar rays PA and PB is here zero. The centre of curvature $C_0$ of any point C of the coupler is the one point of the polar ray PC having no instantaneous velocity, and it is determined in this figure by means of the velocity vector $v_C$ and the component $v_{PC}$ of the pole velocity $v_P$ (Hartmann's method). $\rho_C = C_0C$ is the 'radius of curvature' of the path of point C.

Instantaneous poles must not be confused with centres of curvature! An instantaneous pole defines the state of motion of a planar body, and the tangents of the paths of its points represent the first-order derivative of the paths. A centre of curvature, however, represents the second-order derivative of the paths or the state of accelerations (so the radial acceleration of the point C in Fig. 3.2a is $a_C = v_C^2/\rho_C$). To help clarify this situation, the joints $A_0$ and $B_0$ (the centres of curvature of the joints A and B) were deliberately chosen to lie on opposite sides of the pole P.

Variable instantaneous poles (and 'instantaneous axes' in space) may therefore always be used as reference points for static force analyses, but never for the solution of dynamic problems.

Fig. 3.2b shows Bobillier's method of determining radii of curvature, without giving its theoretical background. The intersection point $D_{AB}$ of the

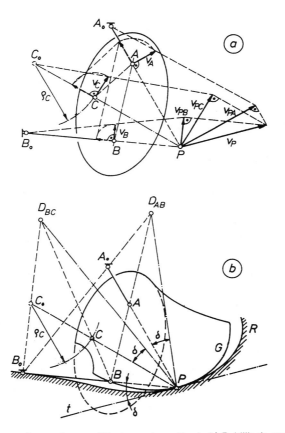

**Fig. 3.2** Centres of curvature: a) Hartmann's method b) Bobillier's method

lines $A_0B_0$ and AB is the instantaneous pole of the relative motion of the links $A_0A$ and $B_0B$. The angle $\delta$ between the line $D_{AB}P$ and the polar ray AP is equal to the angle between the the polar ray BP and the tangent t to the polar path or 'polode' (or the direction of $v_P$ in Fig. 3.2a). This situation is valid for any points or polar rays. Consequently the centre of curvature $C_0$ of a point C is to be found by adding the angle $\delta$ to the polar ray CP, the determination of $D_{BC}$ via B and C, and last of $C_0$ via $D_{BC}$ and $B_0$ (it is obviously not necessary here to know the polode tangent t!).

The instantaneous pole P travels along the 'fixed polode' R in relation to the 'fixed' environment and along the 'moving polode' G with respect to the 'coupler' of the four-joint chain. Both curves are easily designed by drawing different positions of the points A and B or - assuming these two to be fixed - of the points $A_0$ and $B_0$. The moving polode G rolls off on the fixed polode R without any slip and generates the same motion of the coupler as the real pair of the links $A_0A$ and $B_0B$. All mechanisms that show the same fixed and moving polodes are equivalent.

As already mentioned in Chapter 2, wheel suspensions (except for rare cases) form statically constrained systems which consist of kinematic mechanisms with as many degrees of freedom as the number of their wheels. Normally the number of spring elements is the same also. Changes of wheel attitude are caused by external forces, the suspension mechanism then being displaced according to the characteristics of the springs.

Suspension systems must withstand various combinations of external forces which relate to particular driving manoeuvres and which are usually treated separately for theoretical analysis, but may be combined too, as when vertical wheel load, longitudinal accelerating/braking or impact forces are accompanied by cornering forces.

During the evolution of automotive technology, a number of technical terms or 'characteristics' have been introduced which give essential information about the properties of wheel suspensions in connection with the typical load cases mentioned above. These characteristics mainly represent effective lever arms or proportions of forces, and examples of them are the roll centre, scrub radius, brake-force support angle and spring ratio. Some of these characteristics have been standardized and are easily determined for simple suspensions, even by graphical methods. On more sophisticated spatially arranged suspensions, however, they are often difficult to determine, so it is necessary to give them improved definitions to ensure compatibility of the physical statements with the traditional definitions.

For kinematic mechanisms with one degree of freedom - e. g. independent wheel suspensions - a clear interdependence exists between the forces at the suspension members and the geometry of motion, as will be explained later. Mechanisms with two degrees of freedom are represented by the rigid-axle and compound suspensions; here clear relationships between forces and geometry show up if the two degrees of freedom ('parallel wheel travel' and 'rolling') are treated separately.

The geometry of motion of a mechanism can be analysed by quite simple methods. In what follows, it will be shown that all the suspension and steering characteristics that are used in automotive technology can be exactly and compatibly determined by applying the rules of kinematics. To demonstrate the basic idea of this method, the different procedures of checking the balance of forces and moments relating to the state of motion will be compared on a simple planar mechanism - **Fig. 3.3**.

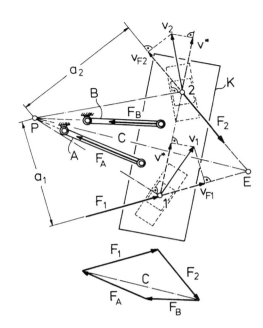

**Fig. 3.3**
Relationship between the balance of forces and the geometry of motion

A rigid body K is controlled in a plane by two links A and B. A force $F_1$ acts on the point 1 of the body. As the mechanism has one degree of freedom, a counteracting force $F_2$ is necessary to secure the balance of forces. $F_2$ acts on the point 2 of the body, and its line of action is given.

The links A and B can transfer forces $F_A$ and $F_B$ only in the directions of their longitudinal axes. Four forces $F_1$, $F_A$, $F_B$ and $F_2$ have to be balanced, while the amount of $F_1$ and the lines of action of the other forces are given. In Fig. 3.3 this problem has been solved by the Culmann method using an 'intermediate resulting force' C, its line of action running between the intersection point E of the lines of action of $F_1$ and $F_2$ and the intersection point P (the pole of the mechanism) of the lines of action of $F_A$ and $F_B$. C is therefore the resulting force of $F_A$ and $F_B$, or, with the inverse sign, also of $F_1$ and $F_2$. C and $F_2$ can immediately be determined by a triangular diagram of forces using the given force $F_1$, and subsequently $F_A$ and $F_B$ with the aid of force C.

In the mechanism of Fig. 3.3, the point P represents the instantaneous pole of the body K in relation to the fixed environment. The forces $F_A$ and $F_B$ of the links A and B cannot exert a moment around that pole, so all other forces acting on K must be in balance with respect to P. An instantaneous pole may therefore be used as reference point for the calculation of a static balance of moments. In the case shown, the forces $F_1$ and $F_2$ act around P by the effective lever arms $a_1$ and $a_2$, and $F_2$ follows from the condition of balance $F_1 a_1 = F_2 a_2$ or $F_1/F_2 = a_2/a_1$.

Moreover, the pole P is also the centre of the actual velocity vector diagram of the mechanism. The velocity vectors $v_1$ and $v_2$ of the force application points 1 and 2 are directed rectangularly to the respective polar rays 1-P and 2-P and must show equal components $v^*$ in the direction 1-2; thus the proportion of $v_1$ to $v_2$ is determined. If the body K rotates around the pole P, the application point 1 of the force $F_1$ travels with velocity $v_1$ which has a component $v_{F1}$ in the direction of $F_1$. The 'power' effected at the velocity $v_{F1}$ by the force $F_1$ is then $F_1 v_{F1}$. To achieve a balance, the same power must be effected at the application point 2, the velocity component of which in the direction of $F_2$ is $v_{F2}$. The balance of power is therefore $F_1 v_{F1} = F_2 v_{F2}$ and leads to the force $F_2$.

Obviously the relationship of the forces is reciprocal to that of the velocities: $F_1/F_2 = v_{F2}/v_{F1}$. With this agreement, and the relationship $a_2/a_1 = F_1/F_2$, the close agreement of the analysis of forces and the analysis of velocities is demonstrated; this agreement is very important for the analysis of more complicated spatial suspension mechanisms where instantaneous axes can no longer be found by simple means.

## 3.2 Basic rules of vector calculation

To analyse the motion of wheel suspension systems, two principal methods are available – 'analytical geometry' and vector calculation. While attempts to use analytical geometry soon lead to complicated equations, the vector calculation method proves easily understandable and extraordinarily well suited to computer application. In addition it enables very simple definitions of suspension and steering 'characteristics', even on spatial mechanisms, as will be shown later.

It is appropriate to repeat here the few basic rules of vector calculation that will be needed hereafter for the kinematic analysis of suspensions.

All laws and equations of vector calculation are defined in a 'right-handed' coordinate system with rectangularly arranged axes – e.g. the x-, y- and z-axes in **Fig. 3.4**, where the x-axis trails by 90° anticlockwise behind the y-axis when one looks in the positive direction of the z-axis. The same is valid for the sequences y-z-x and z-x-y (the rule of 'circular permutation').

A vector is a spatially orientated physical quantity – e.g. a length, a velocity, a force, an angular velocity or a torque. In an x-y-z-system, a vector **a** has the components $a_x$, $a_y$ and $a_z$.

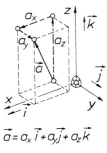

$$\vec{a} = a_x \vec{i} + a_y \vec{j} + a_z \vec{k}$$

**Fig. 3.4** Vector with components

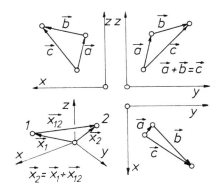

**Fig. 3.5** Vector addition

A vector of the amount '1' is called a 'unit vector'. The unit vector **i** of the x-axis has the components $i_x = 1$, $i_y = 0$ and $i_z = 0$; accordingly the components of the unit vectors **j** and **k** of the y-axis and the z-axis are 0, 1, 0 and 0, 0, 1.

The absolute or 'scalar' amount of the vector **a** can be calculated by means of the Pythagorean theorem from the square sum of its components:

$$a = |\mathbf{a}| = \sqrt{a_x^2 + a_y^2 + a_z^2} \qquad (3.1)$$

A unit vector can be got from every vector by dividing it by its amount:

$$\mathbf{e}_a = \mathbf{a}/|\mathbf{a}| \qquad (3.2)$$

The addition of vectors **a** + **b** = **c** is done by the addition of their components, as shown in **Fig. 3.5** in three views:

$$a_x + b_x = c_x \qquad a_y + b_y = c_y \qquad a_z + b_z = c_z \qquad (3.3a,b,c)$$

The lower left part of this figure shows that a vector $\mathbf{x_2}$ from the origin to a point 2 — the 'position vector' of point 2 — can just as well be found by addition of the position vector $\mathbf{x_1}$ of a point 1 and the connecting vector $\mathbf{x_{12}}$ from point 1 to point 2.

In the multiplication of vectors there are three variants with different physical meanings — **Fig. 3.6**. The 'scalar' product of two vectors **a** and **b** is the product of the component of one vector in the direction of the second and the amount of the latter, and it results from the products of the respective x-, y- and z-components (see Fig. 3.6a):

# Kinematic Analysis of Wheel Suspensions

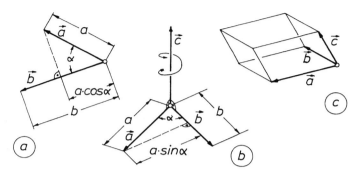

**Fig. 3.6** Multiplication of vectors:
a) scalar product   b) vector product   c) parallelepipedial product

$$\mathbf{a} \cdot \mathbf{b} = a_x b_x + a_y b_y + a_z b_z = c \tag{3.4}$$

Here c is no longer a vector but a scalar, a physical quantity (e.g. the amount of a length or a velocity).

Another interpretation can be derived from Fig. 3.6a:

$$\mathbf{a} \cdot \mathbf{b} = |\mathbf{a}||\mathbf{b}|\cos\alpha \tag{3.5}$$

and an application of the scalar product is, for example, the calculation of the product of a force and a velocity (i.e. a 'power'), if their vectors are skewed.

In contrast with the scalar product, the result of the 'vector product'

$$\mathbf{a} \times \mathbf{b} = \mathbf{c} \tag{3.6}$$

is a vector. The vector **c** is rectangular to the plane that is defined by the vectors **a** and **b**; when rotating the vector **a** into the vector **b** clockwise by the shortest possible angle – that is ≤ 180° – the vector **c** points away from the observer (according to this, the vector product of the unit vectors in Fig. 3.4 is $\mathbf{i} \times \mathbf{j} = \mathbf{k}$ and the swivelling angle in the x-y-plane is 90°). The scalar amount of the resulting vector **c** can be calculated from equations (3.1) and (3.2), or by the geometrical interpretation shown in Fig. 3.6b:

$$c = |\mathbf{a}||\mathbf{b}|\sin\alpha \tag{3.7}$$

Regarding the previously mentioned rule for the definition of the attitude of the vector **c**, or also equation 3.7, it is evident that $\mathbf{a} \times \mathbf{b} = -\mathbf{b} \times \mathbf{a}$.

In what follows, the vector product is applied mainly to form 'normal vectors' (as, for instance, a velocity vector that is perpendicular to a plane defined by an axis of rotation and a radius vector).

For the components of the vectors, a vector product can be represented *formally* by the determinant

$$\begin{vmatrix} \mathbf{i} & \mathbf{j} & \mathbf{k} \\ a_x & a_y & a_z \\ b_x & b_y & b_z \end{vmatrix} = \mathbf{c} \qquad (3.8)$$

(formally for the reason that the first row contains vectors instead of quantities!), and the components of **c** follow as

$$\begin{aligned} c_x &= a_y b_z - a_z b_y \\ c_y &= a_z b_x - a_x b_z \\ c_z &= a_x b_y - a_y b_x \end{aligned} \qquad (3.9\,a,b,c)$$

while here once more the effect of circular permutation may be noticed.

Less important in relation to this book is the 'parallelepipedial product' –

$$V = \mathbf{a} \cdot (\mathbf{b} \times \mathbf{c}) \qquad (3.10)$$

– a scalar product of the vector **a** and the resultant of the vector product **b**×**c**. Assuming the vectors to represent spatial length elements, the resultant V of the product, a scalar, gives the volume of the parallelepiped whose edges are defined by these three vectors, Fig. 3.6c. A possible application of the parallelepipedial product might be to ensure that three vectors (e. g. suspension links) are positioned in three parallel planes or aligned in a common plane; in both cases the result must be V = 0.

## 3.3 Considerations of wheel-suspension systematics

Several ways are available for classifying the various mechanisms used in wheel suspension.

The two most frequently encountered types are, of course, **independent** and **rigid-axle suspensions**. An independent suspension is a mechanism that controls one wheel carrier with one wheel, while on a rigid axle two wheels are associated with one wheel carrier.

There was some uncertainty earlier about the correct classification of those suspensions controlling two opposite wheels of an 'axle' neither independently of each other nor by direct connection; such systems are known nowadays as **compound suspensions** – a term unlikely to cause any confusion with the familiar tandem-axle layouts (or 'dual axles') of trucks.

A rigid-axle suspension is almost always arranged symmetrically to the median plane of the vehicle. Classification of such mechanisms in relation to the types of links or joints is of little help because of the limited variational possibilities. In addition, there are some axle suspensions that are not designed in a kinematically 'exact' manner but are over- or under-constrained, as already mentioned in Section 2.3.3.

It is even more difficult to find a good classification of compound suspensions since these may show a close relationship with either the independent or the rigid-axle type. In addition, there are so many design possibilities for compound systems.

At best, a classification can be set up for independent suspensions based on their similarities to familiar kinematic systems such as double-crank or inverted-slider crank mechanisms. However, classification of wheel suspensions with their spatial geometry by a plan that originates from 'planar' kinematics must be contemplated only with caution. Naming according to external appearance or to design features (e.g. 'double-wishbone layouts', 'strut suspensions' and so on) gives us little guidance as to kinematic properties.

In practice, a strut suspension with its long telescopic damper and its relatively short transverse link cannot provide the same advantageous camber change with wheel travel as a double-wishbone design, while an inverted-slider crank mechanism (the basic model of the strut system) theoretically needs not be inferior kinematically to a double-crank mechanism (the basic model of double-wishbone suspension). In practice, much more important than the kinematic classification of strut suspension is the fact that the damper is used as a suspension link which may cause undesired friction on the one hand but saves a lot of interior space on the other.

Another, somewhat generalized, classification of independent wheel suspensions can be established on the basis of **kinematic potential** levels, as will be explained in what follows.

Except in the case of a direct connection of the wheel carrier to the vehicle body by a joint, the carrier of an independent suspension system normally is the coupler of a kinematic mechanism 'chain'. If the four-joint chain of Fig. 3.3 is assumed to be three-dimensional, it is clear that several of the joints must be realized as the turning variety to avoid forcing the coupler K out of the drawing plane. The three-dimensional coupler will then move in a view towards the drawing plane in the same manner as did the

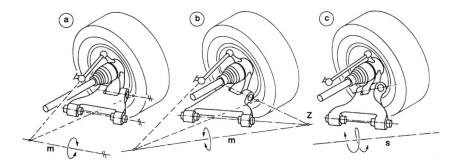

**Fig. 3.7** Planar, spherical and spatial suspensions with trapezoidal links

two-dimensional body, the 'pole' P in the plane changing into an 'instantaneous axis' perpendicular to the plane.

The same is valid for the independent suspension consisting of a trapezoidal link and a rod link shown in **Fig. 3.7**a. As the axes of the two turning joints of the trapezoidal link are parallel, the instantaneous axis m of the wheel carrier must also be parallel with the joints' axes and is determined by the intersection point of the line of action of the rod and the plane of the axes of revolution. With wheel travel, the instantaneous axis remains parallel to the joint axes. In a drawing plane that is perpendicular to the axes, the displacement of the wheel carrier and the links can even be precisely investigated graphically by means of compasses and ruler! Here the wheel carrier follows a **planar path**.

If the axes of the turning joints of the trapezoidal link are inclined towards each other in their plane, Fig. 3.7b, the intersection point Z is fixed at the link and therefore also at the wheel carrier and the vehicle body. All points of the wheel carrier now move on concentric spherical surfaces about the fixed point Z, the wheel carrier itself having a **spherical motion** about Z. In contrast with the previous planar motion, the instantaneous axis m is not displaced by parallel translation with wheel travel but swings about the point Z.

Comparing Fig. 3.7a and b it is easy to see that planar motion is only a special case of spherical motion with the central point Z shifted to infinity. On the other hand, the suspension layouts of Fig. 2.9 in Chapter 2 are now shown to be spherical mechanisms with corporeal central points Z.

A spherical chain is characterized by a fixed central point Z which is the focus of the axes of the links and the instantaneous axis - **Fig. 3.8**. The real joints can be displaced along these axes without changing the kinematic functioning of the mechanism, easily visualized as a pyramid of paper with flexible edges, and the two mechanisms drawn respectively by full and

# Kinematic Analysis of Wheel Suspensions

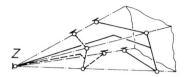

**Fig. 3.8** Spherical four-link mechanisms

dashed lines are therefore kinematically equivalent. This is, however, not valid for the analysis of forces or for the elastic behaviour (elasto-kinematics).

The four linkage axes in Fig. 3.8 can be realized by turning joints (true for the 'cardan joint', the best-known spherical mechanism), but that requires very precise and rigid components to avoid constraint. It is therefore usual in practice to provide no more turning joints than necessary — see Fig. 3.7b — sufficient only to comply with the correct degree of freedom predicated by equation (2.1) of Chapter 2. This results in a 'statically defined' mechanism which is insensitive to tolerances and elastic compliance.

In a mechanism, the basic form of spherical motion (and planar, for that matter) is, as stated earlier, the rotation of a coupler about an instantaneous axis.

All points on the coupler which at the particular moment coincide with the instantaneous axis have zero velocity, so this model cannot represent the 'common' case of three-dimensional motion in space! The points on the instantaneous axis can move too only by superimposing a shift along that axis. Hence, the instantaneous rotation with velocity vectors that are all perpendicular to the axis of rotation changes into an 'instantaneous screw motion' (**4**), and the instantaneous axis becomes an 'instantaneous screw axis'. The resulting velocity vectors of all points of the coupler are now composed of their circumferential velocity vectors (the amounts of which are proportional to their radii from the screw axis) and of the shift-velocity vector along the screw axis.

The turning joints of the trapezoidal-link suspension in Fig. 3.7c have skew axes which (in contrast with the spherical suspension of Fig. 3.7b) do not intersect. Here the wheel carrier executes a path in space that is characterized by instantaneous rotations about the two turning joints and at the same time an overlaid spherical motion about the rod's ball joint at the vehicle body. The wheel carrier in Fig. 3.7c thus performs a **spatial motion**, which can be described as an 'instantaneous screw'.

In space, an instantaneous axis is uniquely defined by four parameters — e.g. its inclination angles in two views, its distance from a reference point and the inclination of the distance vector relative to a plane. However, an instantaneous screw axis is not defined uniquely until the axial shift is specified, which represents a fifth parameter. A particular shift applies to each attitude of the screw axis, and both normally vary with the motion of the mechanism.

The basic properties of a wheel suspension are described by their suspension/steering 'characteristics', as already mentioned. These characteristics reveal the functioning of the suspension in relation to specific driving manoeuvres. For example, the **roll centre** (see Chapter 7) tells about the mode of lateral-force transfer between wheel and vehicle, while the **wheel-travel angle** and the **support angle** (see Chapters 5 and 6) influence the transfer of longitudinal forces. Very important for the vehicle's driving stability are the **self-steering properties** ('bump-steer', see Chapter 7) of the suspension, and the **camber change** with wheel travel; this change can improve the lateral grip of the tyres, especially as the stability limits are approached. In suspensions with wheels driven by transmission shafts, but without hub-reduction gears, the wheel-travel angle is identical with the traction-force support angle, too.

For all these five characteristics to be freely dimensioned, the suspension must be based on a mechanism that is defined by five parameters, and this is the 'spatial' mechanism.

Spherical or planar wheel suspensions allow the free dimensioning of only four characteristics. Since a traction-force support angle is of no interest on non-driven or 'dead' axles, and since the wheel-travel angle is today of considerably less importance than elasto-kinematics, spherical or planar suspensions are adequate for dead axles, unless elasto-kinematic intentions justify the application of more sophisticated spatial systems.

## 3.4 Motion analysis of the wheel carrier

To analyse the motion of a planar or spherical wheel suspension, it would be sufficient to determine its instantaneous axes and the angular velocity of the wheel carrier. For a spatial mechanism the instantaneous shift would be required in addition.

With spatial mechanisms, work involving the instantaneous axes is rather complicated and not very helpful. It is much simpler to determine the instantaneous vector of the angular velocity $\omega_K$ of the wheel carrier and the velocity vector of a reference point on it. In the following, it is preferable to take the wheel centre M for the reference point, and hence its velocity vector $v_M$ for the reference velocity.

Kinematic Analysis of Wheel Suspensions   33

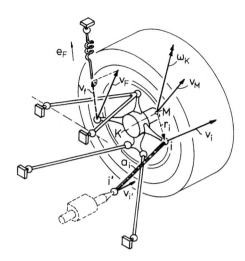

**Fig. 3.9**
State of motion of an
independent wheel suspension

With given vectors $\boldsymbol{\omega}_K$ and $\mathbf{v}_M$, the state of motion of an independent wheel suspension can be defined - **Fig. 3.9**. The velocity vector $\mathbf{v}_i$ of any point i of the wheel carrier K then follows using the connecting radius $\mathbf{r}_i$ from the wheel centre M to the point i as

$$\mathbf{v}_i = \mathbf{v}_M + \boldsymbol{\omega}_K \times \mathbf{r}_i \qquad (3.11)$$

If i and i' are the joints of a rod link, and, if the rod is assumed to be rigid, the distance i–i' must remain constant. The joint i' may have a velocity too - for instance because it is operated via a steering mechanism. Then the velocity vectors of both joints must have equal components in the direction of the rod vector $\mathbf{a}_i$,

$$\mathbf{v}_i \cdot \mathbf{a}_i = \mathbf{v}_{i'} \cdot \mathbf{a}_i \qquad (3.12)$$

From the equations (3.11) and (3.12) it follows that

$$(\mathbf{v}_M + \boldsymbol{\omega}_K \times \mathbf{r}_i) \cdot \mathbf{a}_i = \mathbf{v}_{i'} \cdot \mathbf{a}_i \qquad (3.13a)$$

or, expressed in terms of the vector components of the velocities and of the radius $\mathbf{r}_i$ (while the components of the radius are the differences between the position vectors of the points i and the wheel centre M)

$$\begin{aligned}
& \{v_{Mx} + \omega_{Ky}(z_i - z_M) - \omega_{Kz}(y_i - y_M)\}(x_i - x_{i'}) \\
+ & \{v_{My} + \omega_{Kz}(x_i - x_M) - \omega_{Kx}(z_i - z_M)\}(y_i - y_{i'}) \\
+ & \{v_{Mz} + \omega_{Kx}(y_i - y_M) - \omega_{Ky}(x_i - x_M)\}(z_i - z_{i'}) = \\
& = v_{i'x}(x_i - x_{i'}) + v_{i'y}(y_i - y_{i'}) + v_{i'z}(z_i - z_{i'}) \qquad (3.13b)
\end{aligned}$$

Since the suspension in Fig. 3.9 shows five rod links, two of which are combined to form a triangular link (with coinciding ball joints at the wheel carrier), each rod generates an equation of the type (3.13a or b).

The spring is a 'highly compliant' sixth link which completes the suspension as a statically determined system. Under wheel travel it is compressed by a velocity $v_f$ which is a component of the velocity $\mathbf{v}_F$ of its pivot point at the wheel carrier in the direction of the unit vector $\mathbf{e}_F$ of the spring axis. If $\mathbf{r}_F$ is the radius vector (not drawn) between the wheel centre M and the pivot point, the spring condition is:

$$(\mathbf{v}_M + \boldsymbol{\omega}_K \times \mathbf{r}_F) \cdot \mathbf{e}_F = v_f \tag{3.14}$$

or

$$(v_{Mx} + \omega_{Ky} r_{Fz} - \omega_{Kz} r_{Fy}) e_{Fx}$$
$$+ (v_{My} + \omega_{Kz} r_{Fx} - \omega_{Kx} r_{Fz}) e_{Fy}$$
$$+ (v_{Mz} + \omega_{Kx} r_{Fy} - \omega_{Ky} r_{Fx}) e_{Fz} = v_f \tag{3.15}$$

Five linear equations of the type (3.13) and the spring equation (3.15) form a system for the calculation of the respective three components of the vectors $\mathbf{v}_M$ and $\boldsymbol{\omega}_K$ of the wheel carrier. It is better to carry out the kinematic analysis separately for the bump-and-rebound wheel travel processes ($v_f \neq 0$, while all $v_{i'} = 0$) and for the steering process ($v_f = 0$ while $v_{i'} \neq 0$ at the track rod joint and $= 0$ at all the other joints at the vehicle body). The separation of these processes is very suitable for the easy determination of the suspension/steering characteristics - If elastic compliance is taken into account, all suspension joints may of course show displacements.

The motion analysis can also be done given one of the six wanted components of $\mathbf{v}_M$ and $\boldsymbol{\omega}_K$ - e.g. the velocity component $v_z$. Then only five equations for five components are to be solved. If the suspension contains a triangular link, and if the joint in its apex is taken as the reference point, the system of equations splits into two systems of two and three equations, the first of which can immediately be solved. An exact graphical solution of the problem **must** then be possible, because three linear equations correspond to the intersection point of three planes in space, and two equations to the intersecting line of two planes. Both problems can be solved graphically in a simple manner, as is well known. In practice, though, the graphical analysis of a spatial wheel suspension system requires the skilful application of descriptive geometry, but it was the only method of designing suspensions until the early 1970s.

With a true five-link suspension (as, for example, in Fig. 2.13 in Chapter 2) those five equations remain connected, and a graphical analysis would need to be executed in a five-dimensional space - which is somewhat impossible!

Kinematic Analysis of Wheel Suspensions    35

This is a primary reason why the development of true spatial five-link suspensions was so problematic before efficient computers became available.

The point was made in Section 3.1 that, in a statically defined system such as a wheel suspension with its spring, there is a close agreement between the balance of forces and the state of motion. In a plane, an instantaneous pole can be regarded as the instantaneous centre of rotation, which means that it is both the centre of the various velocities and the reference point for the balance of moments or forces.

The same can be said with respect to planar or spherical mechanisms as shown in **Fig. 3.10**. The instantaneous axis m may serve as a reference axis for calculating the balance of forces acting on the wheel carrier. The intersection point $P_1$ of the axis with a plane $\Pi_1$ (which here is the medial plane of the wheel) is stationary at the moment and forms the centre or 'pole' of the velocities of all points that lie in that plane. The projection $v_{M1}$ of the velocity vector $v_M$ of the wheel centre M in the plane $\Pi_1$ is therefore rectangular to the polar ray M–$P_1$. All forces in $\Pi_1$ are balanced if the sum of their moments about the pole $P_1$ is zero. A spatial force with a line of action intersecting the instantaneous axis m does not generate any moment at the wheel carrier, and the same is valid for a force in $\Pi_1$ with a line of action running through $P_1$.

In contrast with the instantaneous axes of planar and spherical mechanisms, the screw axis of a spatial mechanism is not suitable for a reference axis for moments and forces, or for a geometrical locus for 'poles'. Since the screw shows an axial shift coupled with rotation, forces that act on the screw axis but are not directed rectangularly to it will cause torques in the mechanism. Moreover, intersection points of a screw axis with a plane

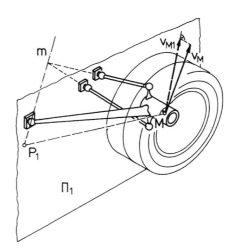

**Fig. 3.10**
Spherical suspension
with its instantaneous axis
and an instantaneous pole
in a particular plane

not perpendicular to that axis are not static but move in the plane according to the shift velocity of the screw and cannot therefore serve as 'poles'.

**Fig. 3.11** shows a screw axis s which intersects an x-y-plane $\Pi_{xy}$ at a point $D_S$. The point of a spatial body (e.g. the coupler of a mechanism or the wheel carrier of a suspension) that coincides with $D_S$ has a shift velocity **t** along s and, since the screw axis is not perpendicular to the plane, a velocity in $\Pi_{xy}$ – namely the projection or component of the shift **t** – and so cannot be regarded as a 'pole' in $\Pi_{xy}$. However, there is always (and in every plane) one point of the spatial body that is stationary at the particular moment – i.e. that point $P_{xy}$ showing opposing components (but of the same amount) of the circumferential velocity **u** of the screw and the shift velocity **t**. As **t** is parallel to the screw axis s in space, and therefore also the projections of both in the plane $\Pi_{xy}$, the connecting radius from the point $D_S$ to the wanted point $P_{xy}$ must appear rectangularly to the projection of the screw axis s. In Fig. 3.11 this point $P_{xy}$ is shown true-to-scale for a screw having an axis s, and its path is extended to a full helix H to facili-

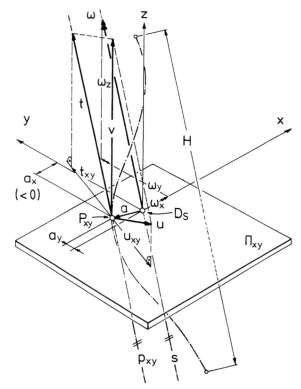

**Fig. 3.11** Instantaneous screw and an 'instantaneous pole' in a particular plane

Kinematic Analysis of Wheel Suspensions                                         37

tate visualization (while it should be recalled here that an instantaneous screw will normally not show a constant shift!). The components $u_{xy}$ of the circumferential velocity **u** and $t_{xy}$ of the shift velocity **t** cancel each other. A vertical velocity vector **v** is left which has no influence on the events in the plane $\Pi_{xy}$. The point $P_{xy}$ is the singular point in the plane $\Pi_{xy}$, the path of which is a rectangular trajectory of the plane: it may therefore serve as an 'instantaneous pole' in the plane. A force in $\Pi_{xy}$ whose line of action runs through $P_{xy}$ does not cause any torque on the spatial body. This becomes evident, too, in a three-dimensional view: the velocity vector **v** of the point $P_{xy}$ is rectangular to the plane $\Pi_{xy}$. A force in a particular plane, however, cannot generate a 'power' with a velocity vector rectangular to the plane.

In all planes that are parallel to $\Pi_{xy}$, corresponding 'poles' can be determined. Their geometrical locus is a line $p_{xy}$ parallel to the screw axis s.

The coordinates of $P_{xy}$ with reference to the intersection point $D_S$ of the screw axis are easy to calculate: with a (wanted) connecting vector **a** in the plane $\Pi_{xy}$, the velocity of $P_{xy}$ is $\mathbf{v}_P = \boldsymbol{\omega} \times \mathbf{a} + \mathbf{t}$. The premise $a_z = 0$ and the conditions $v_{Px} = 0$ and $v_{Py} = 0$ lead to

$$a_x = -t_y/\omega_z \qquad a_y = t_x/\omega_z \qquad (3.16\,a,b)$$

The equations for the pole coordinates in the y-z-and z-x planes derive from equation (3.16) by circular permutation. It should be reiterated that 'poles' defined in such a way may be applied only for the analysis of velocities or forces that act in their reference plane.

At the end of the analysis of a wheel suspension system the question might arise as to whether the mechanism is of the spatial type or not. A screw motion will always exist if the vector $\mathbf{v}_M$ shows a component in the direction of $\boldsymbol{\omega}_K$ (which component will then be the shift **t**). With the unit vector of the screw axis $\mathbf{e}_\omega = \boldsymbol{\omega}_K/|\boldsymbol{\omega}_K|$, the amount of the shift is $t = \mathbf{v}_M \cdot \mathbf{e}_\omega$. The circumferential velocity is then the difference between the total velocity and the shift, $\mathbf{u} = \mathbf{v}_M - t\mathbf{e}_\omega$, and its amount is $u = |\mathbf{u}|$. The helix angle follows from $\tan\alpha = t/u$. With the amounts u and $\omega_K$, the radius between the screw axis and the wheel centre is $r_M = u/\omega_K$ and, correspondingly, the pitch of the screw – which is equal at **all** points of the wheel carrier – is given by

$$H = 2\pi\, r_M \tan\alpha$$

$H < 0$ means a 'left-hand helix' – often the case with spatial rear-wheel suspensions. Even in a spatial mechanism the pitch may be zero in certain positions; the motion of spherical or planar mechanisms shows no pitch at all.

## 3.5 External and internal forces

In practice with spatial mechanisms, working with 'poles' and 'instantaneous axes' is not very rewarding. For the analysis of motion and forces it is easier to 'return to the origins' – i.e. to the instantaneous screw and the equation of energy.

If the velocity vector **v** of a point P of a mechanism is given, that means:
a) all external forces acting on P that are directed rectangularly to **v** do not generate any 'power' and are rigidly transferred by the mechanism to the 'fixed part' of the system.
b) all external forces acting on P that show a component in the direction of **v** generate a 'power' and must be balanced by the power of counter-acting forces.

In what follows, both criteria will frequently be applied to determine the 'characteristics' of the suspension or of the steering geometry. Moreover, the law of energy balance (b) leads to a simple method of calculating the reaction forces caused by external forces even on complicated spatial mechanisms.

**Fig. 3.12** shows an independent wheel suspension consisting of two triangular links, a rod link and a spring. The wheel carrier receives any forces $\mathbf{F}_j$ which act on any points j. A desired result may be the axial component $F_{i\alpha}$ of the reaction force $\mathbf{F}_i$ on the vehicle-side cylindrical rubber bush i of the lower triangular link – i.e. the force in the direction of the unit vector $\mathbf{e}_\alpha$ of the bush axis.

If all joints at the vehicle body are assumed to be held fast and the spring to be incompressible, but the joint i to be displaced in the direction of $\mathbf{e}_\alpha$ by a (virtual) 'velocity' $\mathbf{v}_{i\alpha}$, the resulting special state of motion of the suspension is clearly related to the displacement $\mathbf{v}_{i\alpha}$ and results in virtual velocities $\mathbf{v}_{j\alpha}$ of the working points j of the forces $\mathbf{F}_j$. Since at the particular moment a 'power' is possible only at the points j and the joint i, an equation of energy balance (or of 'virtual work' too) can be applied to the system: the overall sum of the scalar products of the force vectors and the corresponding velocity vectors must be zero.

$$F_{i\alpha} v_{i\alpha} + \sum_j (\mathbf{F}_j \cdot \mathbf{v}_{j\alpha}) = 0 \qquad (3.17a)$$

Given the components of the forces and the velocities, the axial force $F_{i\alpha}$ of the joint i follows from

$$F_{i\alpha} = - \sum_j (F_{jx} v_{j\alpha x} + F_{jy} v_{j\alpha y} + F_{jz} v_{j\alpha z}) / v_{i\alpha} \qquad (3.17b)$$

# Kinematic Analysis of Wheel Suspensions

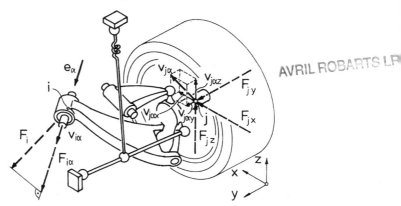

**Fig. 3.12** Determination of a reaction force by 'virtual work'

By a suitable choice of virtual velocity vectors, all internal forces of the suspension can be determined in the same manner.

This method has the advantages that the equations serving for the kinematic analysis can be applied also for the analysis of forces, and that the equations for static analysis no longer need to be established. Incidentally, work with 'velocities' makes it easy to define the suspension and steering 'characteristics', most of which are derivatives of forces or velocities. To demonstrate this, the method may be applied to a simple planar front-wheel suspension consisting of two longitudinal links and subjected to a braking force, **Fig. 3.13**.

The deceleration force causes a pitching moment which depends on the height of the vehicle's centre of gravity and its wheelbase, and which subjects the front wheel to an increase of loading at the expense of that on the rear wheel.

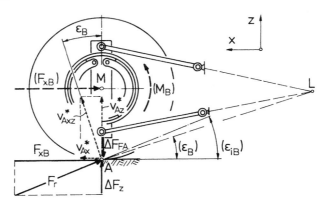

**Fig. 3.13** Simple planar front suspension with an imposed braking force

As is usual today, the friction brake, here shown as a shoe-type drum brake, acts between the wheel carrier and the wheel.

With the deceleration of the vehicle, the additional forces acting on the tyre contact point A are the 'dynamic' wheel-load transfer $\Delta F_z$ and the braking force $F_{xB}$, the resultant of the two being the skew force $F_r$. This resultant must be balanced by the spring-force change and the forces in the suspension links.

The spring force may already have been 'reduced' to the tyre contact point A - i.e. multiplied by the spring ratio - and is therefore indicated here as $\Delta F_{FA}$ ('A' for tyre contact point).

This suspension has a 'longitudinal pole' L which is the intersection point of the links; with wheel travel, the wheel carrier rotates about this pole.

The brake is the 'torque support' that transfers the braking torque $M_B$ (the product of braking force and tyre radius) to the wheel carrier by friction.

Since the braking torque is approximately independent of the rotary frequency of the wheel, it is possible as a simplification to imagine the same process with the vehicle at a standstill, with a 'locked' brake. In this case, the wheel and the tyre contact point A can be assumed to be fixed to the wheel carrier and, under wheel travel, to rotate with it about the pole L. The (virtual) velocity of the tyre contact point resulting from assuming a locked torque support (here the brake) may hereafter be indicated by a star (*). In a spatial mechanism, such a 'velocity' is then to be determined according to equation (3.11) using the radius $r_A$ from the wheel centre M to the contact point A:

$$\mathbf{v}_A^* = \mathbf{v}_M + \boldsymbol{\omega}_K \times \mathbf{r}_A \qquad (3.18)$$

In the actual example, the virtual velocity $v_A^*$ can simply be found as a rectangular vector to the polar ray AL. Its components allow calculation of the spring-force change $\Delta F_{FA}$ caused by both the braking force $F_{xB}$ and the wheel-load transfer $\Delta F_z$, and the 'balance of power' is $(\Delta F_z - \Delta F_{FA})v_{Az}^* - F_{xB}v_{Ax}^* = 0$, or, with $\tan \varepsilon_B = v_{Ax}^*/v_{Az}^*$,

$$\Delta F_{FA} = \Delta F_z - F_{xB} \tan \varepsilon_B$$

The angle $\varepsilon_B$ between the polar ray AL and the road-surface line is the 'braking-force support angle' (**3**) - a suspension characteristic of 'longitudinal dynamics' - and allows the actual 'anti-dive' property of the suspension to be evaluated. Obviously the spring-force change $\Delta F_{FA}$ diminishes as the support angle $\varepsilon_B$ increases.

Kinematic Analysis of Wheel Suspensions

In Fig. 3.13, the brake mechanism acts between the wheel and the wheel carrier, as already mentioned. The considerations made before are naturally no longer valid in the simple form, if the brake (or, more generally, the torque support) is fixed to the vehicle body and the torque is transferred to the wheel by a driveshaft. However, the basic approach used for the example in Fig. 3.13 gives a 'recipe' for the accurate calculation of a suspension characteristic in conformity with the virtual wheel travel or steering input: **Determine the velocity vector of the tyre contact point during virtual wheel travel or steering movement while assuming a locked torque support which is involved in the torque transfer, and apply the theorem of virtual work**.

The justification for assuming a 'locked' torque support on the basis of constant friction torque is surely not very satisfactory, especially with respect to traction torque, and was brought in mainly to introduce the method.

More credible, perhaps, will be the argument that, for the analysis of a quasi-static driving situation, the quasi-constant forward velocity of the vehicle can be subtracted from all internal velocities in order to get the residual relative velocities. This method sidesteps the problem of applying the rules of statics to a mechanism that contains elements rotating at high frequency. The 'fiction' of a locked torque support offers extraordinary simplifications for the use of the theory of virtual work; this is not obvious in the example given in Fig. 3.13 but will show up in more complicated mechanisms.

On nearly all independent suspensions, the torque support for the traction situation is mounted **not** on the wheel carrier but on the vehicle body, the traction torque being transferred to the wheel by a driveshaft with universal joints; moreover, an additional reduction gear may perhaps be installed at the carrier to modify the output torque.

Clearly the analysis of the torque transmitted from engine to wheel by articulating driveshafts (and which may be reduced by gearing in the wheel carrier) and the determination of the forces acting on wheel and suspension links, will afford a substantial application of the laws of statics. However, the rule of the 'locked' torque support points to a relatively simple solution.

To achieve a virtual velocity vector of the tyre contact point, assuming a locked torque support mounted to the vehicle body, it is of course necessary to bring the possible driveshafts and reduction gears into the frame.

## 3.6 Influence of driveshafts and reduction gears

To estimate the influence of a driveshaft on a wheel-suspension mechanism, the properties of driveshaft joints should first be considered.

The best-known universal joint is the 'cardan joint' - **Fig. 3.14**; it is a spherical device with four rotational axes which intersect at a central point Z and are arranged mutually at right angles. It has been used in clocks since the 16$^{th}$ Century. Fig. 3.14a shows the normal version, a 'fixed' joint, and 3.14b the extensible version, the 'pot joint'. If the shafts are deflected by an angle $\alpha$, the angular velocity and the torque vary twice per revolution. In a coordinate system that refers to shaft 1 - Fig. 3.14c - the unit vector $\mathbf{e_1}$ of yoke 1 has the components $e_{1x} = 0$, $e_{1y} = -\cos\varphi_1$ and $e_{1z} = \sin\varphi_1$, where $\varphi_1$ is the angle of rotation of the yoke, while the unit vector $\mathbf{e_2}$ of yoke 2, with its angle of rotation $\varphi_2$, has the components $e_{2x} = -\cos\varphi_2\sin\alpha$, $e_{2y} = \sin\varphi_2$ and $e_{2z} = \cos\varphi_2\cos\alpha$. As the two rotational axes of the coupler of the joint always form a right angle, the scalar product of their unit vectors must be zero, $\mathbf{e_1}\cdot\mathbf{e_2} = 0$, which leads at once to the ratio of the rotational angles:

$$\tan\varphi_2 = \tan\varphi_1 \cos\alpha \qquad (3.19)$$

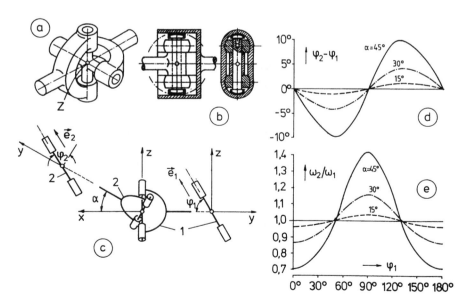

**Fig. 3.14** The cardan joint

# Kinematic Analysis of Wheel Suspensions

The ratio of the angular velocities is given by differentiation of equation (3.19):

$$\omega_2/\omega_1 = \cos\alpha/(1 - \sin^2\varphi_1 \sin^2\alpha) \tag{3.20}$$

The diagrams of Fig. 3.14 d and e show the difference $\varphi_2 - \varphi_1$, the 'gimbal error', and the ratio $\omega_2/\omega_1$ versus the rotational angle $\varphi_1$ for different deflection angles $\alpha$. According to the balance of power, the torque ratio derives from $M_{D1}/M_{D2} = \omega_2/\omega_1$.

Considering the significant deviation of the rotational angles of the shafts with increasing deflection, the cardan joint is not very well suited for duties as the wheel-side joint for steered wheels. In fast vehicles, too, the oscillations of torque and of angular velocity adversely affect the smoothness even on non-steered wheels. That is why so-called 'homokinetic (or constant-velocity) joints' have found favour on wheel driveshafts, except for special cases.

One of the few designs that have proved successful is the 'tripod joint', **Fig. 3.15**. It looks similar to the 'pot joint' (Fig. 3.14b) but has three radial arms displaced by 120° to one another. With a deflection angle $\alpha$, the shaft bearing the arms can adjust itself without any constraint in the housing of the other shaft, while its centre point circulates (with respect to the housing centre) in the opposite direction to the rotation of the shafts, by the eccentricity

$$e = \frac{r}{2}\left(\frac{1}{\cos\alpha} - 1\right) \tag{3.21}$$

with three times the angular velocity of the shafts. For a deflection angle of 20°, the eccentricity is approximately 3% of the radius r.

As the deflection angle of a tripod joint is limited, at least in its normal form, its application to the wheel-side joint of a steered suspension is rare. On the vehicle side of the assembly, though, it is used more frequently because its needle-roller bearings allow an easy axial motion even when carrying a torque, and thus softening the transfer of oscillations from the

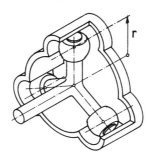

**Fig. 3.15** The tripod joint

engine to the wheel (and so to the steering, too). This characteristic is of special advantage in a vehicle with automatic transmission if this is held at rest by the brake with the drive engaged and the torque converter therefore slipping.

Constant angular velocity requires a symmetrical operation of the system. With cardan shafts a compensation of the angular deviations can be achieved by a 'Z' or 'W' arrangement, **Fig. 3.16**a or b, although the shaft tube itself rotates with an oscillating velocity. The 'double cardan joint' (Fig. 3.16c) is derived from the W arrangement for use as a wheel-side joint for steerable wheels. Compensation of the angular errors of course requires an exactly symmetrical disposition of the shafts (as for the rigid axles of trucks). If this is not possible (as in the case of independent suspensions), the joint needs a centring device.

True constant-velocity joints normally incorporate balls to transmit the torque, these being forced to stay in the plane of symmetry by the appropriate design of their slideways. Such an arrangement is shown in Fig. 3.16d. The six balls are inserted into slideways of the housing and of the driver of the two shafts. For a given deflection angle $\alpha$, the (virtual) cylinder surfaces that contain the ball centres intersect in the plane of symmetry, and, aided by opposing threads in the slideways, the balls line up along the intersection curve of the cylinders – an ellipse.

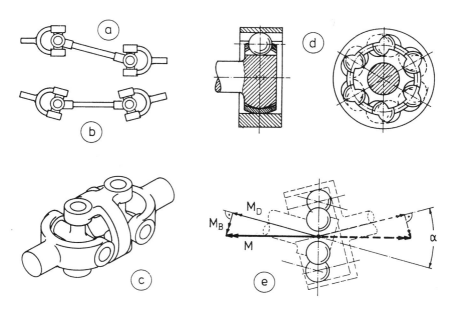

**Fig. 3.16** Constant-velocity arrangements

Kinematic Analysis of Wheel Suspensions 45

Torque deviation on a deflected joint causes reaction moments similar to those of a bevel gear. The resulting moment M of the transmission balls is a rectangular or 'normal' vector of the plane of symmetry - Fig. 3.16e. The torque $M_D$ on the shafts is one component of M, and the second is a bending moment $M_B$:

$$M = M_D/\cos(\alpha/2) \tag{3.22}$$

$$M_B = M_D \tan(\alpha/2) \tag{3.23}$$

For a constant-velocity joint, the direction and amount of the vector of the bending moment are constant in space for a constant deflection angle $\alpha$; the shaft therefore carries a rotating bending moment.

On a cardan joint, however, the vector of the resulting moment is directed rectangularly to the crossing axes of the spider, its amount and direction therefore changing twice per revolution and causing bending vibrations having double the rotational frequency of the shaft. These bending vibrations cause humming noises in the vehicle structure and, much more than the torque oscillations, are the main reason for the decreasing popularity of cardan joints in passenger cars.

A very simple and obvious constant-velocity arrangement is shown in **Fig. 3.17**a. The two shafts are connected by a central ball joint and, at equal distances from the ball joint, they carry turning joints on which equal-length triangular links are mounted; these triangular links are connected by a further ball joint. Both ball joints always remain in the plane of symmetry of the shafts and, since a moment can be transmitted only by a couple of forces acting on the joints, the vector of the resulting moment is always directed rectangularly to the plane of symmetry.

Clearly, this mechanism is not suited for high revolutions because of its unbalanced masses, but it does represent the basic principle of the well-known 'fixed' constant-velocity joint of Fig. 3.17b. Six balls, two of them seen in the section, are held in the plane of symmetry by means of a cage (not drawn). In contrast with the joint of Fig. 3.16d, the slideways of the balls are arranged not on cylinders but on spheres. The centre points of these spheres are displaced relative to the centre of the joint and correspond to the turning joints of Fig. 3.17a.

Fig. 3.17c shows a joint with parallel slideways in the medial section, an arrangement that allows axial movement between the shafts. On a deflected joint, the balls line up in the plane of symmetry $\Pi_S$, Fig. 3.17d. If one of the shafts is assumed to be fixed, the other can first swivel about the line of symmetry of the shafts with an angular velocity $\omega'$, while retaining the deflection angle of the shafts, and/or can change the deflection angle by

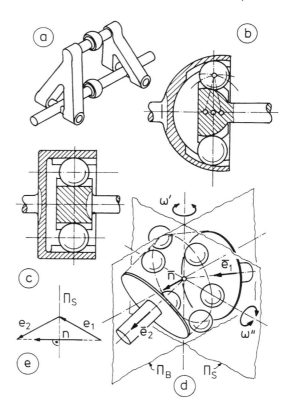

**Fig. 3.17**
Constant-velocity joints – functional description and degrees of freedom

an angular velocity $\omega''$. Since $\omega'$ and $\omega''$ lie in the plane of symmetry $\Pi_S$, the axis of the relative rotation of both shafts must also lie in $\Pi_S$.

The 'normal' vector **n** of the plane of symmetry $\Pi_S$ lies in the plane of deflection $\Pi_B$ which is defined by the unit vectors $\mathbf{e}_1$ and $\mathbf{e}_2$ of the shafts. A unit normal vector can be calculated by adding the unit vectors of the shafts and dividing the resulting vector by its amount:

$$\mathbf{n} = \frac{\mathbf{e}_1 + \mathbf{e}_2 \, \text{sgn}(\mathbf{e}_1 \cdot \mathbf{e}_2)}{|\mathbf{e}_1 + \mathbf{e}_2 \, \text{sgn}(\mathbf{e}_1 \cdot \mathbf{e}_2)|} \tag{3.24}$$

as indicated in Fig. 3.17e. The factor $\text{sgn}(\mathbf{e}_1 \cdot \mathbf{e}_2)$ shall ensure that the component of one unit vector in the direction of the other has a positive sign, inverting, if necessary, the sign of the vector $\mathbf{e}_2$.

On a constant-velocity joint, the vector of the resulting moment (see Fig. 3.16e) acts in the direction of the normal vector. The moments at the joints of a driveline cause reaction moments at the wheel suspension and the steering, as is generally appreciated.

Kinematic Analysis of Wheel Suspensions

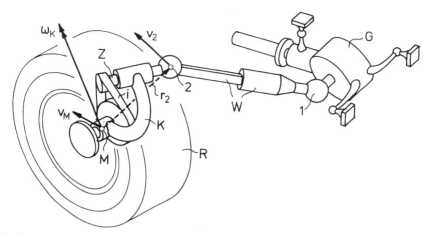

**Fig. 3.18** Driveline with a driveshaft and a reduction gear

**Fig. 3.18** shows schematically a suspension system and a driveline with a final-drive gearbox G (assumed to be fixed to the vehicle body), a driveshaft W with two constant-velocity joints 1 and 2 and the wheel carrier K of the wheel suspension – type is immaterial. The wheel carrier may in addition be provided with a reduction gear (represented here by a belt drive) which reduces the angular velocity of the wheel relative to that of the gearbox input shaft Z by a ratio i. If there is no reduction gear, the shaft Z coincides with the wheel stub-axle (at least for kinematic considerations).

For the analysis of the effects of the driveshaft on the suspension using the theorem of virtual work, the (constant) angular velocity of the final-drive output shaft will be eliminated, as already explained in section 3.5, assuming the engine, and with it the final-drive gear, to be 'locked'.

Provided that the motion of the wheel carrier K is already known and defined by the parameters $\boldsymbol{\omega}_K$ and $\mathbf{v}_M$, the velocity of the wheel-side driveshaft joint 2 follows by equation (3.11) as

$$\mathbf{v}_2 = \mathbf{v}_M + \boldsymbol{\omega}_K \times \mathbf{r}_2 \tag{3.25}$$

with the radius vector $\mathbf{r}_2$ connecting the wheel centre M and the joint 2.

Normally the vector $\mathbf{v}_2$ does not act rectangularly to the axis of the shaft W but shows an axial component in the latter's direction, which causes a lengthening or shortening of the shaft with a shift velocity $\mathbf{v}_s$, **Fig. 3.19**. Where the shaft's unit vector is $\mathbf{e}_W$, the shift velocity is

$$\mathbf{v}_s = (\mathbf{v}_2 \cdot \mathbf{e}_W) \mathbf{e}_W \tag{3.26}$$

**Fig. 3.19**

Relative velocities at the inboard driveshaft joint

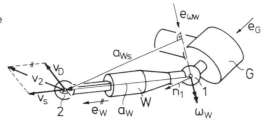

The circumferential velocity $v_D$ of the joint 2 that swivels the shaft W about the inboard joint 1 is the vector difference of the joint velocity $v_2$ and the shift velocity $v_S$:

$$v_D = v_2 - v_S \tag{3.27}$$

As mentioned above, the joint halves can have relative rotation about an axis that lies in the plane of symmetry of the joint – that is arranged rectangularly to the normal vector of the plane. The angular velocity vector $\boldsymbol{\omega}_W$ of the shaft W when swivelling about the joint 1 must therefore be rectangular to the normal vector $n_1$ of the joint's plane of symmetry. As the joint 1 cannot have any translational velocity, the vector $\boldsymbol{\omega}_W$ must also be rectangular to the vector $v_D$. The normal vector $n_1$ can be calculated by equation (3.24) using the unit vectors $e_G$ of the final-drive output shaft and $e_W$ of the driveshaft; then the unit vector $e_{\omega W}$ of the angular velocity $\boldsymbol{\omega}_W$ of the shaft W about joint 1, which must be rectangular to $n_1$ and $v_D$, can be determined by the vector products

$$e_{\omega W} = (n_1 \times v_D)/(|n_1 \times v_D|) \tag{3.28}$$

If $a_W$ is the connecting vector of the shaft joints 1 and 2, its component $a_{Ws}$ rectangular to $e_{\omega W}$ is the effective radius of the shaft joint 2 about the line of action of $e_{\omega W}$:

$$a_{Ws} = a_W - (a_W \cdot e_{\omega W})e_{\omega W} \tag{3.29}$$

and the amount of $\boldsymbol{\omega}_W$ follows from the amounts of $a_{Ws}$ and $v_D$:

$$\omega_W = v_D/a_{Ws} \tag{3.30}$$

The vector of the angular velocity of the shaft W is therefore

$$\boldsymbol{\omega}_W = \omega_W e_{\omega W} \tag{3.31}$$

Kinematic Analysis of Wheel Suspensions                                   49

The normal vector $n_2$ of the plane of symmetry of the shaft joint 2 at the wheel side can be determined from the unit vector $e_Z$ of the reduction-gear input shaft Z and the unit vector $e_W$ of the driveshaft W using equation (3.24), **Fig. 3.20**. Vector $\omega_{Z,W}$ of the relative angular velocity of the input shaft Z and the driveshaft W must lie rectangularly to the normal vector $n_2$. The resultant angular velocity $\omega_Z$ of the input shaft is on the one hand the vector sum of the angular velocities $\omega_W$ of the shaft W and $\omega_{Z,W}$ of the input shaft Z in relation to the drive shaft W, and on the other hand the vector sum of the angular velocities $\omega_K$ of the wheel carrier K and $\omega_{Z,K}$ of the input shaft Z with respect to the wheel carrier. However, the direction of the vector $\omega_{Z,K}$ is already known from the design of the wheel carrier and the unit vector $e_Z$, so $\omega_{Z,K}$ can be expressed by its amount and its unit vector:

$$\omega_{Z,K} = \omega_{Z,K} e_Z$$

and only the amount $\omega_{Z,K}$ is still unknown. These conditions lead to the vector equation $\omega_W + \omega_{Z,W} = \omega_K + \omega_{Z,K} = \omega_K + \omega_{Z,K} e_Z$ or $\omega_{Z,W} = \omega_K + \omega_{Z,K} e_Z - \omega_W$, and since $\omega_{Z,W}$ must be perpendicular to $n_2$ - i.e. $\omega_{Z,W} \cdot n_2 = 0$ - the amount of $\omega_{Z,K}$ follows as

$$\omega_{Z,K} = \frac{(\omega_W - \omega_K) \cdot n_2}{e_Z \cdot n_2} \tag{3.32}$$

If there is no reduction gear at the wheel carrier, the input shaft Z and the wheel stub-axle coincide; the angular velocity vector of the wheel R (regarded as a spatial body) is then $\omega_R = \omega_Z = \omega_K + \omega_{Z,K} = \omega_K + \omega_{Z,K} e_Z$.

In the case of a reduction gear with a ratio i (i > 0 for equal direction of rotation of shaft Z and the wheel, as in the belt drive in Fig. 3.20), the relative angular velocity of the wheel body R relative to the wheel carrier K

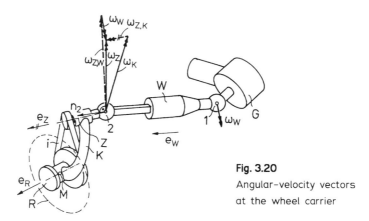

Fig. 3.20
Angular-velocity vectors at the wheel carrier

is, however, reduced by the ratio i, $\omega_{R,K} = \omega_{Z,K}/i$. With the unit vector $\mathbf{e}_R$ of the wheel axle, the absolute angular velocity of the wheel for a suspension with a driveshaft and a reduction gear (assuming the torque support to be mounted to the vehicle and to be locked) becomes

$$\boldsymbol{\omega}_R = \boldsymbol{\omega}_K + (\omega_{Z,K}/i)\mathbf{e}_R \qquad (3.33)$$

Via the velocity vectors $\boldsymbol{\omega}_R$ and $\mathbf{v}_M$, a virtual velocity vector of the tyre contact point A can be defined according to equation (3.18). A virtual velocity of a suspension with a driveline consisting of driveshafts and a torque support at the vehicle body may be characterized in what follows by a double star (**) in order to distinguish it from the case of a torque support at the wheel carrier. The virtual velocity of the tyre contact point then results in analogy to equation (3.18)

$$\mathbf{v}_A^{**} = \mathbf{v}_M + \boldsymbol{\omega}_R \times \mathbf{r}_A \qquad (3.34)$$

This vector $\mathbf{v}_A^{**}$ enables analysis of the forces of a wheel suspension in respect of the influence of driveshafts and possible hub reduction gears and to determine (for instance) a support angle $\varepsilon^{**}$.

The investigations of this section can be effected also through a computer-aided design program. In that case it is better to make sure that a 'universal joint' offered by the software really is a constant-velocity and not a cardan joint, since the latter would cause errors depending on the positions of its yokes. For the purpose of kinematic analysis, a constant-velocity joint can, on the other hand, easily be generated using the set-up given in Fig. 3.17a.

## 3.7 Attitude and motion of the wheel

Nearly all external forces act on the tyre contact point, and several of these are influenced to some extent by the wheel attitude (e.g. wheel camber or toe-in angle). Hence, the coordinates and the angular parameters of the wheel are needed for every position of the suspension.

The wheel axis will hereafter be defined by the wheel centre M and an auxiliary fixed point H, related to the wheel carrier - see **Fig. 3.21**. If the steering angle $\delta$ and the wheel camber $\gamma$ are given, and also the y-coordinate of the point H, its other two coordinates become

$$x_H = x_M + (y_M - y_H)\tan\delta \qquad (3.35)$$
$$z_H = z_M + (y_M - y_H)\tan\gamma/\cos\delta \qquad (3.36)$$

Kinematic Analysis of Wheel Suspensions 51

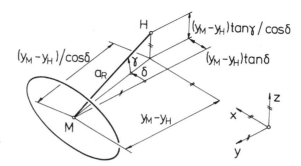

**Fig. 3.21**
Definition of the wheel axis
by an auxiliary point H

The auxiliary point H is treated like every point of the wheel carrier, and its velocity and displacement vectors are calculated according to equation (3.11).

Regarding the definition of the steering angle $\delta$ – see Fig. 3.21, or also Fig. 1.2 in Chapter 1 – it will be recalled that a positive toe-in angle corresponds to a negative steering angle.

For every position of the wheel suspension, the wheel camber $\gamma$ and the steering angle $\delta$ can be calculated from the coordinates of the wheel centre M and the auxiliary point H:

$$\sin\delta = (x_H - x_M)/\sqrt{(x_H - x_M)^2 + (y_H - y_M)^2} \qquad (3.37)$$

$$\sin\gamma = (z_H - z_M)/\sqrt{(x_H - x_M)^2 + (y_H - y_M)^2 + (z_H - z_M)^2} \qquad (3.38)$$

The tyre contact point A is *not* normally fixed to the wheel carrier, and its coordinates must be determined anew for every position of the suspension, **Fig. 3.22**:

$$\begin{aligned}x_A &= x_M + R\sin\gamma\sin\delta \\ y_A &= y_M - R\sin\gamma\cos\delta \\ z_A &= z_M - R\cos\gamma\end{aligned} \qquad (3.39\,a,b,c)$$

where R is the tyre radius. If the tyre contact point A is assumed to be temporarily fixed to the wheel carrier (as, for instance, in the case of a 'locked' brake on the wheel carrier – see Section 3.5), its position must always first be calculated from equation (3.39).

The displacement of the suspension mechanism may be investigated with the aid of software for the analysis of multi-body systems. To ascertain the suspension or steering characteristics, too, access to the velocity vectors is needed.

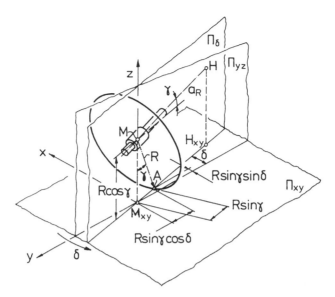

**Fig. 3.22** Determination of steering and camber angles, and of the coordinates of the tyre contact point

If no software is available, the equations given in Sections 3.4 to 3.6 can be used to calculate the suspension travel step by step. For a five-link suspension with its spring element, six equations of the type (3.13) allow determination of the vectors of the wheel-centre velocity and the wheel-carrier angular velocity, and through them the velocities of all other points of the suspension, using equation (3.11). With the vertical wheel travel or the displacement of a track rod divided into 'infinitesimal' steps of about 0.1–0.01 mm, very adequate results are possible. The displacements of all the points will then be added to their former positions to give the required figures.

This method not only is easy to program but has the further advantage of delivering precise velocity and displacement vectors of all points of the suspension for every step in the calculation; these vectors are of course needed for the analysis of forces (see Section 3.5) and for the determination of the suspension and steering 'characteristics' (Chapters 5 to 8). To get these characteristics it is better to carry out separate analyses of the bump-and-rebound wheel travel and the steering process.

# Chapter 4

# The Tyre

The transfer of forces and moments from the road surface to the vehicle is done by the tyre, an elastic torus made of natural and artificial rubber, reinforced by a synthetic-fibre or steel-wire envelope and filled with air under pressure, **Fig. 4.1**. The 'bead' 2 of the tyre is mounted on the shoulder of the rim 1, where the 'bead core' 3 which is usually made of steel wires holds it in position. The 'carcass' or cord casing 4 consists of plies of cords which are moulded around the bead core and stiffen the tyre to withstand the air pressure. On a 'bias-ply' or 'cross-ply' tyre (a), successive cord plies 4 cross at an angle of about 45° to the direction of rotation, while on a 'belted' or 'radial-ply' tyre they run directly from bead to bead, approximately at right angles to the direction of rotation, and a circumferentially stiff belt 6 reinforces the tyre under the tread 5.

The most important properties of a tyre are discussed in what follows. With a vertical wheel load $F_z$ the tyre complies by a deflection f, and its contact area to the road is enlarged, Fig. 4.1 (right). The balance of forces is preferably achieved by the product of the contact area and the air pressure (which remains nearly constant) yet involves the sidewalls as little as possible to avoid flexural energy which results in thermal stress and rolling resistance. The 'static effective radius' $R_{St}$ - the distance of the wheel

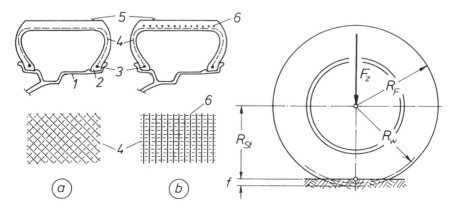

**Fig. 4.1** Tyre types (schematically)

centre from the road surface – is the difference between the 'designed radius' $R_F$ and the static deflection f.

Owing to the mode of load transfer, the tyre initially shows a mildly progressive 'spring characteristic' which then becomes almost linear with a spring rate which, on a passenger car, is about ten to twenty times that of a suspension spring.

When rolling along, the contact area shrinks longitudinally, so the forward velocity of the wheel centre is lower than the circumferential velocity of the tyre tread. The (virtual) effective rolling radius $R_W$ lies between the designed radius $R_F$ and the static radius $R_{St}$. On a radial-ply tyre, the deflection f is higher than that of a cross-ply tyre but, since its circumferential belt offers more resistance to shrinking, its rolling radius $R_W$ is nearer the designed radius $R_F$ than the static radius $R_{St}$, in contrast with a cross-ply tyre. The rolling radius is unrelated to the 'dynamic reinforcement' of the tyre tread at high speed, caused by centrifugal forces; that reinforcement leads only to an almost negligible lifting of the wheel centre. For this reason, and to avoid confusion, it is usual today to specify a 'rolling circumference' instead of a rolling radius.

Owing to the higher circumferential stiffness of its belt, a radial-ply tyre shows less contact-area shrinkage and thus lower longitudinal slip, with the advantages of lower rolling resistance, better tread durability and improved road grip especially where friction coefficients are low – e.g. on a wet surface or on snow or ice.

The load capacity of a tyre depends essentially on the enclosed air volume; 'low-profile' tyres therefore need an appropriately greater width than more ordinary ones with a height-to-tread-width ratio of 80%.

When a tyre is subjected to a lateral force $F_y$, the contact area is deflected sideways according to the lateral compliance of the sidewalls; it therefore assumes a 'slip angle' $\alpha$ with respect to the medial plane of the wheel – **Fig. 4.2** a – because the tread elements arriving in the contact area from the front are not yet fully influenced by the lateral deflection and try to align themselves in the centre plane, so the lateral deflection of the contact area increases towards the rear. The effective plane of the tyre is therefore deflected from the medial plane of the wheel by the slip angle $\alpha$. The line of action of the resulting force $F_y$ of the lateral shear diagram ($\tau$) (which has a triangular or trapezoidal shape) is displaced behind the wheel centre by the 'pneumatic trail' $n_R$. Consequently the lateral force $F_y$ generates a 'restoring moment' (or self-aligning moment) $M = F_y n_R$ which tries to swivel the medial plane of the wheel into its true direction of motion.

As the lateral deflection grows, the rearward elements of the contact area that are subjected to the greatest shear will exceed the friction limits

The Tyre 55

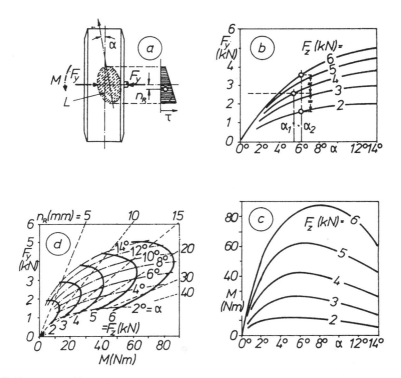

**Fig. 4.2** The tyre subjected to a lateral force:
a) lateral slip angle  b) lateral force  c) restoring moment  d) Gough diagram

and so will begin to slide back to the medial plane, thus causing energy losses and wear on the one hand and limiting the transferable lateral force on the other. Consequently, the lateral shear diagram begins to collapse at the rear end of the contact area and the restoring moment and the pneumatic trail decrease. The characteristic diagram of the lateral force $F_y$ versus the tyre slip angle $\alpha$ – Fig. 4.2b – therefore shows a degressive shape. An increase of wheel load does not lead to a proportional increase of the transferable lateral force, as the tyre begins to 'roll off' sideways over its shoulder and sidewall. The diagram relates to a rather high friction coefficient (13 in low-profile tyre).

The degressive dependence of the lateral force upon the slip angle as well as upon the wheel load is very important for the driving behaviour and for the tuning of the suspension and springing parameters, as will be clarified by the following numerical example, Fig. 4.2b. When a tyre receives a wheel load of 4 kN – i.e. an axle load of 8 kN – it generates a lateral force of 2.6 kN at a slip angle $\alpha_1$ of 5.3°, and the axle is able to balance a lateral force of 5.2 kN, or a lateral acceleration of 5.2/8 = 0.65 g in total. However, on a realistic vehicle with its centre of gravity above the road surface, the lateral acceleration causes a rolling moment which tries to tilt the vehicle's body sideways and leads to different wheel loads on the two sides. Assuming a centre of gravity height of 550 mm and a track width of 1400 mm, the wheel load transfer amounts to about 2 kN (unless the vehicle's other axle gains a share of the rolling moment - e.g. through an anti-roll bar). Thus the load on the outer wheel increases to about 6 kN while that on the inner wheel is only about 2 kN. Obviously a greater slip angle - $\alpha_2 = 6.2°$ - is now necessary in order to achieve a total lateral force $F_{ya} + F_{yi} = 5.2$ kN. For a given lateral acceleration, then, the resulting slip angle of an axle increases with increasing difference between the wheel loads. That difference is determined primarily by the height of the vehicle's centre of gravity and by the axle's track width, and it can be influenced by means of the suspension geometry (roll centre - see Chapter 7) and the springing system (anti-roll bars - see Chapter 5).

From the diagrams of the lateral force and the restoring moment versus lateral slip angle - Fig. 4.2b and c - we can derive the Gough diagram (Fig. 4.2d) which shows the lateral force versus the restoring moment, using the slip angle and the wheel load for parameters. The pneumatic trail can also be derived ($n_R$ = const. means $M/F_y$ = const.).

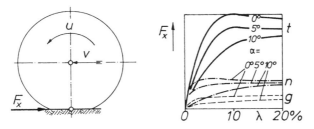

**Fig. 4.3** Longitudinal force and circumferential slip

A longitudinal force $F_x$ in the contact area (e.g. a braking or a traction force) causes circumferential slip, as the elements of the tread that enter the contact area are stretched by a braking force or shrunk by a traction force. The longitudinal force shows a degressive growth with the circumferential slip - **Fig. 4.3** (t = dry, n = wet and g = slippery road). As the maximum grip utilization by a lateral or a longitudinal force is limited in accordance with Newton's law of friction, combinations of lateral and longitudinal forces, or of slip angle and circumferential slip, can be calculated together in a diagram enveloped by the circle of the adhesion limit - **Fig. 4.4**a.

The Tyre                                                                                          57

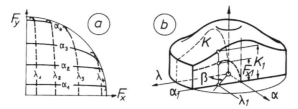

**Fig. 4.4** Lateral and circumferential forces

Fig. 4.4b combines lateral and longitudinal forces in a spatial diagram (**18**) where the sideways slip angle $\alpha$ and the circumferential slip $\lambda$ form the base area and the adhesion limit forms the height (with equal valuation at $\alpha = 90°$ and $\lambda = 100\%$).

The inclination of the wheel centre plane by a camber angle $\gamma$ - **Fig. 4.5** - generates a lateral 'camber force', as the tyre tread is deflected, shrinking differently on the two sides, and tries to roll off on a curved path similar to that of a cone; this effect is more pronounced on a cross-ply than on a radial-ply tyre because of the latter's greater circumferential stiffness. One degree of camber results in about one-sixth of the lateral force corresponding to one degree of slip angle for a cross-ply tyre, and in about one-twelfth for a radial-ply tyre. The adhesion limit is not increased by a camber force, but - for a given lateral force - the slip angle is reduced.

Wheel camber causes the tyre material to undergo thermal stress which rises with wheel load and driving velocity. The static camber for different loading conditions (driver alone or fully laden) therefore has to fall between narrow limits. For optimum handling behaviour, a slightly negative camber (combined perhaps with a toe-in angle) is advantageous since the tyres will then run preloaded by a small lateral force and so will respond more quickly to steering inputs. For optimum traction on ice, zero camber and toe-in angles are, of course, ideal, and such optimum conditions are naturally wanted for cornering, too. Here the rigid axle proves best in that it maintains a virtually constant camber angle. However, on a driven rigid axle (or 'live' axle) with integrated final-drive gear, the camber and toe-in angles must be zero with respect to the axle's half-shafts.

Compound suspensions are generally designed for appropriate camber characteristics. On the other hand, an independent suspension always

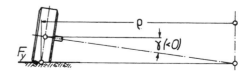

**Fig. 4.5** Camber force

needs a compromise between the demands of straight-ahead driving and of cornering. A progressive increase of camber angle with wheel bump travel might seem the ideal because it allows favourable negative camber at the outer wheel during extreme cornering in spite of providing moderate camber around static load. On double-wishbone suspensions, for example, this is easily realized, but it will normally lead to a rather high 'positive' camber (with respect to the road surface) at the inner wheel and, moreover, to a roll-centre geometry that provokes the well-known 'jacking-up' effect (see Chapter 7). The hoped-for advantages may thus be neutralized.

An effect similar to that of camber force can be obtained by a non-symmetrical construction of the tyre belt resulting in a constant lateral force (so-called 'conicity'). Conicity may be caused by tolerances but can also be the result of deliberate design in order to achieve a lateral preload force similar to a toe-in force.

Due to the arrangement of the under-tread cords one upon another, the top cord has the greatest influence and every tyre shows a 'ply-steer' effect - i.e. a constant sideslip angle without generating a lateral force; the opposite effect occurs during driving backwards.

Ply steer and conicity are not noticeable in driving unless their values change with variations in the traction or braking forces.

In the foregoing, only static tyre properties have been discussed. According to rule-of-thumb, a transient state becomes a steady state after approximately one revolution of the wheel.

Of all the constructional elements of the chassis, the tyre is more affected by production tolerances than any other - hardly surprising when one considers the materials used. Discrepancies of shape such as radial or lateral runout are quite easily discovered, but irregularities in the distribution of material or stiffness ('tyre non-uniformity') can be determined only with special testing machines. All these variations cause fluctuation of radial, lateral and circumferential forces, with amplitudes that vary with driving speeds and conditions. Contrary to the experience of normal mechanical engineering, amplitudes may be very substantial, even for high-order harmonics. The resulting high-frequency oscillations can embrace the audibility range, and an essential function of the rubber mounts in wheel suspension systems is to insulate the vehicle body against vibration and noise.

# Chapter 5

# Springs and Dampers

## 5.1 Purpose of springing and damping

The primary purpose of springing and damping is to protect the vehicle body from impact forces and high accelerations resulting from road-surface irregularities. How the spring force increases with wheel travel depends on the spring rate: the lower the spring rate, the lower the force caused when a wheel is lifted by a bump in the road. When being compressed or extended, a spring temporarily stores or releases energy; the vehicle body receives only the spring-force fluctuation.

Between the vehicle, the suspension and the road surface are the 'unsprung masses' - i.e. the wheel with the wheel carrier and a proportion of the suspension linkage. These unsprung masses would undergo high accelerations from road irregularities if the tyre did not itself represent a spring. To avoid continuing oscillations of the vehicle, excited by force or energy transfer, 'shock absorbers' or 'dampers' are provided which cause the oscillations to diminish. The dampers also serve to improve the road-holding since they prevent the wheels from bouncing off the road surface after bumps.

On the other hand, the dampers' resistance to sudden wheel movements does result in impact forces. Harmonization of springs and dampers therefore always requires a compromise between the demands of handling and riding comfort.

As comprehensive literature is available dealing with vehicle springing theory (**16**), only the most important modes of vibration will be discussed in what follows. Since the same is valid for springs, the main focus here will be on the special features of their interaction with the suspension mechanism.

## 5.2 Vehicle vibrations

### 5.2.1 Basic system

The simplest model of an oscillating system is shown in **Fig. 5.1** - a vehicle mass m supported by a spring. The restoring force F of the spring increases with wheel travel s by a linear function according to the spring rate c:

$$F(s) = cs \qquad (5.1)$$

see Fig. 5.1b. When carrying the mass m, the spring reacts by a static compression

$$s_0 = mg/c \qquad (5.2)$$

where g is the acceleration due to gravity. The static position of equilibrium represents, of course, the initial point of any vibration. When displaced from the static position by a travel z, the spring generates a restoring force which accelerates the mass m:

$$m\ddot{z} = -cz$$

The solution of this differential equation is a 'harmonic' function, for example

$$z(t) = a\cos(\omega_0 t)$$

and with the amplitude a, the time t and a (still unknown) parameter $\omega_0$ this results in

$$-m\omega_0^2 a\cos(\omega_0 t) = -ca\cos(\omega_0 t)$$

or

$$\omega_0^2 = c/m \qquad (5.3)$$

If T is the time of vibration, one cycle of vibration is completed for $\omega_0 T = 2\pi$, and

$$\omega_0 = 2\pi/T \qquad (5.4)$$

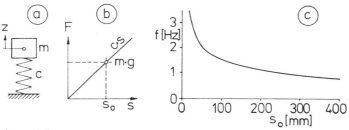

**Fig. 5.1** Undamped linear vibrating system

# Springs and Dampers

is the 'angular frequency' of the 'natural' vibration. The natural frequency in 'hertz' – i.e. cycles/s – or Hz is

$$f_0 = 1/T \tag{5.5a}$$

and the number of oscillations in [min$^{-1}$] is

$$n_0 = 60 f_0 \approx 10 \omega_0 \tag{5.5b}$$

From the equations (5.4) and (5.5) we get

$$\omega_0 = 2\pi f_0 \tag{5.6}$$

or, with the static compression, see equation (5.2),

$$\omega_0 = \sqrt{g/s_0} \tag{5.7}$$

With a linear spring, there is consequently an immediate relationship between the static compression $s_0$ caused by the carried weight and the system's natural frequency, Fig. 5.1c. Most acceptable frequencies for a human being range between about 0.7 and 2.0 Hz which correspond to static compressions of around 500 to 60 mm. On 'comfortable' vehicles, wheel travel of about ±100 mm and more must be provided; if the natural frequency of the springing system is high and the static compression small, this may lead to a total unloading of the springs (i.e. lifting off the mounting if not fixed to it) or of the wheel (i.e. jumping off the road surface).

Vibrations are excited either by an oscillating force $F_E(t)$ acting on the mass (direct excitation), **Fig. 5.2**a, or by a forced displacement $z_E(t)$ of the spring mounting (indirect excitation), Fig. 5.2b. The frequency $\omega$ of excitation is usually different from the natural frequency $\omega_0$.

For a 'harmonic' excitation force $F_E \cos(\omega t)$, the equation of motion of the directly excited mass is $m\ddot{z} + cz = F_E \cos(\omega t)$, and with the statement $z(t) = a \cos(\omega t)$ this leads to

$$-m\omega^2 a + ca = F_E \tag{5.8}$$

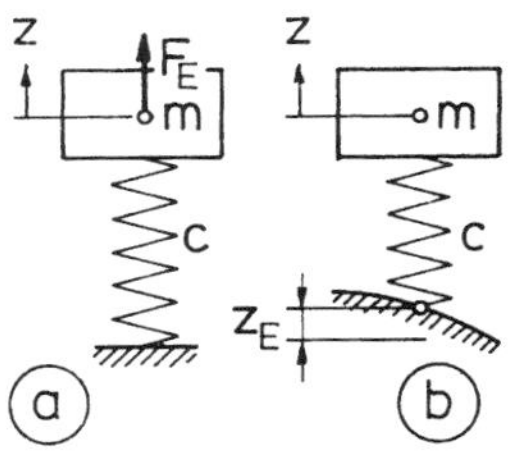

**Fig. 5.2**
Linear system
a) with direct excitation
b) with indirect excitation

For an exciting displacement $z_E(t) = h\cos(\omega t)$ with amplitude h, the equation of the indirectly excited vibration is $m\ddot{z} + c[z - h\cos(\omega t)] = 0$ or, with the statement $z(t) = a\cos(\omega t)$,

$$-m\omega^2 a + c(a - h) = 0 \tag{5.9}$$

After dividing the equations (5.8) and (5.9) by the mass m and the spring rate c, replacing the mass and the spring rate by the natural frequency $\omega_0$ according to equation (5.3), and introducing the 'frequency ratio'

$$\eta = \omega/\omega_0 \tag{5.10}$$

the amplitudes of the excited mass are given by

$$a = (F_E/c)/(1 - \eta^2) \tag{5.11}$$

for the directly excited and

$$a = h/(1 - \eta^2) \tag{5.12}$$

for the indirectly excited vibration.

For a linear system, the 'magnifying rates' or 'magnifiers' $ac/F_E$ and $a/h$ of the amplitudes of response and of excitation obviously depend on the frequency ratio $\eta$ only. For $\eta < 1$ the amplitudes of response and of excitation show the same phase position or the same sign ('subcritical state'); for $\eta = 1$ the response amplitude increases to infinity ('resonance') and for $\eta > 1$ the vibrating mass moves in opposition to the excitation ('supercritical state').

As shown by Fig. 5.1c, a linear springing system on a vehicle with varying load has the disadvantage of a decreasing natural frequency with growing static deflection or load. If a constant natural frequency with varying load is desired, the quotient $c/m = \omega_0^2$ must be constant. The spring rate c is the derivative of the spring force F and the spring deflection s, $c = dF/ds$, and the spring characteristic follows from the integral $F(s) = \int c(s)\,ds$. With the (varying) static load $F = mg$, the spring rate can be written $c = m\omega_0^2 = F\omega_0^2/g$; consequently the differential equation of the spring characteristic is $dF/ds = \omega_0^2 F/g$ and its integral $\ln F - \ln F_0 = \ln(F/F_0) = \omega_0^2 s/g$ (with the 'integration constant' $F_0$). The spring characteristic with a constant natural frequency at any load is therefore defined by the equation

$$F(s) = F_0 e^{(\omega_0^2 s/g)} \tag{5.13}$$

and is shown in **Fig. 5.3**.

# Springs and Dampers

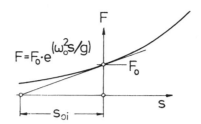

**Fig. 5.3**
Spring characteristic for constant natural frequency

Representing the local derivative of the spring force F versus the deflection s, the rate c of a non-linear spring is valid only for a very small amplitude where a linearization is permissible. The static compression $s_0$ in equation (5.7) has to be replaced here by the subtangent $s_{oi}$ (which in the case of Fig. 5.3 is constant for any travel s owing to the constant natural frequency). For large amplitudes, the spring rate is no longer sufficient to determine the vibration frequency. The positive or negative amplitudes meet the condition of equal energy absorption and are, for instance, to be calculated by integrating the vibration energy U versus spring deflection using the spring force F: $dU = F ds$ or $U = \int F \, ds$.

In practice, 'progressive' spring characteristics are generally provided (though not exactly according to Fig. 5.3) to avoid a sensible decrease of the natural frequency with increasing load as well as heavy bump-stop contact on bad roads, and are normally realized by the superimposition of the characteristics of a primary spring and an auxiliary spring (preferably of the rubber-elastic type). Another possibility is to provide a variable 'kinematic springing rate' by appropriate design of the wheel suspension (see Section 5.4).

Theoretically, undamped vibrations will never fade away entirely. Since this is not desirable on vehicles, vibration dampers are applied. However, such dampers introduce additional forces and may also affect the riding comfort adversely. Dampers convert viscous energy into thermal energy, and this may result in slightly increased rolling resistance on bad roads.

The oldest damping measure, normally undesired, is friction. The friction force works in the same sense as the decelerating force of the spring, opposite to the accelerating force, thus distorting the 'harmonic' vibration and increasing the acceleration; moreover, impact forces lower than the friction force are directly transmitted to the vehicle body without any reaction of the spring. It follows that friction should be kept as low as possible. The optimal damping device for harmonic vibrations is a viscous damper with damping forces proportional to the damper velocity. As the velocity and acceleration maxima are displaced by a phase angle of 90°, the

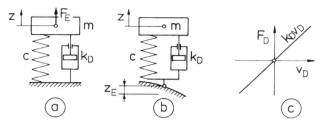

**Fig. 5.4** Damped linear vibration system

maximum damper force and the maximum accelerating force (i. e. the spring force!) are displaced by the same phase angle and thus cannot add together.

**Fig. 5.4** shows schematically a simple vibrating system with a mass, a linear spring and a viscous damper. The damping force $F_D$ is assumed to be proportional to the velocity $v_D$ of the damper piston,

$$F_D = k_D v_D$$

with the 'damper rate' $k_D$. The respective equations of motion for the directly and indirectly excited vibrations (Fig. 5.4a and b) are now extended by a 'velocity term':

$$m\ddot{z} + k_D \dot{z} + cz = F_E \cos(\omega t)$$
and
$$m\ddot{z} + k_D[\dot{z} + h\omega \sin(\omega t)] + c[z - h\cos(\omega t)] = 0$$

Since these two equations contain sine and cosine terms, a statement $z(t) = a_1 \sin(\omega t) + a_2 \cos(\omega t)$ enables them to be solved. With $\omega_0^2 = c/m$ according to equation (5.3) and the 'degree of damping'

$$D = k_D/(2m\omega_0) = k_D/(2\sqrt{mc}) \tag{5.14}$$

(a non-dimensional characteristic for the properties of the damped system which amounts to about 0.2 - 0.3 on road vehicles) each of the equations of motion results in two equations for the points of time $\omega t = 0$ and $\omega t = \pi/2$. The two amplitudes $a_1$ and $a_2$ are displaced in phase by 90° and therefore can be added geometrically. This leads to the magnifying rates (**16**)

$$a = (F_E/c)/\sqrt{(1 - \eta^2)^2 + 4D^2\eta^2} \tag{5.15}$$

for the directly excited and

$$a = h\sqrt{1 + 4D^2\eta^2}/\sqrt{(1 - \eta^2)^2 + 4D^2\eta^2} \tag{5.16}$$

for the indirectly excited system.

Springs and Dampers

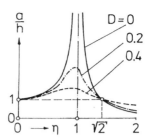

**Fig. 5.5** Magnifying rates of a linear system

**Fig. 5.5** shows the magnifying rate a/h or 'resonance curve' of an indirectly excited system (the more interesting for wheel suspension design) versus the frequency ratio η with the degree of damping D as parameter. For $\eta = \sqrt{2}$, the magnifier a/h holds the value '1' independently of D. In the region of $0 < \eta \leq \sqrt{2}$ the responding amplitude a is higher than the exciting amplitude h, while beyond $\eta = \sqrt{2}$ it is lower. The amplitude of resonance (η = 1) decreases with increasing degree of damping D. In the region of $0 < \eta \leq \sqrt{2}$ the amplitude a decreases and beyond $\eta = \sqrt{2}$ increases with D. A degree of damping D = 0 means 'no damping', in which case equation (5.16) changes into equation (5.12).

Genuinely speed-sensitive damping is an ideal case that is never achieved in practice. **Fig. 5.6** illustrates the effects of different damping mechanisms by examples of the attenuating process of a system consisting of a mass of 300 kg supported by a spring with a rate of 12.1 N/mm. The mass is assumed to be lifted by 100 mm and then released. For D = 0, the system vibrates continuously; with viscous damping (D = 0.3), however, the vibration starts to fade quite quickly and thereafter fades more slowly (as indicated, it will theoretically never finish, and its amplitude decreases exponentially). Damping by an amplitude-sensitive friction force r generates a behaviour similar to that of viscous damping, but causes higher velocity and acceleration amplitudes. It is even more disadvantageous than damping by a constant friction force $F_R$. Typical of the latter, though, is a final 'sticking'

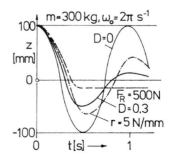

**Fig. 5.6**
Attenuating processes with different damping mechanisms

within an amplitude range $\pm F_R/c$, which means that exciting forces lower than $F_R$ will be transmitted immediately to the mass.

### 5.2.2 System of two masses

The 'unsprung' masses (i.e. primarily the wheels and wheel carriers) come up to about 8-10% of the vehicle mass on passenger cars and in effect are sprung with a spring rate that results from the parallel arrangement of the suspension spring and the tyre; the latter is the dominant factor and its rate is about tenfold the higher of the two. According to equation (5.3), the natural frequency of the unsprung mass may therefore be assumed about tenfold higher than that of the vehicle as a whole.

**Fig. 5.7** shows a system of two masses which simulate the vehicle and unsprung masses. The damping properties of the tyre being almost negligible, only the wheel mass $m_1$, the tyre spring rate $c_1$, the vehicle mass $m_2$, the suspension spring rate $c_2$ and the damper rate $k_2$ are regarded here. The magnifying rate $a_2/h$ of the vehicle mass shows two resonance frequencies, one of them for $\eta_2 = 1$ (vehicle resonance) and one for $\eta_2 \approx 10$ (which means, too, $\eta_1 = 1$ or wheel resonance). The amplitude $a_P/(c_1 h)$ of the wheel-load fluctuation is also drawn. The amplitude of vehicle movement decreases remarkably with increasing degree of damping D while the wheel-load amplitude is less influenced and the resonance frequency area is enlarged.

The popular dictum that stiffer damping results in better roadholding (though at the expense of ride comfort) is not entirely valid. For a given spring rate, an increasing degree of damping D leads of course to a transitory reduction of wheel-load oscillations while adversely affecting ride comfort; as indicated schematically in **Fig. 5.8** (according to (12)), however, good roadholding properties reach a maximum and finally drop, in line with

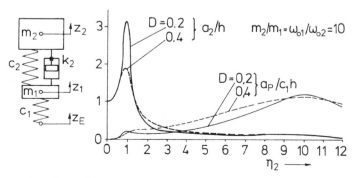

**Fig. 5.7** Damped system of two masses

Springs and Dampers

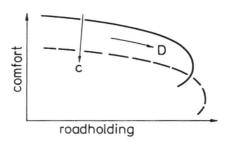

**Fig. 5.8**
Schematic conflict diagram of ride comfort and roadholding

a drastic loss of comfort. The reason is easy to envisage: with excessive damping the spring can no longer force the wheel to follow the road irregularities, so the wheel begins to jump from one peak to the next. 'Sporting' (stiffer) tuning of the springs (dashed curve) offers only a small improvement in roadholding, but with further loss of comfort.

According to equation (5.14), the degree of damping D is proportional to the damper rate $k_D$ and the reciprocal of the square-root of the spring rate c, $D \sim k_D/\sqrt{c}$. For a given damping level or vibration behaviour of the vehicle, improved riding comfort (or a reduction of the comfort-damaging damping forces) can best be achieved by lowering the spring rate and thus the natural frequency. A 'soft' and comfortable setting of the springing system thus allows the desired degree of damping to be attained in spite of low damping forces, with significant advantages in comfort, noise isolation and - last but not least - fuel consumption.

The evaluation of vibration comfort in vehicles has long been an object of intensive research. Impairment of comfort is noticed according to the frequency band: from low to high frequencies, first amplitudes, then velocities and lastly accelerations are cited as the main criteria. Acoustic disturbances are of course inseparably involved, and visual recognition of the road surface state (and thus of the cause of disturbances) may soften the subjective impression of discomfort.

### 5.2.3 Pitching and rolling

The considerations in the preceding section of the vehicle's vibrational behaviour were based on the simplifying assumption that the mass of the vehicle body could be split into partial masses above the axles or the wheels - something that is not always correct.

Therefore some investigations may finally have to be made into the vibrational characteristics of the complete vehicle while ignoring the influence of the unsprung masses (i.e. the wheels etc.), the tyre spring rates and the dampers.

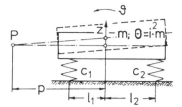

**Fig. 5.9** Pitching vibration

In the side view of a vehicle, **Fig. 5.9**, both the springs of an axle can be treated in combination (i.e. as a 'parallel arrangement' of two springs), since the vehicle's design is usually symmetrical in cross-section.

The vehicle body has two degrees of freedom in side view - bouncing in the z direction and pitching about a transverse axis by the pitch angle $\vartheta$. With the distances $l_1$ and $l_2$ of the axles from the centre of gravity, the wheel travels of a generalized and combined pitching and bouncing movement are $s_1 = z - \vartheta l_1$ at the front (1) and $s_2 = z + \vartheta l_2$ at the rear axle (2). If m is the vehicle mass, i the radius of gyration and thus $\Theta = i^2 m$ the moment of inertia, the equations of motion follow as

$$m\ddot{z} + z(c_1 + c_2) - \vartheta(c_1 l_1 - c_2 l_2) = 0 \qquad (5.17)$$
$$i^2 m \ddot{\vartheta} + \vartheta(c_1 l_1^2 + c_2 l_2^2) - z(c_1 l_1 - c_2 l_2) = 0 \qquad (5.18)$$

The statements $z = a\cos(\omega t)$ and $\vartheta = \vartheta_0 \cos(\omega t)$ lead to the equation for the two natural frequencies:

$$i^2 m^2 \omega^4 - m\omega^2 [c_1 l_1^2 + c_2 l_2^2 + i^2(c_1 + c_2)] + c_1 c_2 (l_1 + l_2)^2 = 0 \qquad (5.19)$$

In the equations (5.17) and (5.18), there occur identical coefficients $(c_1 l_1 - c_2 l_2)$ of the variables $\vartheta$ and z. If these 'coupling terms' were to be zero, the two equations would become independent of each other and would immediately deliver the natural bounce and pitch frequencies. Accordingly, the condition for the 'decoupling' of pitch and bounce is

$$c_1 l_1 = c_2 l_2 \qquad (5.20)$$

and looks similar to the condition of balance of two forces or masses. The front and rear springs may then be assumed to be combined into a 'substitute spring' with a rate $c = c_1 + c_2$ acting on an 'elastic centre' which here coincides with the vehicle's centre of gravity. In this special case of decoupled springs, the natural bounce and pitch frequencies follow from

$$\omega_z^2 = (c_1 + c_2)/m \qquad (5.21)$$
$$\omega_\vartheta^2 = (c_1 l_1^2 + c_2 l_2^2)/(i^2 m) \qquad (5.22)$$

Springs and Dampers

If $\omega_z$ and $\omega_\vartheta$ were to be equal, only one natural frequency would appear in the side view of the vehicle, and tuning of the vehicle's vibrational behaviour would be easier. The condition $\omega_z = \omega_\vartheta$ leads to

$$i^2 = l_1 l_2 \tag{5.23}$$

well known from the 'physical' or 'reversible' pendulum which can be suspended either way round at two points, meeting the equation (5.23) and showing the same natural frequency in both cases.

Normally the solution of the equations (5.17) and (5.18) will result, however, in two forms of combined pitching and bouncing, one of them dominated by the bouncing motion and the other by the pitching one. The quotient of the amplitudes of both vibrations is the distance $p = z(t)/\vartheta(t)$ of a 'pole' P about which the vehicle body pivots in side view. As the accelerations are proportional to the amplitudes (except for their signs), the pole distance p can be directly calculated using the two equations. The special case $i^2 = l_1 l_2$ according to equation (5.23) leads to

$$p^2 - p(l_1 - l_2) - l_1 l_2 = 0$$

with the two solutions $p_1 = l_1$ and $p_2 = -l_2$, meaning that the vehicle body pivots about either the front or the rear axle. With the partial masses $m_1 = m l_2/(l_1 + l_2)$ and $m_2 = m l_1/(l_1 + l_2)$ of the vehicle at the front and rear axles, their contribution to the pitching moment of inertia is $\Theta = m_1 l_1^2 + m_2 l_2^2 = m l_1 l_2$, and this is the total moment of inertia of the vehicle - see equation (5.23)! The 'real' vehicle characterized by a mass and a moment of inertia can then be replaced by two mass points at the front and rear axles. Equation (5.23) defines the 'decoupling of masses', and the two points meeting this equation are 'percussion points'. An impact on axle 1 does not cause any reaction force on axle 2 and *vice versa* (**1**)(**16**).

Only in the case of 'decoupled masses' is it physically correct to carry out the simplification of splitting the vehicle mass into two mass points at the axles.

Pitching is excited by acceleration or braking, but above all by road irregularities. These act on the rear axle later than on the front axle by a delay which follows from the wheelbase l and the driving velocity v: $\Delta t = l/v$. **Fig. 5.10** shows the vibrating behaviour of a vehicle which runs over an obstacle shaped like half a sine-wave. The vehicle is assumed to have 'decoupled masses' according to equation (5.23), as being appropriate for passenger cars of medium size. The pitch angle $\vartheta$ follows from the amplitudes $z_v$ and $z_h$ of the front and rear axles and from the wheelbase. If the two axles are sprung with equal natural frequencies (here 80 min$^{-1}$),

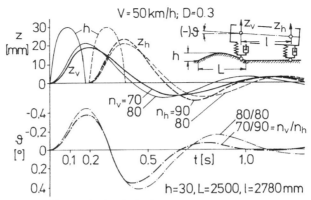

**Fig. 5.10** Pitching caused by a single obstacle

the delay $\Delta t$ will last as long as the combined bounce-and-pitch vibration. However, if the front axle has a lower frequency than the rear one (here 70 to 90 min$^{-1}$), the 'quicker' vibration of the rear axle enables it to catch up on the front axle and achieve much the same phase position at the beginning of the second half-period of the vibration, where the pitch angle has almost faded away.

In practice, a compromise between the demands of decoupled pitch and bounce vibrations and of minimized pitch on single obstacles is usual by arranging for the natural frequency of the front axle to be about 5-20% lower than that of the rear axle.

Small and light passenger cars often have quite a long wheelbase for the vehicle length, in order to gain a greater cabin length, and so do not meet the equation (5.23). This leads to a pitch frequency higher than the bounce frequency, which is very disagreeable for passengers. By means of a longitudinal compound springing system, though, the pitch frequency can be lowered to an acceptable level while retaining a suitable bounce frequency. The torsion bar in **Fig. 5.11** connects the wheel suspensions of one side like a scale-beam, while the effective spring forces at the wheels can be fixed – even differently – by the choice of the lever radii. The vehicle of course needs at least one 'directive' spring, $c_1$, and perhaps another, $c_2$, to ensure a definitive position for each external load. The bounce rates of the front and rear axles follow from their directive-spring rates and from their share of the compound spring rate; the latter does not influence the pitch rate as long as it acts on both axles with equal force, like a true scale-beam.

Because the pitch spring rate is lowered by a longitudinal compound springing system of this kind, anti-dive and/or anti-squat devices at the suspensions will be advantageous (see Chapter 6).

Springs and Dampers

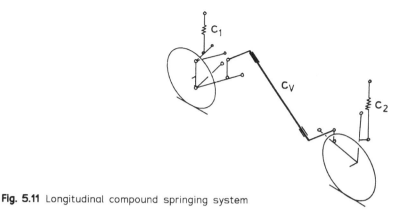

**Fig. 5.11** Longitudinal compound springing system

During a 'roll vibration', the vehicle can be assumed to be symmetrical in cross-section. With the track width b, **Fig. 5.12**a, the opposed wheel travels of an axle at a roll angle $\varphi$ are $s = (b/2)\varphi$ and the spring force changes are $F = c(b/2)\varphi$; thus, the restoring moment is $M = Fb = c(b^2/2)\varphi$. The roll spring rate $c_\varphi = M/\varphi$ of an axle is therefore

$$c_\varphi = c b^2/2 \tag{5.24}$$

When calculating the overall roll spring rate or 'roll stiffness' of a vehicle, it can be assumed that the front and rear axles normally show different roll spring rates, quite apart from having different suspension systems. These measures serve to transform the roll moment during cornering into desirable wheel load transfers between the front and rear wheels, and thus influence the driving behaviour. The roll rates of the axles are, so to speak, used for a deliberate distribution of the roll moments. As will be shown in Chapter 7, the front axle of a passenger car is normally given a higher roll moment than the rear axle.

This requirement is contrary to the usual specifying of the springs for optimal pitching behaviour - see Fig. 5.10 - where the rate of the front springs is lower than that of the rear.

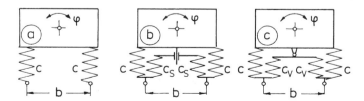

**Fig. 5.12** The roll spring rate
    a) bounce spring   b) stabilizer spring   c) transverse compound spring

To achieve a higher spring rate in roll at the front axle, despite the softer bounce rate, stabilizer springs ('anti-roll bars') are adopted; they react to antimetric travel of the two wheels only and have no effect on symmetrical travel. If the front roll stiffness should become intolerable in respect of comfort when set up according to the demands of driving behaviour, an alternative solution is to reduce the rear roll rate by incorporating a transverse compound spring which takes over a share of the bounce rate but does not affect the roll rate. Figures 5.12b and c respectively show schematically the functions of these two solutions ($c_S$, $c_V$ = stabilizer and compound spring rates per wheel).

In practice, stabilizer springs are best realized by torsion bars which are easily installed and attuned, **Fig. 5.13**. In Fig. 5.13b the bar serves also as a suspension link on the MacPherson principle (**22**). Three examples of transverse compound springs are shown in **Fig. 5.14**, namely a transverse leaf spring (a) which acts like a scale-beam, a coil spring mounted between two wheel suspensions (b) and a z-shaped torsion bar (c), the counterpart of a stabilizer bar.

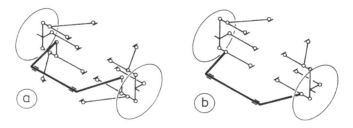

**Fig. 5.13** Stabilizer (anti-roll) bars

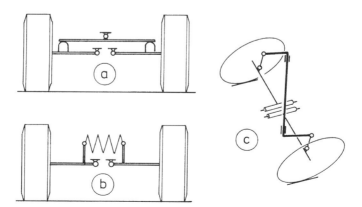

**Fig. 5.14** Transverse compound springs

Springs and Dampers

## 5.3 Interaction of springs

Vehicle springs act generally in conjunction with other constructional parts, as for example auxiliary springs, flexible mountings or suspension links. In many cases, springs with different properties are combined to achieve a desired overall effect, as shown in **Fig. 5.15** by three examples.

The hydropneumatic strut seen in Fig. 5.15a has a spring rate $c_2$ depending on the cross-sectional area of its piston rod, the pressure of the liquid or gas in the accumulator, and the gas volume. An additional coil spring with the rate $c_1$ carries part of the total load F, and the hydropneumatic spring serves only to compensate for the static load change (the so-called 'partially charged' hydropneumatic spring). With a displacement f of the working point of F, the coil spring is compressed by f and the piston rod moves in the cylinder by the same amount. The force increment is therefore $\Delta F = f(c_1 + c_2)$. This 'parallel arrangement' of the springs shows an effective spring rate $c_{res} = \Delta F/f$ or

$$c_{res} = c_1 + c_2 \qquad (5.25)$$

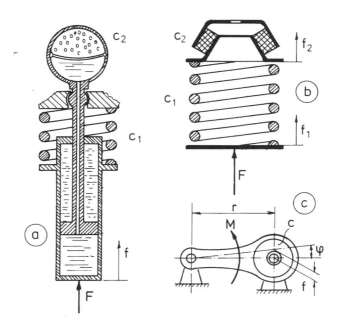

**Fig. 5.15** Interaction of springs
a) parallel arrangement  b) series arrangement  c) torque support

In Fig. 5.15b the coil spring with the rate $c_1$ is attached to its support by an intermediate rubber mount with the rate $c_2$ – e.g. for better noise isolation. The external force F compresses the spring by a deflection $f_1 = F/c_1$ and also acts through it on the rubber mount which is deflected by $f_2 = F/c_2$. The working point of the force F is therefore displaced by a resulting travel $f_{res} = f_1 + f_2 = F/c_1 + F/c_2 = F(1/c_1 + 1/c_2)$, and the resulting spring rate of this 'series arrangement' is $c_{res} = F/f_{res}$ or

$$c_{res} = c_1 c_2 / (c_1 + c_2) \tag{5.26}$$

In Fig. 5.15c, a rubber bush with the spring rate c supports a rotatably mounted lever with the radius r. A moment M acting on the lever swivels it by an angle $\varphi$ and deflects the rubber bush by a travel $f = r\varphi$ which causes a restoring force $F = cf = cr\varphi$. This force exerts a restoring moment $M = Fr = cr^2\varphi$ on the lever. The effective rotational spring rate of the lever arrangement is therefore $c_\varphi = M/\varphi$ or

$$c_\varphi = cr^2 \tag{5.27}$$

The rubber mounting in **Fig. 5.16** can be regarded as a combination of two springs with perpendicular lines of action. The spring in the direction 1 has been weakened in comparison with that in the direction 2 by kidney-shape apertures A. The rubber mounting thus yields two different spring rates $c_1$ and $c_2$ in perpendicular directions. If a skew force F acts on the mounting, its components $F\sin\alpha$ and $F\cos\alpha$ respectively in the directions 1 and 2 cause displacements $f_1 = F\sin\alpha/c_1$ and $f_2 = F\cos\alpha/c_2$ in these directions. The resulting displacement f is not aligned with the force F but shows a deviation towards the axis 1, caused by the lower spring rate $c_1$. This process looks very similar to that of the 'skew bending' of a beam with different geometrical moments about the axes of its cross-section. The mathematical treatment is identical if the geometrical moment about the axis 1 is replaced by the spring rate $c_2$ and *vice versa*. Hence, the well-

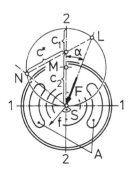

**Fig. 5.16**
Rubber mounting with an applied skew force

Springs and Dampers

known graphical method of determining the deflection of a skew-loaded beam is applicable, too, for the rubber mounting: beginning at the geometrical centre S, the spring rate $c_2$ is drawn on the axis 2 and the rate $c_1$ is added. The sum of the rates is the diameter of a circle that is intersected by the line of action of the force F at a point L. The line from L through the point on the vertical diameter S–2 where the spring rate $c_2$ adjoins $c_1$ intersects the circle at N. On a real bent beam, the line N–S would define the 'neutral axis' and is here to be interpreted as the 'normal line' of the deflection f. The value of f follows from the force F and the 'spring rate' $c^*$ which can be read from the drawing: $f = F/c^*$.

On a rubber mounting – as well as on a beam – it would be useless to attempt to generate more than two principal axes or spring rates in one section plane (possibly by non-symmetrical distribution of material).

With vehicle chassis, complete assemblies are frequently mounted through rubber springs – e.g. subframes that carry wheel suspensions or parts of them for better noise isolation, easier pre-assembly or elasto-kinematic coordination. Assuming these subframes to be rigid, it is useful to estimate the properties of such a system by a simple method.

**Fig. 5.17** shows an arrangement of rubber mountings supporting an aggregate (an engine, a gearbox or the subframe of a wheel suspension) in a single plane. Each of the mountings has a compression-spring rate $c_1$ and a shear rate $c_2$. If all mountings or springs of the system are assumed to be free of load in a reference position, and all spring rates to be linear, the system of springs can be replaced by two principal rates $c_I$ and $c_{II}$ and a rotational rate $c_\varphi$ acting about the elastic centre $S_F$ of the system.

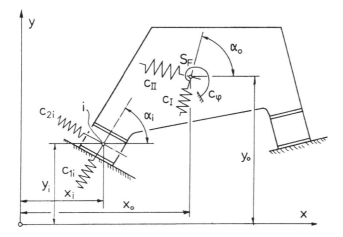

**Fig. 5.17** Planar system of springs with elastic centre and principal spring rates

The effective levers of the spring components at a working point i about the elastic centre $S_F$, with its (still unknown) coordinates $x_0$ and $y_0$, are

$$r_{1i} = (y_0 - y_i)\cos\alpha_i - (x_0 - x_i)\sin\alpha_i \qquad (5.28a)$$

and $\qquad r_{2i} = (y_0 - y_i)\sin\alpha_i + (x_0 - x_i)\cos\alpha_i \qquad (5.28b)$

and with these levers the resultant rotational rate of the system follows according to equation (5.27): $c_\varphi = \Sigma(c_{1i}r_{1i}^2 + c_{2i}r_{2i}^2)$ or

$$c_\varphi = \sum_i \{c_{1i}[(y_0 - y_i)\cos\alpha_i - (x_0 - x_i)\sin\alpha_i]^2 \\ + c_{2i}[(y_0 - y_i)\sin\alpha_i + (x_0 - x_i)\cos\alpha_i]^2\} \qquad (5.29)$$

The elastic centre $S_F$ is the one point in the plane about which the system shows the minimum rotational spring rate, because when rotating about any other point the principal springs $c_I$ and $c_{II}$ would have effective lever arms about this point too, and would thus increase the effective rotational rate. The conditions $\partial c_\varphi/\partial x_0 = 0$ and $\partial c_\varphi/\partial y_0 = 0$ for the minimum rotational rate result in the following equations for the coordinates $x_0$ and $y_0$ of the elastic centre $S_F$:

$$x_0 \sum_i (c_{1i}\sin^2\alpha_i + c_{2i}\cos^2\alpha_i) + y_0 \sum_i (c_{2i} - c_{1i})\sin\alpha_i\cos\alpha_i \\ - \sum_i [x_i(c_{1i}\sin^2\alpha_i + c_{2i}\cos^2\alpha_i) + y_i(c_{2i} - c_{1i})\sin\alpha_i\cos\alpha_i] = 0 \qquad (5.30a)$$

$$x_0 \sum_i (c_{2i} - c_{1i})\sin\alpha_i\cos\alpha_i + y_0 \sum_i (c_{1i}\cos^2\alpha_i + c_{2i}\sin^2\alpha_i) \\ - \sum_i [x_i(c_{2i} - c_{1i})\sin\alpha_i\cos\alpha_i + y_i(c_{1i}\cos^2\alpha_i + c_{2i}\sin^2\alpha_i)] = 0 \qquad (5.30b)$$

If the point of action i of a spring with rate $c_{1i}$ is displaced in the x direction by a travel $f_x$, the spring will be compressed by a deflection $f_x\cos\alpha_i$ and will exert a force $F_{1i} = c_{1i}f_x\cos\alpha_i$ with a component $F_{1ix} = c_{1i}f_x\cos^2\alpha_i$. Accordingly, the force component in the y direction caused by a travel $f_y$ is $F_{1iy} = c_{1i}f_y\sin^2\alpha_i$. The coefficients $\cos\alpha_i$ and $\sin\alpha_i$ are, so to speak, the 'spring ratios' in the x and y directions. Corresponding ratios can be derived for a perpendicular spring with the rate $c_{2i}$. The resulting principal spring rates $c_I$ and $c_{II}$ of the system are the overall sums of the effective spring rate components:

$$c_I = \sum_i [c_{1i}\cos^2(\alpha_0 - \alpha_i) + c_{2i}\sin^2(\alpha_0 - \alpha_i)] \qquad (5.31a)$$

$$c_{II} = \sum_i [c_{1i}\sin^2(\alpha_0 - \alpha_i) + c_{2i}\cos^2(\alpha_0 - \alpha_i)] \qquad (5.31b)$$

# Springs and Dampers

The angle $\alpha_0$ of the principal system follows from the condition that the principal spring rates are the maximum or minimum values of all resulting spring rates of the system in any particular direction; $dc_I/d\alpha_0 = 0$ leads to

$$\tan(2\alpha_0) = \frac{\sum_i (c_{2i} - c_{1i}) \sin(2\alpha_i)}{\sum_i (c_{2i} - c_{1i}) \cos(2\alpha_i)} \qquad (5.31c)$$

$\alpha_0$ and the coordinates $x_0$ and $y_0$ follow from the equations (5.31c) and (5.30), and consequently $c_\varphi$ and the principal spring rates $c_I$ and $c_{II}$ from (5.29) and (5.31).

Knowing the elastic centre $S_F$ and the principal spring rates, the displacement of a body carrying an external load is easy to determine – **Fig. 5.18**. The resulting travel f may be found graphically by the method shown in Fig. 5.16 or by geometrical addition of the deflection vectors $f_I = F_I/c_I$ and $f_{II} = F_{II}/c_{II}$. As the working line of the force F passes the elastic centre $S_F$ at a distance $r_F$, the force F generates a moment $M_D = F r_F$ and a clockwise rotational angle $\varphi = M_D/c_\varphi$.

Knowledge of the elastic centre and principal spring rates enables the properties of an elastically mounted assembly to be judged. However, it should be remembered that the above-mentioned relationships and methods are valid only so long as the system consists of linear springs that are all free of load at the same reference position. In all other cases, non-linear methods must be applied, especially if the sprung assembly itself cannot be assumed to be rigid.

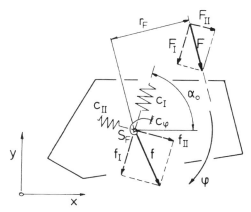

**Fig. 5.18**
Displacement of a sprung system carrying external forces

## 5.4 Interaction of springs and suspension

Except in special cases, the springs of wheel suspensions cannot be positioned exactly in the vertical working line of the wheel load but must be displaced laterally and perhaps inclined. The force acting between the road surface and the tyre is the sum of the relevant share of the 'sprung weight' of the vehicle body and the 'unsprung weight' of the wheel and wheel carrier. Hence, the suspension springs have to deal with only the sprung weight.

The 'spring ratio' is the derivative of the sprung wheel load and the force or the moment of the installed spring. **Fig. 5.19** shows an independent suspension (here a simple 'swing axle') with a coil spring and a telescopic damper which are both connected to the wheel carrier (i. e. the axle tube) as well as to the vehicle body by ball joints.

With a differential travel of the wheel carrier (here a rotation about the axis of its turning joint at the vehicle body), the tyre contact point A moves with a velocity $v_A$, and the spring's lower joint is compressed by a velocity component $v_f$ acting along its axis. The (virtual) 'power' exerted by the sprung wheel load (or 'reduced' spring force) $F_{FA}$ and the vertical component $v_{Az}$ of the velocity $v_A$ must be balanced by the 'power' of the spring force $F_F$ and the spring compression $v_f$: $F_F v_f = F_{FA} v_{Az}$ or

$$F_{FA} = F_F (v_f / v_{Az}) \tag{5.32}$$

The spring ratio is

$$i_F = v_f / v_{Az} \tag{5.33a}$$

or $\quad i_F = F_{FA} / F_F \tag{5.33b}$

Because of the identical z-component, a virtual velocity $v_A^*$ (see Chapter 3) may also be applied here.

By multiplication of the 'velocities' with the time differential dt, the spring ratio results, too, as the derivative of the spring travel and the vertical wheel travel:

$$i_F = df/ds \tag{5.33c}$$

This definition is valid independently of the position of the spring in the suspension system, and it may not be mounted between the suspension and the vehicle body but between two suspension links, and thus will be compressed from both ends.

Looking at Fig. 5.19, it is easy to envisage that the direction of the vector $v_A$ and the inclination of the spring will change with wheel travel, so

Springs and Dampers

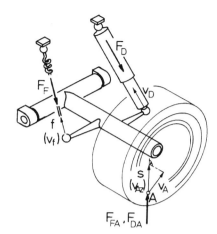

**Fig. 5.19** Spring and damper ratios

the spring ratio is variable – something that is true of nearly all wheel suspensions.

The 'spring rate' is the derivative of the spring force and the compression

$$c_F = dF_F/df \qquad (5.34)$$

and may itself be variable, for instance increasing with wheel bump travel ('progressive spring').

For independent suspensions, because of the definite relationship between the spring force and the sprung wheel load, general considerations are not affected by how and in which position the spring is mounted. To simplify practical work and theoretical investigations, a representative 'spring characteristic' $F_{FA}(s)$ reduced to the tyre contact point A is therefore preferably utilized. The 'reduced' spring rate then follows from

$$c_{FA} = dF_{FA}/ds \qquad (5.35)$$

In the term $F_{FA} = F_F i_F$ – see equation (5.33b) – the spring ratio $i_F$ is usually variable with wheel travel, as mentioned above. The reduced spring rate has therefore to be calculated (**2**) by the rule

$$c_{FA} = (\partial F_{FA}/\partial F_F)(dF_F/ds) + (\partial F_{FA}/\partial i_F)(di_F/ds)$$

and, with $\partial F_{FA}/\partial F_F = i_F$, $dF_F/ds = (dF_F/df)(df/ds) = c_F i_F$ and $\partial F_{FA}/\partial i_F = F_F$, this rate results in

$$c_{FA} = c_F i_F^2 + F_F (di_F/ds) \qquad (5.36)$$

The spring rate $c_{FA}$ reduced to the tyre contact point A is obviously determined not only by the well-known first term of equation (5.36), but also shows in the case of a variable spring ratio $i_F$ a second term - the so-called 'kinematic spring rate'. The two terms may be interpreted as follows: the first term is valid for any spring-force transmission ratio $i_F$, even if it is constant (as, for example, with every normal gear drive), while the second term is imaginable for a spring of constant length only if its transmission ratio is changed with wheel travel (e.g. by shifting the spring mount along a suspension link).

The realization of a 'kinematic spring rate' is easily recognized on a simple 'planar' combination of a trailing link and spring, as is usual on motor-cycle rear suspensions - **Fig. 5.20**:

The spring force $F_F$ acts on the link at a radius b and the (sprung) wheel load $F_z$ at a radius a, so the balance of moments is therefore $F_z a = F_F b$ with the spring ratio $i_F = b/a$. Both radii will obviously increase with wheel travel on bump. The second term of equation (5.36) can now be written, using the variables a and b: $di_F/ds = d(b/a)/ds$ or

$$di_F/ds = (\partial i_F/\partial a)(da/ds) + (\partial i_F/\partial b)(db/ds)$$

where $\partial i_F/\partial a = -b/a^2$ and $\partial i_F/\partial b = 1/a$. The derivatives $da/ds$ and $db/ds$ (i.e. the increases of the radii a and b with wheel travel s) can easily be determined graphically using Fig. 5.20: the vertical component $v_{Mz}$ of the wheel-centre velocity $v_M$ is equal to the wheel-travel velocity, and the radius a is increased by the horizontal component $v_a$. Hence $da/ds = v_a/v_{Mz}$. From $v_M$ we can derive (e.g. graphically) the velocity $v_B$ of the lower spring joint, and its component $v_f$ is the compression velocity of the spring while its component $v_d$ swivels the spring about its joint at the vehicle, thus

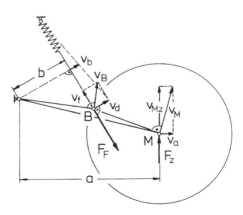

**Fig. 5.20**
Kinematic influence on the reduced spring rate

# Springs and Dampers

increasing the radius b by $v_b$. With $db/ds = v_b/v_{Mz}$, the reduced spring rate follows according to equation (5.36) as

$$c_{FA} = c_F(b/a)^2 + F_F[-(b/a^2)(v_a/v_{Mz}) + (1/a)(v_b/v_{Mz})]$$

Torsional springs in the form of torsion bars or cylindrical rubber bushes can exert a spring force by means of a lever arm only, so naturally the spring ratio is variable. Even on a 'classic' trailing-arm layout, **Fig. 5.21**, the variation of the effective wheel-load lever a with wheel travel has a significant effect. The torsional spring with spring rate $c_\varphi$ may be assumed to be already twisted by an angle $\varphi$ in the position shown – i.e. 'preloaded' – while the trailing arm is inclined to the horizontal by an angle $\alpha$ (in Fig. 5.21, $\alpha < 0$ at the particular moment). The 'sprung' wheel load follows from the horizontal component a of the radius r and the torque $M_d$ as

$$F_{Az} = M_d/a = c_\varphi \varphi / a$$

and the 'spring ratio' $i_F = 1/a$ is here a dimensional quantity – i.e. a reciprocal length. According to equation (5.36) the 'reduced' spring rate is

$$c_{FA} = dF_{Az}/ds = (\partial F_{Az}/\partial \varphi)(d\varphi/ds) + (\partial F_{Az}/\partial a)(da/ds)$$

where $\partial F_{Az}/\partial \varphi = c_\varphi/a$, $d\varphi/ds = 1/a$, $\partial F_{Az}/\partial a = -\varphi c_\varphi/a^2$ and $da/ds = v_a/v_{Mz}$. This results in

$$c_{FA} = (c_\varphi/a^2)[1 - \varphi(v_a/v_{Mz})]$$

and with $a = r\cos\alpha$ and $v_a/v_{Mz} = -\tan\alpha$ comes the well-known equation

$$c_{FA} = c_\varphi(1 + \varphi \tan\alpha)/(r^2 \cos^2\alpha)$$

It is noteworthy that the minimum spring rate of the slightly s-shaped characteristic does not occur in the horizontal position of the lever arm but

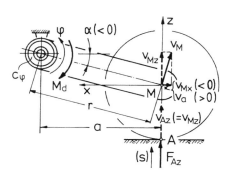

**Fig. 5.21**
Trailing arm and torsional spring

– depending on the preload angle φ – more or less below it. The shorter the lever, the more marked the s-shape of the spring characteristic.

On modern wheel suspensions which are deliberately designed for elasto-kinematic function, kinematic tricks to influence the spring characteristic are not normally recommendable, since they usually lead to marked changes of the lines of actions of forces with wheel travel and thus generate fluctuating horizontal force components at the suspension mechanism. When installed between the suspension and the vehicle body, the spring should preferably retain an approximately vertical position to avoid horizontal force components. In particular, longitudinal forces that load-up the 'longitudinal compliance' of the suspension may render a good elasto-kinematic balance difficult or impossible. It is, of course, possible to deal with these disturbing forces by elasto-kinematic means also – e.g. by suitable choice of the angles of incidence of suspension links (see Chapter 9) – but probably in theory rather than in practice, because the high spring forces and the considerable tolerances of the elastic properties of rubber mountings could lead to ongoing problems during series production. If a skew arrangement of springs is inevitable, these should be integrated into the suspension or mounted between it and a subframe connecting the whole assembly to the vehicle body.

On rigid-axle and compound suspensions, with their two degrees of freedom, there is no longer a definite relationship between the wheel attitude and the deflection of the relevant spring. The position of the spring in the assembly is influential, too. **Fig. 5.22** shows a rigid axle in cross-section. The two springs are mounted between the tyres with a 'spring track' $b_F$ which is considerably smaller than the wheel track b. Assuming these springs to be in the cross-sectional plane and not mounted to suspension links, parallel wheel travel s will compress both springs by a deflection f = s, and the spring rate $c_p$ (reduced to the tyre contact point) is equal to the rate $c_F$ of the installed spring. With vehicle roll or anti-

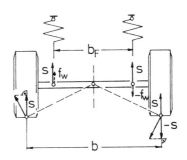

**Fig. 5.22** Spring track of a rigid axle

# Springs and Dampers

metric wheel travel $\pm s$, however, the springs will be deflected by a minor travel $f_W = \pm s\,(b_F/b)$, with $b_F/b$ as the 'antimetric spring transmission ratio'. Consequently the effective spring rate per wheel during vehicle roll is

$$c_W = c_F(b_F/b)^2$$

and is significantly lower than the parallel rate $c_p$. Moreover, the spring rates $c_p$ and $c_W$ can be influenced by inclining the springs or mounting them to suspension links.

The damper shown in Fig. 5.19 is basically subject to the same conditions as the spring. In contrast to a spring, though, the damper force $F_D$ does not depend on the damper travel but on the velocity $v_D$ of the damper piston:

$$F_D = k_D v_D \tag{5.37}$$

and the 'damper rate' $k_D$ is usually chosen differently for different velocity ranges. With the 'damper ratio'

$$i_D = v_D/v_{Az} \tag{5.38}$$

the 'reduced' damper force at the tyre contact point A results in

$$F_{DA} = F_D\, i_D \tag{5.39}$$

or with a 'reduced' damper rate $k_{DA}$

$$F_{DA} = k_{DA} v_{Az} \tag{5.40}$$

and, according to equations (5.40) and (5.39), the reduced damper rate is

$$k_{DA} = k_D\, i_D^2 \tag{5.41}$$

The damper rate represents the instantaneous relationship of the damper velocity and the damper force and is, unlike the spring rate, no 'derivative' of the force (here: over the velocity and not the piston travel). Hence, a 'kinematic damper rate' similar to that of equation (5.36) does not exist.

## 5.5 Vehicle springs

### 5.5.1 General remarks

Fast vehicles require a wheel travel of ±100 mm and more, and, of the 'technical' springs, the bending and torsional types are the most suitable for them. Of the bending springs the 'leaf spring' is frequently used, not least because of its ability to form part of a suspension linkage by performing a wheel-locational function, and it maintains its popularity on commercial vehicles beside the gas spring. Torsional springs – i.e. the torsion-bar and coil types – are not directly suitable for wheel-suspension duties but require linkage mechanisms; this is why their preferred application is on passenger cars. Rubber springs are rarely favoured today for 'whole' vehicle systems, but the numerous rubber mountings in wheel suspensions are becoming increasingly sophisticated within the field of 'elasto-kinematic' harmonization.

### 5.5.2 Leaf springs

The types of leaf springs used in vehicles are grouped in **Fig. 5.23**. The multi-leaf spring has a long history dating from the days of horse-drawn carriages: Fig. 5.23a shows a 'quarter-elliptic' spring; Fig. 5.23b a parallel arrangement of two quarter-elliptic springs, as frequently used in earlier times as an alternative to double-wishbone suspension; Fig. 5.23c a 'semi-elliptic' spring (the most frequently used type); Fig. 5.23d a non-symmetrical variant which might perhaps be chosen to save space or for kinematic reasons; Fig. 5.23e a parallel arrangement of a primary spring and an auxiliary one that becomes operative with high wheel load; Fig. 5.23f the

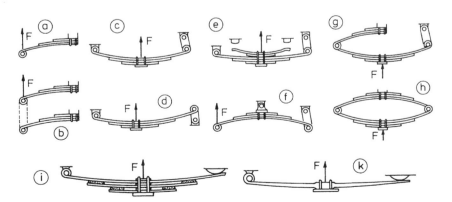

**Fig. 5.23** Leaf spring types

'cantilever spring' for greater deflection but lower load; g and h the 'three-quarter-elliptic' and 'fully-elliptic' springs which originated on horse-drawn carriages and represent series arrangements - see Section 5.3. On a 'spacing spring' (i), the leaves are separated by elastomeric spacers to minimize friction. The frictionless 'parabolic spring' (k) is milled as a single leaf, shaped in accordance with the bending moment and the section modulus for even stressing.

A leaf spring is a bending beam with a very low geometrical moment of inertia and a high deflection. The basic equation of bending deflection follows from the elastic modulus E, the geometrical moment of inertia $I_B$, the bending moment $M_B$ and the bending radius

$$\rho = E I_B / M_B \qquad (5.42)$$

which is the reciprocal of the second-order derivative of the bending curve.

For the single leaf of **Fig. 5.24**a with constant cross-section, the effective spring rate is

$$c = 3 E I_B / l^3 \qquad (5.43)$$

and minimum bending radius and maximum tensile stress occur at its fixed end. For the optimum material utilization, the section modulus of the leaves of the examples (b) and (d) - i.e. 'parabolic' and 'triangular' springs - increases proportionally to the length of the leaf and to the bending moment. In practice, the triangular spring (d) becomes a multi-leaf spring with leaves of equal width and increasing length, today preferably with elastomeric spacers (Fig. 5.23i) for minimum friction. If $I_{B0}$ is the (maximum) geometrical moment of inertia at the fixed end, the spring rate at the free end of an 'ideal' triangular spring results in

$$c = 2 E I_{B0} / l^3 \qquad (5.44)$$

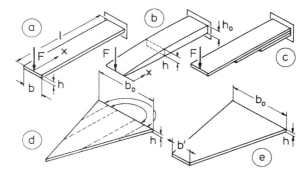

**Fig. 5.24** Cross-sections and shapes of leaf springs

and the constant bending tension and bending radius allow a greater deflection combined with better material utilization. For more precise investigations into multi-leaf springs (c), however, the model of a 'trapezoidal' spring (e) with slightly lower deflection is the better suited.

A symmetrical semi-elliptic spring as in Fig. 5.23c is a parallel arrangement of two quarter-elliptic springs, and their spring rates are additive.

For a non-symmetrical spring as in Fig. 5.23d, however, no simple parallel arrangement applies since the reaction forces of its halves are reciprocal to their lengths - **Fig. 5.25** - $F_1/F_2 = l_2/l_1$, while the respective spring rates follow according to equation (5.44) as

$$c_1/c_2 = (l_2/l_1)^3$$

The deflections $f_1$ and $f_2$ of the free ends lead to an inclination of the spring section at the point of action of the external force F by an angle

$$\alpha = (f_2 - f_1)/(l_1 + l_2)$$

and to a resulting spring travel

$$f = f_1 + l_1 \alpha$$

With the simplifying assumption that the spring has truly triangular leaves and that their cross-section in the middle is equal on both sides (i.e. that the leaves run through the applicational point of F), the resulting spring rate follows according to equation (5.44)

$$c = 2EI_{B0}(l_1 + l_2)/(l_1^2 l_2^2) \tag{5.45}$$

and the angle of inclination of the medial area is

$$\alpha = \frac{F}{2EI_{B0}} \frac{l_1 l_2 (l_2 - l_1)}{l_1 + l_2} \tag{5.46}$$

An idealized triangular spring (but nearly a trapezoidal spring, too) shows a constant bending radius, and the radius of curvature of the path of its free end amounts to about $7/9$ of the free length - **Fig. 5.26**. If the spring is serving as a suspension link, it can be kinematically defined by this radius.

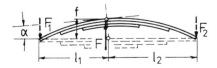

**Fig. 5.25** Non-symmetrical leaf spring

Springs and Dampers

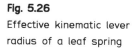

**Fig. 5.26**
Effective kinematic lever radius of a leaf spring

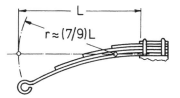

A semi-elliptic spring which also has a wheel suspension function is normally connected to the chassis by a 'fixed' joint and a displaceable mounting – **Fig. 5.27**. The centre of curvature K of the mounting point of a rigid-axle beam or the like is then to be found on the 'chord' of the contour line of the spring near to the 'fixed' point, and in a flattened state of the leaves its distance from the mounting point 1 is $\rho = (7/9) l_1$.

The mounting point of the axle beam swivels with wheel travel about a 'pole' P at a distance p to be calculated from the inclination angle $\alpha$ according to equation (5.46) and the spring deflection f that follows from the force F and the spring rate c – see equation (5.45): $p = f/\alpha$ with $f = F/c$ or

$$p = l_1 l_2 / (l_2 - l_1) \qquad (5.47)$$

The centre of curvature K remains more or less fixed with wheel travel, but the pole P moves with its polar ray (positions 1 and 2 in Fig. 5.27). The pole P corresponds exactly with the 'instantaneous pole' of a linkage mechanism. On a leaf-sprung rigid axle which rotates about a pole in side view with wheel travel (as for instance the thrust-ball axle shown in Fig. 2.14), and which is fixed to the spring, it will be helpful if the poles of the axle suspension and the spring coincide in order to achieve adequate loading and equal tension in both halves of the spring. The pole is significant, too, for the suspension geometry – as, for example, for the determination of the 'support angle' (see Chapter 3, Fig. 3.13). This function is not affected by a superimposed and perhaps considerable elastic wind-up caused by a traction or a braking torque.

A leaf spring that serves as the only axle suspension link has to withstand the traction or braking torques, which cause opposed reaction forces

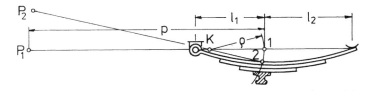

**Fig. 5.27** The 'instantaneous pole' of a non-symmetrical leaf-spring suspension

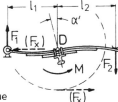

**Fig. 5.28** Leaf-spring wind-up caused by an applied torque

$F_1$ and $F_2$ on the leaf ends – **Fig. 5.28** – whereby the axle winds up by an angle $\alpha'$. With the spring rates of the two leaf-spring arms, the resulting rotational rate of the spring is according to equation (5.27):

$$c_{\alpha'} = c_1 l_1^2 + c_2 l_2^2$$

and with the spring rates according to equation (5.44) $c_1 = 2EI_{B_0}/l_1^3$ and $c_2 = 2EI_{B_0}/l_2^3$ the wind-up rate follows as

$$c_{\alpha'} = 2EI_{B_0}(l_1 + l_2)/(l_1 l_2)$$

Dividing this by equation (5.45) results in

$$c_{\alpha'} = c\, l_1 l_2 \tag{5.48}$$

On a symmetrical spring ($l_1 = l_2 = l$), the wind-up rate is therefore $c_{\alpha'} = c l^2$, which means that for a given spring rate c the resistance against wind-up increases proportionally to the square of the length. This, apart from the lower number of leaves and therefore friction, is a decisive advantage of long leaf springs over shorter ones.

The position of the centre of curvature K of free spring movement as in Fig. 5.27 is important also in relation to the suitable mounting of a longitudinal 'drag rod' or 'steering rod' of a steered rigid axle (see also Chapter 8). If elastically wound up by a torque (Fig. 5.28), the axle will swivel about a point D which lies approximately beside the upper leaf and which may be determined by calculation or by testing. In side view, a drag rod should run through this point to prevent braking or traction torques from causing undesired steering angles.

Passenger cars sometimes have a transverse leaf spring that is supported twice at the vehicle body as well as at the wheel suspensions, **Fig. 5.29**. With parallel wheel travel, Fig. 5.29a, the bending moment M is constant in the medial area of the spring, and a constant leaf cross-section will be appropriate at least in this area. However, with antimetric wheel travel – e.g. with vehicle roll – the deformation of the spring is s-shaped, as shown in Fig. 5.29b. For a constant cross-section of the leaf or leaves,

Springs and Dampers

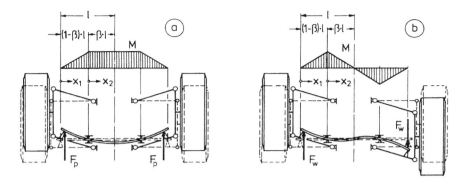

**Fig. 5.29** Twice-supported leaf spring with parallel (a) and antimetric (b) wheel travel

and consequently a constant geometrical moment of inertia $I_B$, the parallel spring rate $c_p$ and the antimetric rate $c_w$ per wheel result, using the auxiliary measurements given in the drawing

$$c_p = 3EI_B/[l^3(1-\beta)^2(1+2\beta)] \quad (5.49a)$$

$$c_w = 3EI_B/[l^3(1-\beta)^2] \quad (5.49b)$$

or $c_w/c_p = 1 + 2\beta$; the antimetric or roll rate is higher than the parallel rate – not surprising in view of the deformation lines. Hence, this type of spring has an integral 'stabilizer' or anti-roll device. The stabilizer rate per wheel is $c_S \doteq c_w - c_p$ or

$$c_S = 6EI_B\beta/[l^3(1-\beta)^2(1+2\beta)] \quad (5.49c)$$

### 5.5.3 Torsion bars

A torsion bar needs a lever arm to act on a wheel suspension, **Fig. 5.30**. With the shear modulus G, the polar geometrical moment of inertia $I_D$ and the length l, the angular spring rate is

$$c_\varphi = GI_D/l \quad (5.50)$$

For a circular section of diameter d, the polar geometrical moment of inertia is $I_D = \pi d^4/32$ and consequently the angular spring rate

$$c_\varphi = G\pi d^4/(32\,l) \quad (5.51)$$

The effective spring rate at the point of action of an external force F depends on the preload angle and the position of the lever arm, see Section 5.4.

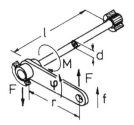 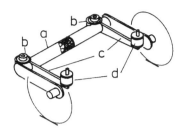

**Fig. 5.30** Simple torsion-bar springing

**Fig. 5.31** Torsion-bar springing and flexibly mounted subframe

The spring force F causes a reaction force at the mounting of the torsion bar which hinders noise isolation. If the torsion bar is mounted on a subframe, **Fig. 5.31**, the reaction force is better transmitted back into the cross-sectional plane of the wheels – e.g. by extension arms c – to conserve the flexibility of the subframe's resilient mountings.

The considerable length of torsion bars works against their installation in modern vehicles. A series arrangement of a torsion bar and a coaxial torsion tube, **Fig. 5.32**, is of little benefit because of the latter's high torsional stiffness. However, an elegant solution is shown in **Fig. 5.33** (21): torsion bars with rates $c_{\varphi 1}$ and $c_{\varphi 2}$ are installed on both sides of the true axis of rotation d of a link 1. The lower torsion bar is fixed to the chassis at one end and rotationally mounted at its other end, which bears a lever arm 2 of the length $a_1 + a_2$ and, at its opposite end, the second (upper) torsion bar which acts at a radius $a_1$ on the link. A moment M twists the upper torsion bar by an angle $\varphi_1 - \varphi_2 = M/c_{\varphi 1}$ and the lower bar by an angle $\varphi_2 = M/c_{\varphi 2}$. For $\varphi_1 a_1 = \varphi_2 (a_1 + a_2)$, the upper torsion bar is actually translated parallel in space without any constraint. This requires $c_{\varphi 1}/(c_{\varphi 1} + c_{\varphi 2}) = a_1/(a_1 + a_2)$ and, for identical torsion bars, $a_1 = a_2$. The mounting g is not necessary but may be advantageous to avoid oscillations.

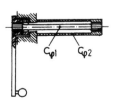 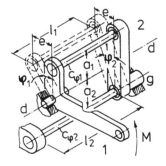

**Fig. 5.32** Torsion bar and coaxial torsion tube

**Fig. 5.33** Series arrangement of two torsion bars

Springs and Dampers

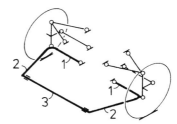

**Fig. 5.34** Multi-functional torsion bar

In the same way as the twice-supported leaf spring of Fig. 5.29, a torsion bar can be adapted for multifunctional use. **Fig. 5.34** shows a wheel suspension in which the lower transverse links 1, tension rods 2 and stabilizer bar 3 are made from a single piece of material (**20**). The tension rods are twisted with wheel bump and rebound and represent the suspension springs.

Similar inventions that try to reduce the number of components have not survived, primarily because of harmonization problems (which may affect suspension geometry), bulky construction, complicated finishing procedures and difficult integration into an elasto-kinematic system. The only exception is the MacPherson layout (see Fig. 5.13b) which is still quite popular today.

The preferred use of torsion bars today is for anti-roll stabilizers. Especially on front wheels, a skewed arrangement of the arms of the bar is almost unavoidable with respect to the wheels' steering articulation, and the bar is thus subjected to bending even over its medial section - **Fig. 5.35**. The axial section modulus of a beam with a circular cross-section is only half its polar one, so the material utilization is reduced and the tension peak at the edges is increased. The diagram shows the share of the bending energy $A_B$ of the overall deformation energy dependent on the angle $\alpha$ for a numerical example; the bending energy for $\alpha = 0$ is caused by the lateral arms. Cranking of the medial section increases the share of bending.

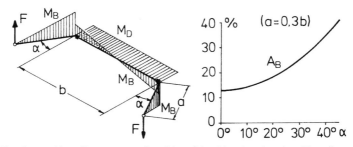

**Fig. 5.35** Torsion and bending on an anti-roll bar ($M_D$, $M_B$ = torsional and bending moments)

## 5.5.4 Coil springs

A coil spring is a 'wound-up' torsion bar with the advantage of needing no lever arm to exert a spring force. Several examples are given in **Fig. 5.36**. The basic type (a) is cylindrical with constant coil and wire diameters and constant pitch, giving linear loading characteristics. The ends of the spring may be ground flat (an expensive process) - Fig. 5.36a (bottom) - or an 'inactive' final coil of about 270° is incorporated with a pitch slightly greater than the wire diameter, which sits in a suitably shaped washer - Fig. 5.36a (top). If there is danger of tilt, e.g. with considerable angular movement of the support, the end may perhaps be fixed by a final coil with reduced diameter and a bolt, Fig. 5.36b (top), or may be mounted on a ball joint (bottom).

For constant coil and wire diameters, a non-linear but progressive spring characteristic can be achieved by variable pitch - Fig. 5.36c - which causes the coils to close one after another. A progressive characteristic results, too, from a series arrangement of two springs, as in Fig. 5.36d. If both springs are linear, the effective spring rate follows from equation (5.26) until one spring is 'coil-bound' and then is equal to the rate of the remaining spring; the characteristic therefore shows two linear sections with a kink. A 'barrel spring', Fig. 5.36e, with varying coil and wire diameters, and varying pitch, enables almost any desired progressive characteristic combined with minimum installed height since the coils can close right up, one within another, on the spring seat.

The length of the (straightened) wire of a coil spring with w coils and respective coil and wire diameters of D and d is $l = w\pi D$, and its torsional rate is, according to equation (5.51), $c_\varphi = Gd^4/(32wD)$; the coil radius $D/2$ acts as the effective lever of an external force F, **Fig. 5.37**. The balance of the torques exerted by the force F and by the twisted wire is $FD/2 = c_\varphi \varphi =$

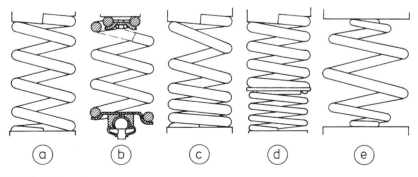

**Fig. 5.36** Coil springs

Springs and Dampers

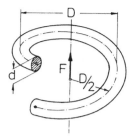

**Fig. 5.37** A coil spring as a 'wound-up' torsion bar

$Gd^4\varphi/(32wD)$ (where G is the shear modulus) and, with the spring deflection $f = \varphi D/2$ per coil, the effective spring rate is

$$c = Gd^4/(8wD^3) \qquad (5.52)$$

Coil springs can be adapted to nearly all spatial conditions. If the available height is limited, a low number of coils and a thicker wire must be adopted. It should be appreciated, however, that the permissible tensile loading decreases slightly with wire diameter and that the 'inactive' coils become a larger proportion of the total number of coils, and therefore of the weight too. In practice, a cylindrical coil spring normally has at least four active coils. Another possibility to save installed height is to mount the spring on a suspension link and thus to achieve a 'spring ratio' below unity. Long and slim springs with a relatively large number of coils are lighter but tend to buckle earlier, necessitating control by proved methods (**11**).

If a coil spring is mounted on a suspension link, it will bend with wheel travel unless remedial measures are taken (see Fig. 5.36b). On a spring that is arranged perpendicular to a link in the normal position – **Fig. 5.38**a – most of the bending occurs at the lower end, and in the relaxed or rebound state there is danger of its tilting off its seating. A long and slim spring is here more advantageous. The smallest possible amount of bending, and equal moments at both ends (hence, a minimal installed length, too),

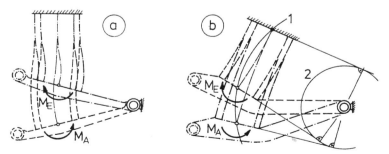

**Fig. 5.38** Bending of a coil spring mounted on a suspension link

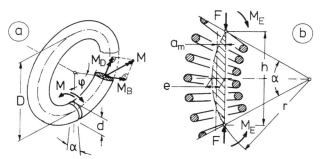

**Fig. 5.39** Approximate determination of the bending moment of a symmetrically deformed coil spring

can be achieved by an arrangement as shown in Fig. 5.38b, where the ends of the spring lie on a common circle 1 and the planes of the supports are tangential to a common cylinder 2 about the inner joint of the suspension link.

For this special case of symmetrical bending, an approximate calculation of bending moments is possible. The components of an external moment M acting on a coil are a bending moment $M_B(\varphi) = M\sin\varphi$ and a torque $M_D(\varphi) = M\cos\varphi$ – **Fig. 5.39**a. The integral of energy over one coil produces the bending angle $\alpha$ between the ends of the coil:

$$\alpha = (M/GI_D)\int_0^{2\pi}(D/2)\cos^2\varphi\,d\varphi + (M/EI_B)\int_0^{2\pi}(D/2)\sin^2\varphi\,d\varphi$$

and with this the bending rate $c_\alpha = M/\alpha$ for w coils becomes

$$c_\alpha = Gd^4/[16wD(1 + 2G/E)]$$

For a given angle $\alpha$ and a height h, Fig. 5.39b, the simplifying (and fairly true) assumption of a circle-shape bending of the central axis with the radius $r = (h/2)/\sin(\alpha/2)$ leads to an arc length $u = r\alpha$ of the central axis of the spring and thus, with the relaxed length $l_0$, to a force $F = c(l_0 - u)$. The medial lever of the force is approximately equal to the quotient of the hatched area and the arc length: $a_m \approx r(\alpha - \sin\alpha)/(2\alpha)$, and therefore the medial bending moment exerted by the force is $M_F \approx Fa_m$, while the reaction moment at the spring seatings is the difference between the bending moment of the spring and the counteracting moment of the force: $M_E \approx c_\alpha \alpha - M_F$. If $M_E > F(D/2)$, there is danger of tilting. This simple but useful assessment does not, however, absolve the designer from guarding against buckling!

Springs and Dampers

### 5.5.5 Rubber springs

Rubber-elastic springs are applied in the suspensions of the faster types of vehicles mainly as bump-stops or auxiliary springs, and such springs are developed in close cooperation with the suppliers. These rubber springs are made not only from natural or synthetic rubber but also from other elastomeric materials. The principal considerations that follow apply for all kinds of elastomeric springs.

Apart from wheel suspension applications, all rubber-elastic mountings of vehicle assemblies such as the engine, gearbox, steering box and suspension links should be regarded as rubber springs which are harmonized with respect to the requirements of acoustics and elasto-kinematics. For geometrically simple types the spring characteristics can be determined approximately (**10**).

The most frequently applied rubber mounting in suspensions is the cylindrical turning joint, **Fig. 5.40**. For a solid-rubber joint, the equations for the axial and the radial spring rates $c_a$ and $c_r$ look similar:

$$c_a = 2\pi h G / \ln(r_2/r_1) \qquad (5.53)$$

$$c_r = 7{,}5 \, k \pi h G / \ln(r_2/r_1) \qquad (5.54)$$

The coefficient $k$ depends on the ratio of the length $h$ and the rubber thickness $s = r_2 - r_1$, and it increases from the value 1 for $h/s = 0$ progressively to 2.1 for $h/s = 5$, while the shear modulus $G$ increases in the region of 53–113 N/mm² with a Shore hardness of HS = 45-65. The relationship of $c_r$ and $c_a$ can be chosen within a wide range, but one below $c_r/c_a = 4$ is difficult to achieve with a solid-rubber joint. The torsional spring rate is

$$c_\varphi = 4\pi h G / (1/r_1^2 - 1/r_2^2) \qquad (5.55)$$

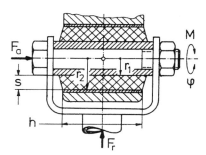

**Fig. 5.40** Cylindrical rubber joint

Since rubber does not like tensile stress, heavy-duty suspension joints are preferably designed to deal with a combination of compressive stress and shear. On simple joints as in Fig. 5.40, a compressive preload is generated after vulcanization by squeezing the outer bush down to a smaller diameter or by enlarging the inner bush. The equations (5.53) to (5.55) are then no longer exactly valid.

A cylindrical rubber joint in a wheel suspension is normally loaded not only with pure rotation about its principal axis but also with tilting of its inner and outer bushes, the 'coning angle'. This may be because it replaces a true ball joint (see Fig. 2.2) or because the suspension mechanism is elastically deformed by external forces; in many cases, however, the coning angle is caused by a deliberately applied skew displacement of the joint axis and the suspension link's effective axis of rotation for elasto-kinematic purposes. For 'stiff' cylindrical rubber joints with a length h at least five times the rubber thickness s, **Fig. 5.41**, the 'coning rate' can be approximated by using an analogy to normal beam bending theory: a beam of length l and with a rectangular cross-section of height h and width b has a tensional rate $c = Ebh/l$ and a bending rate $c_\varphi = E(bh^3/12)/l$, as is known from mechanics; the relationship of the rates is therefore $c_\varphi/c = h^2/12$. In a section through its principal axis, a cylindrical rubber joint shows a comparable shape - i.e. a height h and a resulting width $b = 2s$ of the two rubber layers. If the joint is long, and the rubber layers therefore relatively thin, an approximately linear increase in radial compression or relaxation of the material from the centre of the joint to its open faces can be assumed when a coning moment is applied. This is quite similar to the tension diagram

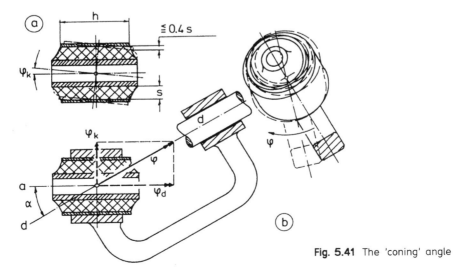

**Fig. 5.41** The 'coning' angle

Springs and Dampers

of the cross-section of a beam under bending, and the ratio $c_k/c_r$ of the coning and radial rates of the joint will therefore be the same as the ratio $c_\varphi/c = h^2/12$ of the bending and the tensional rates of the beam; the coning rate of the joint thus follows from the radial rate as

$$c_k \approx c_r h^2/12 \tag{5.56}$$

As already mentioned, this equation is valid only for relatively stiff joints. On a softer joint with $h < 5b$, however, the compliance of the open faces will cause a drop of tension towards the ends of the joint and will thus prevent a linear tension increase, so the resulting coning rate will be significantly lower than indicated by equation (5.56).

According to rule-of-thumb, the maximum rubber deformation of a joint should not be more than about 40% of the original thickness s - Fig. 5.41a. If a coning deflection of the joint is anticipated, the permissible maximum length h and thus the radial rate $c_r$ (which essentially depends on the h/s ratio) will perhaps meet narrow limits.

The generation of a coning angle from the forced rotation of a rubber joint with an axis a inclined to the effective axis of rotation d of a suspension link by an angle $\alpha$ is visualized in Fig. 5.41b: the suspension link and the outer bush fixed to the link pivot about the axis d by an angle $\varphi$ (which is here symbolized by a 'rotating vector'). The component of the angle $\varphi$ in the direction of the principal axis a of the joint is the joint's rotational angle

$$\varphi_d = \varphi \cos\alpha \tag{5.57a}$$

and the component of $\varphi$ normal to the joint's principal axis a is the coning angle

$$\varphi_k = \varphi \sin\alpha \tag{5.57b}$$

In space, a rubber joint has three translational and three rotational spring rates. For the cylindrical joint shown in **Fig. 5.42**, the radial rate $c_{rad\,1}$

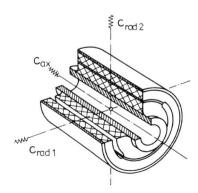

**Fig. 5.42**
Cylindrical rubber joint
and its principal axes

has been 'weakened' by moulded-in slots, while the rectangular rate $c_{rad2}$ has been reinforced by intermediate sheet-metal segments. Measures of this kind are frequently used for rubber joints of subframes provided for elasto-kinematic purposes. For suspension joints with large deflection angles, however, reliability and durability impose design limitations.

The elastic properties of the rubber joints of suspension links and subframes are carefully harmonized for elasto-kinematic behaviour. These properties should, of course, remain unchanged during the vehicle's life. Important conditions here are that the joints are adequately dimensioned for the designed load and that they are protected from extreme stress (e. g. exhaust heat radiation). Moreover, it is very advantageous for all suspension joints to be made from similar rubber compounds with practicable Shore hardness (in a range of about HS = 45-70) to ensure comparable 'aging' throughout.

### 5.5.6 Pneumatic springs

A pneumatic spring contains a defined gas volume which is compressed with spring deflection, **Fig. 5.43**. The spring force F results from the effective working area A and the difference between the internal and ambient air pressures p and $p_a$ respectively:

$$F = A(p - p_a) \tag{5.58}$$

The energy fed to the spring by compression is partly stored by pressure increase and partly lost by thermal dissipation to the environment. For low-velocity spring action (e. g. a static-load change), the isothermal law of pV = constant is valid, while for high velocities (e. g. dynamic spring action over road irregularities), with lack of time for thermal dissipation, the adiabatic law has to be applied - $pV^{\kappa}$ = constant. For realistic investigations, a generalized 'polytropic' law is used:

$$pV^n = \text{constant} \tag{5.59}$$

with the polytropic exponent $1 < n \leq \kappa = 1.4$. This exponent increases with the velocity of compression; high-frequency impacts - e. g. when rolling

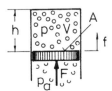

**Fig. 5.43** Schematic gas spring

over the rails of a level-crossing – therefore result in a greater pressure increase (and thus a higher spring rate) than low-frequency disturbances such as the overall unevenness of the crossing.

As any gas tends to diffuse, a pneumatic springing system is always combined with a level-control system. Since the regulating process more or less hides isothermal events, the isothermal law is of little importance for gas springs.

From equation (5.58) and $V = Ah$ we get $dF = A\,dp$ and, using equation (5.59), $dp = -pn\,dV/V$, where $dV = -A\,df$; hence the spring rate $c = dF/df$ results in

$$c = A^2 pn V \qquad (5.60)$$

or with equation (5.58)

$$c = F^2 pn / [(p - p_a)^2 V] \qquad (5.61)$$

for a constant area A.

Two basic types of gas springs are in use – one with 'constant gas volume' and the other with 'constant gas mass'.

Static levelling of the spring with constant gas volume is made by feeding or releasing gas while the volume of its container remains constant. **Fig. 5.44** shows schematically a rolling-lobe spring with a piston 1, the lobe 2, a connecting rod 3 to actuate the control valve (here a three-way component) 4 and a pump 5. The effective working area A is defined by the diameter where the wall of the lobe has its horizontal tangent. The constant-volume spring normally uses ambient air – a feature that makes the system extraordinarily simple. If the static load is changed, the gas volume remains constant, the internal pressure increases according to equation (5.58) proportionally to the load, and the spring rate increases more or less likewise, see equation (5.61). Hence, the constant-volume spring provides a nearly constant natural frequency with changing load. As the volume enclosed by the lobe is normally much too small to give the desired low frequency, an auxiliary volume must be provided – e.g. by enlarging the pot-shaped upper part in Fig. 5.44. The spring rate can be further influenced by contouring the piston to get a variable working area with wheel travel. Air springs operate with maximum pressures of about $7-15 \times 10^5$ Pa and static-load pressures of about $3 \times 10^5$ Pa.

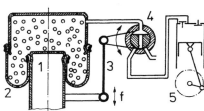

**Fig. 5.44**
Schematic air spring with constant gas volume

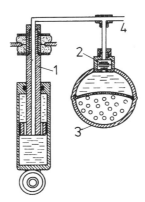

**Fig. 5.45**
Schematic hydropneumatic spring with constant gas mass

In a gas spring with constant mass, hydraulic fluid is the working medium to transfer the pressure to the gas volume, this type of component being therefore known as a 'hydropneumatic spring'. Movement of the fluid with wheel travel is usually controlled by an integral damping device. **Fig. 5.45** shows schematically a hydropneumatic strut, a cylinder with a hollow piston rod 1, a damper valve 2, a gas accumulator 3 and a pipe 4 by which oil can be fed or released. The working area is defined by the diameter of the piston rod, and because of its relatively small cross-section area, higher pressures – about $20$–$50 \times 10^5$ Pa – are necessary. According to the equations (5.51) and (5.58), the influence of the ambient-air pressure $p_a$ on the spring rate is much smaller than with a constant-volume spring. As the gas volume V, in contrast to the case of the constant-volume spring, decreases with growing load, the spring rate increases by a greater amount than the load, and hence the natural frequency also, so a constant-mass gas spring behaves in a non-linear manner.

A comparison of the essential properties of the two gas-spring types is given in **Fig. 5.46**. The upper row of graphs refers to the constant-volume or air spring, and the lower to the constant-mass or hydropneumatic spring.

Both types have in common a decrease of the spring rate $c_0$ in the normal position and of the natural frequency $f_0$ with increasing gas volume (left graphs). The adaptation of the constant-volume spring to growing static load $F_0$ is achieved by regulation of the mass while the volume V remains constant, and only the pressure $p_0$ in the normal position increases. The spring rate does not quite keep abreast of the load, and the natural frequency $f_0$ decreases slightly (upper graph in the middle). In contrast to this, for the constant-mass spring the law $pV = $ constant is valid, and the volume decreases with increasing load and increasing pressure; hence, the

Springs and Dampers 101

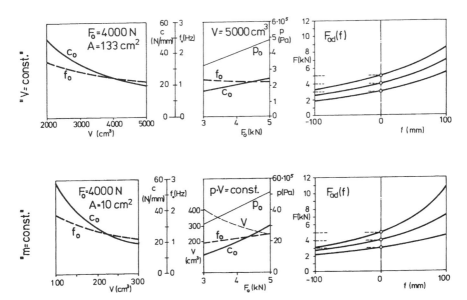

**Fig. 5.46** Comparison of the essential properties of gas springs:
above: constant-volume spring; below: constant-mass spring

spring rate $c_0$ and the natural frequency $f_0$ increase exponentially (lower graph in the middle). The right-hand graphs show examples of dynamic spring characteristics (calculated using the adiabatic coefficient $\kappa = 1.4$) for three static loads. Both spring types show progressive characteristics but the progressiveness of the constant-mass spring is the more pronounced.

## 5.6 Hydraulic dampers

Among the hydraulic damper types, the telescopic damper is favoured because it can be directly connected to the suspension and to the vehicle body, needs no linkage, and, owing to its long stroke and consequently low forces, responds quickly to wheel movements. There are two basic types, shown schematically in **Fig. 5.47**, namely the 'twin-tube' damper (a) and the 'single-tube' or 'gas-pressure' damper (b).

The twin-tube damper consists of a piston rod with piston moving in an inner cylinder which forms an oil reservoir together with an outer cylinder. On compression of the damper, oil (1) flows upward through holes in the piston from the lower to the upper portion of the cylinder. Since the oil is preloaded only by atmospheric pressure, the flow may not be throttled here,

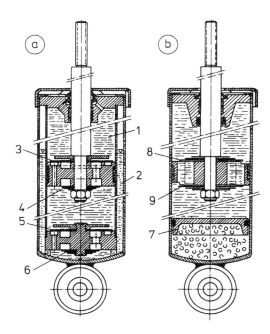

**Fig. 5.47**
Telescopic dampers
a) twin-tube damper
b) single-tube damper

thus avoiding depression above the piston and cavitation which could exert heavy shock loading. The upper side of the piston therefore has only a non-return valve 3, and the damping effect is achieved by a valve 6 in the base of the cylinder restraining the flow of the oil that is displaced by the incoming piston rod and fed to the reservoir 2 between the inner and outer cylinders. With extension of the damper, the oil flows downward through the piston, the damper valve 4 acting as a powerful throttle. The volume vacated by the rising piston rod is filled by the suction of oil from the reservoir 2 with as little resistance as possible via the non-return valve 5 in the base of the cylinder. The twin-tube damper is robust and continues working even after losing a minor quantity of oil, but is preferably mounted in the vertical position.

Alongside the twin-tube damper, the single-tube variant has gained considerable popularity. Compensation for the volume displaced or gained by the piston rod with wheel bump or rebound is achieved by the compression or decompression of a gas volume that is preloaded with a pressure of about $25 \times 10^5$ Pa, and that is separated from the oil by a diaphragm or floating piston 7. The damper piston carries valves for the compression (8) as well as for the decompression stage (9). The high-pressure preload avoids cavitation and helps towards quicker response of the damper. A single-tube unit can be mounted in any desired position, but the piston rod

Springs and Dampers

seal must be able to withstand the high preload pressure and therefore causes generally higher breakaway forces than in a twin-tube damper. Regarding the piston-rod cross-section area, too, the preload pressure generates a 'gas-spring force' that tries to lift the vehicle and so must not be neglected when detailing the springs.

It is worth mentioning that nowadays twin-tube dampers too are preloaded with gas pressure (lower, of course) to gain the advantages of the single-tube type while reducing the risk of failure.

Normally the damper valves are dimensioned for significantly lower compression than extension forces. This can hardly be a tribute to tradition (since in former times there were no base valves and thus no compression force) and it gives rise to ongoing discussion but, since practice still sticks to it there must be genuine reasons. Damper harmonization has to meet a lot of requirements, in respect not only of riding comfort but also of stationary and dynamic stability of the vehicle, and has therefore remained in the province of practical road testing until recently.

On a bad road surface, a damper with the usual non-symmetrical characteristics $k_z$ and $k_d$ of the extension and compression stages should theoretically pull the vehicle body slightly downward because of the stronger extension force. If the damper deflection is assumed equal to the exciting amplitude h - approximately true for a frequency ratio $\eta = \omega/\omega_0 \geq 5$ - the dynamic loss of vehicle height results approximately from the quotient of the average damping force and the spring rate as

$$\Delta h \approx \frac{4 h \eta D}{\pi} \cdot \frac{k_z - k_d}{k_z + k_d}$$

and because of the minute high-frequency amplitudes the loss of vehicle height is negligible.

## 5.7 Controlled suspension systems

A level-control system avoids vehicle height differences with changing static load. The parameters of suspension geometry - e.g. wheel camber or roll-centre height - and of aerodynamics are therefore maintained constant, as are the headlight range, ground clearance and available dynamic wheel travel.

However, conventional levelling systems, which work quasi-statically for minimum power consumption, compensate rather belatedly for acceleration or braking pitching of the vehicle body.

While level control is possible with nearly all types of springs - as for example with mechanical springs by displacement of the spring supports or setting-up of torsion bars - it is generally realized today with gas springs.

Independent regulation of the ride height at all wheels of a vehicle amounts to a statically 'over-constrained' system: because different response times of the actuators, and thus unbalanced wheel-load distribution, can never be excluded, such a system needs control devices which, for instance, limit pressure differences between the springs.

The most common levelling solution is 'three-point control': on a four-wheel vehicle, the springs of one axle are influenced individually and those of the other axle in common (with a pressure-equalization valve between the two gas springs). During cornering, though, the axle with individually controlled springs will try to force the vehicle back into the vertical (no-roll) position, which in the first place amounts to the function of an extremely stiff anti-roll bar (with the usual disadvantages to the vehicle's roadholding properties) and second will cause the vehicle to adopt an oblique attitude at the beginning of subsequent straight-ahead driving - or a disadvantageous starting attitude for a perhaps subsequent opposed corner. For this reason, an automatic switch is normally provided to inhibit the levelling processes in corners.

The simplest levelling method is a 'two-point' control where the gas springs of each axle are connected by equalization pipes, and roll stiffness is achieved by strong anti-roll bars alone.

On vehicles with steel springs in the basic version, a 'partially charged' levelling system may be applied with certain advantages. The steel springs carry the unladen or 'kerb' weight of the vehicle almost entirely on their own, and the gas or hydropneumatic springs compensate for static load changes. Such a system will ensure emergency driving properties even in the case of a failure of the levelling system. **Fig. 5.48** shows a rear suspension with a 'partially charged' levelling system consisting of hydropneumatic struts carrying coil springs. A lever in the middle of the anti-roll bar monitors the medial wheel travel and actuates the control switch by a rod.

Instead of a power-operated pump being necessary, the movements of the vehicle can be used to achieve a pumping effect on the level-control units. Such 'self-pumping' systems do not work, however, if the vehicle is loaded or unloaded while at a standstill.

Early attempts and proposals were made to influence the vehicle or wheel attitude in cornering, partly by purely mechanical means. On an experimental car of the 1930s, **Fig. 5.49** a, the vehicle body was connected to the chassis frame by pendula intersecting at a pole P above the centre of gravity; this allowed the body to swing sideways like a hammock and thus

Springs and Dampers

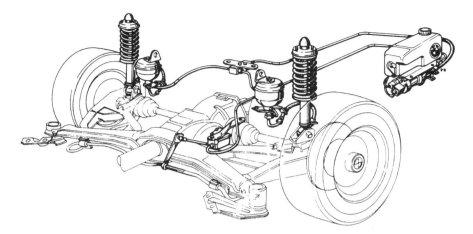

**Fig. 5.48** 'Partially charged' hydropneumatic levelling system (courtesy BMW AG)

to lean inwards in a corner (13), similarly to systems known for railway carriages. While the almost horizontal attitude of the body was found to be comfortable, a lack of sensivity to adhesion limits was criticized.

For the (again experimental) wheel suspensions of a racing car (b), power-control of the wheel camber was actuated by the driver's knees, but his skill proved overtaxed by this task! In the proposal shown in Fig. 5.49c, both the body and the wheels are inclined towards the inner side of a corner by power assistance.

In view of the properties of modern radial-ply tires, camber control needs to be applied only within narrow limits.

By far the most pretentious successors to these optimistic mechanisms are the 'active suspension systems' of today. The term 'active suspension' means springing or damping systems that are activated by sensors and a logical control circuit to conduct the vehicle body over an uneven road surface with minimal roll angles and vibrational accelerations. The ideal state

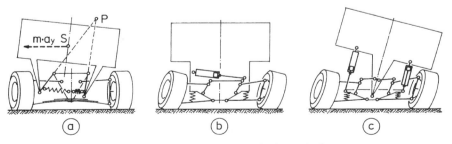

**Fig. 5.49** Early attempts of vehicle and wheel attitude control

would be the ride on a 'skyhook' – i.e. floating of the vehicle orientated to the 'fixed' coordinate system of the globe. Such a system would ideally have to monitor the road surface immediately in front of each wheel. While this measuring process has been until recently hardly practicable with justifiable expense, in respect of the various possible ambient conditions such as ice or flood water, today's quickly responding sensors now make it possible to analyse a wheel deflection at its initial stage and to compensate for it by a control system. The technical content is remarkable, and the safety of the system must be ensured by multiple-circuit control devices because a failure – e.g. by an unexpected load distribution – would immediately and seriously affect the vehicle's driving behaviour.

A less complex variant of 'active suspension' sometimes offered on today's passenger cars compensates only for the roll angle and cedes bounce control to a conventional (and slow) levelling system. The effect is comparable with that of extremely stiff anti-roll bars and may influence the vehicle's driving behaviour (e.g. by reinforcing a 'jacking-up' effect on certain suspension systems – see Chapter 7).

Bearing in mind the present state of conventional 'passive' passenger-car springing systems, with well-balanced pitching behaviour and moderate roll angles even with high lateral acceleration, and their working reliability, active systems *must* show advantages not achievable with these conventional systems if they are to gain widespread acceptability. Among those advantages are the obviation of 'copying vibrations' – an unpleasant rolling movement on uneven roads clearly excited by the irregularities below the wheels of the more-strongly stabilized axle (on passenger cars, the front one) – and, in co-operation with sensors and a vehicle-dynamic processor, on-the-move control of the roll-moment distribution on the axles, to improve driving stability.

## 5.8 The wheel-travel angle

The forces acting on a free-rolling wheel and not caused by braking or traction can be transferred to the wheel carrier only through the wheel bearings and, consequently, the wheel centre M, **Fig. 5.50**. These forces are the wheel load $F_z$, the rolling resistance $F_R$ and other impact or resistance forces such as those induced in aquaplaning. The rolling resistance force $F_R$ acts horizontally on the tyre contact point and is proportional to the wheel load $F_z$. However, if the resulting force from $F_z$ and $F_R$ acted on the tyre contact point, it would miss the wheel centre and would therefore exert a moment at the wheel, which is impossible when wheel-bearing

Springs and Dampers

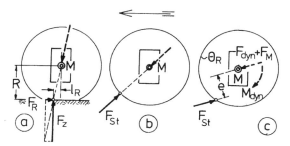

**Fig. 5.50** Forces acting on a free-rolling wheel:
a) rolling resistance  b) concentric and  c) eccentric impact

friction is neglected. In reality, the rolling resistance arises from the deformation energy at the tyre in a complicated manner. The line of action of the resulting force must pass through the wheel centre, Fig. 5.50a, and the condition of equilibrium is $F_R = F_z l_R / R$. With the non-dimensional rolling resistance coefficient $f_R = l_R / R$, the rolling resistance force is defined by

$$F_R = f_R F_z \qquad (5.62)$$

Skewed impact forces, too, can be transferred only by the wheel axis, whether they act concentrically (Fig. 5.50b) or eccentrically (c). An eccentric impact force $F_{St}$ causes a rotational acceleration at the wheel, and the couple formed by the force $F_{St}$ and the sum of the inertia force $F_{dyn}$ and the reaction force at the wheel bearing $F_M$ balances out the moment $M_{dyn}$ of the wheel inertia.

It is clear from Fig. 5.50 b and c that an impact effect on a vehicle can be softened by a suspension designed to provide compliance in the direction of the impact force. However, this can be realized within narrow limits only, owing to other requirements of vehicle dynamics and roadholding.

As a pure kinematic 'wheel suspension characteristic', to describe the reaction of a suspension to impact forces, the 'wheel-travel angle' $\varepsilon$ has been introduced – i.e. the side-view angle by which the wheel centre departs from the vertical with wheel travel. It is very conspicuous for a telescopic fork of a motorcycle – see **Fig. 5.51a** – but can be provided on any wheel suspension, as for instance the trailing arm shown in Fig. 5.51b. The wheel-travel angle is defined as positive on front and rear wheels if the wheel centre moves backward relative to the vehicle with wheel bump, and it follows from the components of the wheel centre velocity $\mathbf{v}_M$:

$$\tan \varepsilon = - v_{Mx} / v_{Mz} \qquad (5.63)$$

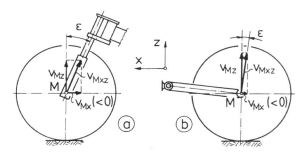

**Fig. 5.51** The wheel-travel angle

Since all high-frequency impacts cause inertial forces at the wheel and on the suspension, the wheel-travel angle is of minor importance in respect of riding comfort and at best offers advantages for low-frequency impacts of high amplitudes, as in off-road driving. Moreover, it does not provide any isolation against the high-frequency disturbing forces exerted by normal road irregularities, such isolation being the purpose of the rubber-elastic joints that are standard in all modern wheel-suspension systems.

# Chapter 6

# Traction and Braking

## 6.1 Steady-state accelerating and braking

Accelerating and braking forces are transmitted from the vehicle to the road surface via the tyre contact areas. The inertia force acts on the vehicle's centre of gravity, generating a pitching moment that respectively increases and decreases the front and rear wheel loadings in case of braking, and conversely for accelerating. The braking and accelerating processes are distinguished only by the sign of the longitudinal acceleration $a_x$. In the following, the braking process of a two-axle vehicle will be analysed first.

The resulting front and rear wheel loads $F'_{zv}$ and $F'_{zh}$ during braking, **Fig. 6.1**, follow from the condition of static balance and the braking deceleration $a_x$ (<0) as

$$F'_{zv} = mg(l - l_v)/l - ma_x h/l \qquad (6.1a)$$
$$F'_{zh} = mgl_v/l + ma_x h/l \qquad (6.1b)$$

Assuming friction coefficients $f_v$ and $f_h$ at the front and rear axles respectively, the braking forces are

$$F_{xv} = F'_{zv} f_v \qquad (6.2a)$$
$$F_{xh} = F'_{zh} f_h \qquad (6.2b)$$

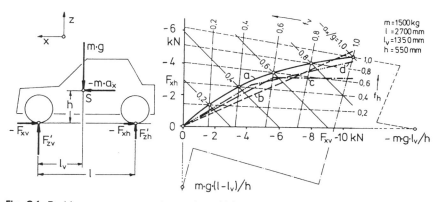

**Fig. 6.1** Braking process on a two-axle vehicle

and their sum is the decelerating force on the vehicle mass:

$$F_{xv} + F_{xh} = ma_x \tag{6.3}$$

For the braking process, $f_v$ and $f_h$ are negative.

If equal grip utilization at both axles is desired ($f_v = f_h$), the equations (6.1) and (6.2) lead to $f_v = f_h = a_x/g$, and the 'ideal' braking-force distribution is

$$F_{xv} = ma_x[g(l - l_v) - a_xh]/(gl) \tag{6.4}$$
$$F_{xh} = ma_x[gl_v + a_xh]/(gl) \tag{6.5}$$

This is the parametric representation of a parabola $F_{xh}(F_{xv})$ with the parameter $a_x$. In the right-hand section of Fig. 6.1 the ideal braking-force distribution of an average passenger car is shown (a). In practice, a diverging distribution is usually chosen to avoid premature overbraking or locking of the rear wheels owing to friction tolerances of the brake linings or pads, because locked rear wheels can exert little lateral grip, and the vehicle will therefore tend to spin. An anti-lock brake system (as common today) will moderate this problem, however.

Each loading state of the vehicle of course requires its own ideal braking-force distribution. If all states of loading have to be met by one 'fixed' distribution, the crucial criterion is normally that of the empty vehicle carrying the driver alone. A fixed distribution, though, prevents optimum braking forces at higher states of loading. The line (b) shows such a braking-force distribution where the rear-axle braking force does not exceed the ideal until a maximum deceleration of 0.8 g. 'Bent' distribution curves can be applied, too, as in case (c) by use of a pressure-limiting valve at the rear axle and in case (d) by a pressure-reducing valve. Vehicles with extreme loading variations - for instance trucks or station wagons as well as many front-drive cars - have pressure-reducing valves with a variable switch-over

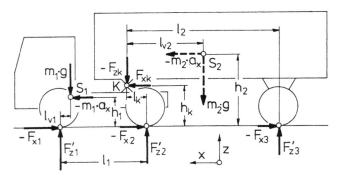

**Fig. 6.2** Braking process on a semi-trailer truck

# Traction and Braking

point depending on the static axle load (the so-called 'brake power regulator').

On a two-axle vehicle, the wheel loads during braking depend on the deceleration $a_x$ only, and not on the installed braking-force distribution. This is no longer valid on vehicles with three or more axles – e.g. a semi-trailer truck, **Fig. 6.2**. The height coordinates $h_K$, $h_1$ and $h_2$ of the semi-trailer coupling, the centres of gravity of tractor and trailer, and the installed braking-force distribution determine the coupling forces $F_{xk}$ and $F_{zk}$ and thus the 'dynamic' load distribution on all axles.

The equations established here for the braking process are valid, too, for accelerating, of course using a positive sign for the acceleration, $a_x > 0$.

## 6.2 The support angle

### 6.2.1 General remarks

Accelerating and braking forces act on the tyre contact points A in the longitudinal direction. **Fig. 6.3** shows schematically an all-wheel-drive vehicle in side view under steady-state straight-ahead acceleration. Assuming the vehicle to have been in a state of equilibrium before accelerating, the 're-duced' spring forces will have corresponded to the static wheel loads. It will, therefore, be sufficient to determine the new state of equilibrium by calculating the additional forces arising from acceleration (assuming an average level of damping $D \approx 0.3$, the vehicle will need approximately one second to assume its new equilibrium state).

The distribution of the front and rear longitudinal forces is normally determined through design, as for instance in the case of all-wheel drive by a power-divider transfer gear, and in the case of braking by the diameters

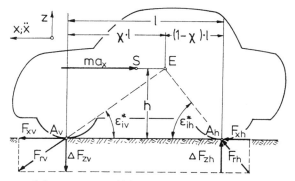

**Fig. 6.3** Diagram of forces on a vehicle under steady-state acceleration

of the brake disc and the brake-caliper pistons. With inverted signs, the diagram of forces in Fig. 6.3 is analogously valid for a steady-state braking process. The situation is simplified insofar as the 'unsprung' and the 'rotating' masses have been combined with the vehicle mass, while aerodynamic forces have been neglected.

The front and rear longitudinal forces $F_{xv}$ and $F_{xh}$ generate the longitudinal acceleration $a_x$ of the vehicle and the inertia force $ma_x$ at the centre of gravity S. The axle-load transfer $\Delta F_{zv}$ at the front and the inverted transfer $\Delta F_{zh}$, of equal amount, at the rear follow from the inertia force $ma_x$, the wheelbase l and the centre-of-gravity height h. The resulting forces $F_{rv}$ and $F_{rh}$ of the longitudinal forces $F_{xv}$ and $F_{xh}$ and the load-transfer forces $\pm \Delta F_{zv,h}$ intersect the horizontal line of action of the inertia force $ma_x$ at a point E, the position of which depends on the longitudinal force distribution. For a longitudinal force share $\chi$ of the front axle, the distance of E from it is $\chi l$. The front-axle share depends on the requirements of optimum friction utilization and of vehicle-dynamics safety criteria, as mentioned in Section 6.1.

The resulting forces $F_{rv}$ and $F_{rh}$ are inclined with respect to the road surface by angles $\varepsilon_{iv}^*$ and $\varepsilon_{ih}^*$. Using Fig. 3.13 in Chapter 3, the reaction of wheel suspension and spring on such forces has already been explained, introducing a 'support angle' $\varepsilon_B$ to characterize the suspension properties under braking force. In accordance with Chapter 3, a support angle that is determined assuming a 'locked' torque support at the wheel carrier may in the following be marked by a star (*). The support angle $\varepsilon^*$ defines in side view the direction of the line of action of a force acting on the tyre contact point A which can be reacted by the suspension mechanism without loading the spring. It was shown, moreover, in Chapter 3 that the support angle follows from a virtual velocity vector $v_A^*$ at the tyre contact point A which arises from a virtual wheel motion assuming a perhaps-engaged torque support to be locked.

If a brake at the wheel carrier like that of Fig. 3.13 is locked, the tyre contact point A can instantaneously be taken as being fixed to the wheel carrier too, and, in terms of longitudinal forces, the suspension mechanism can be replaced in side view by a guiding slot which generates the (virtual) path of the tyre contact point, **Fig. 6.4**. On this guiding slot, the diagram of forces is easily established (see Section 3.5): the longitudinal braking force $F_{xB}$ is balanced by the horizontal component of the reaction force $F_k$ acting on the slot (or, in reality, on the suspension mechanism),

$$F_k \cos \varepsilon^* = F_{xB}$$

Traction and Braking

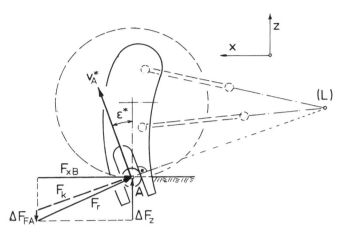

**Fig. 6.4** Braking-force support angle

and the vertical component of $F_k$ helps the spring-force change $\Delta F_{FA}$ to catch up the wheel-load change $\Delta F_z$:

$$F_k \sin \varepsilon^* + \Delta F_{FA} = \Delta F_z$$

From both equations follows

$$\Delta F_{FA} = \Delta F_z - F_{xB} \tan \varepsilon^* \tag{6.6}$$

The greater the support angle, the smaller the spring-force change.

Fig. 6.4 shows a front-suspension system; for a rear one it is clear from the diagram of forces at the vehicle (Fig. 6.3) that the 'guiding slot' or the vector $v_A^*$ would need an opposite inclination to counteract the forces at the rear wheel. Hence, the definition of a support angle must distinguish between front and rear wheels, in contrast with the definition of the wheel-travel angle $\varepsilon$ (see Section 5.8). If $v_A^*$ is the vector of the virtual velocity of the tyre contact point A, assuming a locked torque support at the wheel carrier, the support angle follows from

$$\tan \varepsilon_{v,h}^* = \pm v_{Ax}^* / v_{Az}^* \tag{6.7}$$

where the upper sign is valid for front suspensions, subscript 'v'.

Any wheel travel during an accelerating or braking process would obviously be inhibited if the 'guiding slot' of Fig. 6.4 was perpendicular to the force $F_r$ and this force could, thus, be balanced without loading the spring. From equation (6.6) it follows for $\Delta F_{FA} = 0$ that $\tan \varepsilon^* = \Delta F_z / F_{xB}$, and this defines the 'ideal' support angle to achieve '100% anti-dive' or 'anti-squat' effect. According to the diagram of forces in Fig. 6.3, the ideal support angles at the front and the rear wheels are to be defined by

$$\tan \varepsilon_{iv}^* = h/(\chi l) \qquad (6.8a)$$
$$\tan \varepsilon_{ih}^* = h/[(1 - \chi)l] \qquad (6.8b)$$

If provided with an ideal support angle $\varepsilon_i^*$, the wheel suspension is able to balance the resulting force $F_r$ in a state of (normally 'unstable') equilibrium and without any spring-force change. In reality, the support angles $\varepsilon^*$ are different from the ideal angles and cause spring-force changes and, consequently, wheel travel.

The intersection point E of the forces in Fig. 6.3 is the 'accelerating centre' or 'braking centre' (**19**). Traction-force distribution may be chosen rather differently according to the purpose of the vehicle, while braking-force distribution is determined by the conditions mentioned above. Accordingly, the braking centre is nearer to the rear axle the nearer the centre of gravity is to the front axle.

If only one axle is driven (or braked) - for instance the front axle, **Fig. 6.5** - point E is located above the non-powered or non-braked axle. The corresponding 'ideal' support angle would then be 90°, which is not practicable because it would mean locking of the springing system.

On an average passenger car with a wheelbase of 2700 mm and a centre-of-gravity height of 550 mm, the ideal support angle for single-axle drive/braking is 11° 30', and for all-wheel drive/braking with a front-axle share of 70% the ideal front and rear support angles are 16° 10' and 34° 10'.

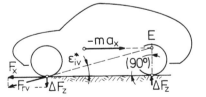

**Fig. 6.5**
Diagram of forces for single-axle drive

## 6.2.2 'Torque support' at the wheel carrier

The suspension characteristic 'support angle', as in Figs 6.3 and 6.4, was introduced on the assumption that the torque support is mounted between the wheel and the wheel carrier. For a locked torque support, the tyre contact point may therefore be regarded as being instantaneously fixed to the wheel carrier. This is true for the suspension types shown in **Fig. 6.6**.

Traction and Braking

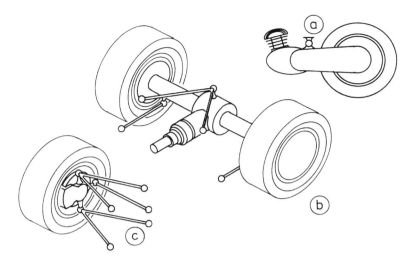

**Fig. 6.6** Examples of torque supports mounted on the wheel carrier

On a drive assembly as frequently used on light motorcycles (a), the engine is mounted at the wheel carrier (here the chain-drive housing is serving as trailing-arm suspension) - so-called 'drive-unit wheel carrier' - and may even be installed inside the wheel ('wheel-hub motor', maybe for hydrostatic drive).

If the final-drive gearbox is integrated within the beam of a rigid axle and the drive torque transferred to the axle via a longitudinal driveshaft (b), and if the motor is assumed to be locked, the final drive will also appear locked with parallel wheel travel. However, the driveshaft generates a 'roll moment' $M_D$ acting between the axle and the vehicle body, **Fig. 6.7**, which causes different wheel loadings $\Delta F_{z1}$ and $\Delta F_{z2}$ and reduces the transferable traction performance unless a differential-lock device is provided. The loading difference can be avoided by a non-central torque-reaction

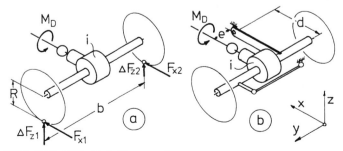

**Fig. 6.7** Rigid axles with final-drive gearboxes

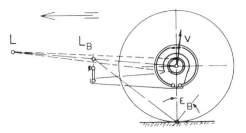

**Fig. 6.8** Schematic of rotatably mounted brake support

arm (Fig. 6.7b); with a final-drive gear ratio i, equal wheel loading is achieved for $e = d/i$. Assuming the brakes also to be mounted on the axle, though, as is normal, that measure would lead to different wheel loadings during braking - an even less desirable situation.

Nowadays for independent suspension the brakes are generally attached to the wheel carriers, as shown in Fig. 6.6c. Inboard disc brakes would be subject to tight dimensional limits and to danger of stone impact; moreover, the driveshafts that transmit the braking torque to the wheels would cause increased resilience of the driveline - a disadvantage in respect of the oscillations excited by anti-lock brake systems.

Depending on the suspension type, it may not be possible for the braking-force support angle to be freely chosen. This situation can be remedied by providing a rotatable arrangement of the torque support (e.g. the brake caliper) at the wheel carrier and to control its movement with wheel travel. **Fig. 6.8** shows the wheel of a rear suspension with an instantaneous pole L, a given wheel-travel direction (v) and a rotatably mounted brake support. An arm connecting the brake support to the chassis defines a 'braking pole' $L_B$ on the polar ray from the wheel centre to L, and with $L_B$ a braking-force support angle $\varepsilon_B$. Mounting of the control arm to a suspension link is also possible. Assuming the brake to be locked, the tyre contact point will swivel about the pole $L_B$ with wheel travel. When applying the methods shown in Chapter 3 to determine the support angle, the rotatably mounted brake support has here transiently to be regarded as the wheel carrier.

The wheel-travel angle $\varepsilon$ (see Section 5.8) and the support angle $\varepsilon^*$ define a 'longitudinal pole' L in the side-view plane, as demonstrated in **Fig. 6.9**: the polar rays to the wheel centre M and to the tyre contact point A, and inclined by $\varepsilon$ and $\varepsilon^*$, intersect at the pole L. With the tyre radius R, the polar distance becomes

$$p = R/(\tan\varepsilon^* \pm \tan\varepsilon) \tag{6.9}$$

Traction and Braking 117

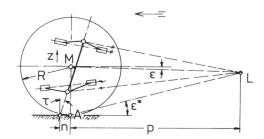

**Fig. 6.9**
Longitudinal pole
and castor change

(upper sign valid for front wheels) and is especially important for steered wheels. This is because the wheel carrier swivels about L in side view with wheel travel and with it the 'kingpin axis', thus changing the castor angle $\tau$ and the castor offset n; this may be felt adversely at the steering wheel in cornering. Because of this, castor-angle change exceeding 2-3° per 100 mm wheel travel should be avoided on the larger passenger cars. The pole distance p may therefore set certain limits on the support angle $\epsilon^*$ on steerable wheels, unless a negative wheel-travel angle $\epsilon$ is accepted.

In contrast to two-track vehicle suspensions, front-wheel forks of motorcycles generally show an opposed range of steering and suspension mechanisms, **Fig. 6.10**. The wheel suspension (here short 'leading links' K) swivels about the kingpin axis when steered, and the 'kingpin' is located between the suspension and the cycle frame at what is called the 'steering head'. This arrangement enables a considerable braking-force support angle $\epsilon^*$ without disturbing change of castor angle and castor offset. A rotatably mounted brake support BA is controlled by the suspension links K (i.e. the wheel carrier) and a torque-reaction link BS, together defining the

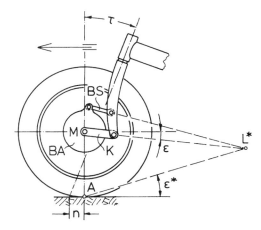

**Fig. 6.10**
Motorcycle fork suspension
with 'sprung' kingpin

118                                                                    Road Vehicle Suspensions

braking pole $L^*$; the instantaneous castor offset change with wheel travel is zero for equal castor and wheel-travel angles: $dn/ds = 0$ for $\tau = \varepsilon$.

The 'Dubonnet' suspension which was frequently utilized in former times (see Chapter 12) shows the same basic arrangement of suspension and steering mechanisms and therefore offers the same functional possibilities.

### 6.2.3 'Torque support' at the vehicle chassis

On nearly all driven independent suspensions, the traction torque is transmitted to the wheels from a chassis-mounted torque support (e.g. the final drive) via driveshafts with universal joints, **Fig. 6.11** a, while, as mentioned earlier, the brakes are nowadays preferably installed at the wheel carriers.

Chassis-mounted final drive is a characteristic, too, of the 'de Dion' axle, Fig. 6.11 b - a light rigid axle that was invented as early as 1893 to reduce unsprung mass and to reduce damage to the road surface. It was frequently adopted for the rear wheels of fast and sporty vehicles, and appeared in the 1920s for the front wheels of a racing car. Its essential advantage is that it avoids the asymmetrical wheel loading resulting from the torque of the longitudinal driveshaft on normal rigid axles, see Fig. 6.7.

Looking at the simple trailing-arm suspension of **Fig. 6.12**, it is clear that the motion is different from that shown in Fig. 6.4 when applying the theorem of virtual work in order to determine the support angle. If the

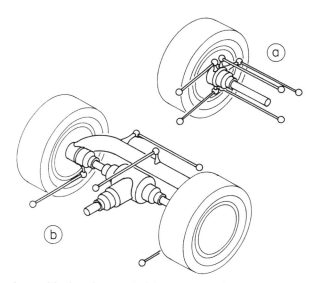

**Fig. 6.11** Suspensions with chassis-mounted torque support

Traction and Braking

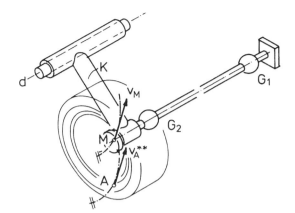

**Fig. 6.12**
Trailing-arm suspension with driveshaft

driveshaft is assumed to be locked at its inner end, it prevents the wheel rotating in side view with bump and rebound, and the wheel performs a translational motion along a circular path given by the trailing arm's rotation about its chassis-side turning-joint axis d. Hence, the tyre contact point A copies the path of the wheel centre M, so its virtual velocity $v_A^*$ is the same as the wheel-centre velocity $v_M$, and the 'longitudinal pole' of the wheel (regarded as a spatial body) lies at infinity.

However, the wheel-centre velocity $v_M$ defines the wheel-travel angle $\varepsilon$ – see Chapter 5, Fig. 5.51 – and it is therefore to be expected that the traction-force support angle will have the same value.

In the following, kinematic parameters that result from the assumption of a 'locked' torque support mounted to the vehicle chassis or body will be marked by a double star (**) to distinguish them from those applying to a wheel-carrier-mounted torque support. The virtual velocity vector of the tyre contact point will therefore be $v_A^{**}$ and the support angle $\varepsilon^{**}$.

Calculation of the support angle is done using the methods and equations given in Chapter 3, Section 3.4, covering the influence of driveshafts according to Section 3.6. The wheel centre velocity $v_M$ and the (virtual) angular wheel velocity $\omega_R$ define the virtual velocity $v_A^{**}$ of the tyre contact point A and, in analogy to equation (6.7), the support angle for driveshaft transmission by

$$\tan \varepsilon_{v,h}^{**} = \pm v_{Ax}^{**} / v_{Az}^{**} \qquad (6.10)$$

the upper sign being valid on front wheels (subscript 'v').

The influence of a possible hub reduction gear is also analysed according to Section 3.6. Hub reduction gears are used on heavy trucks and off-road vehicles to reduce the unsprung weight of the driveline parts and/or to increase the ground clearance.

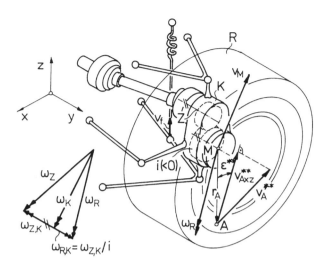

**Fig. 6.13** Schematic diagram of independent suspension with driveshaft and reduction gear

If built into a non-steerable rigid axle which carries the final-drive gearing and is driven by a longitudinal driveshaft, a reduction gear will not affect the support angle, since all driveline parts are 'locked' with the engine.

In contrast, in a suspension driven by transverse shafts from a chassis-mounted final drive, **Fig. 6.13**, relative rotations of the parts will occur as described in Section 3.6; the angular wheel velocity $\boldsymbol{\omega}_R$ and the virtual tyre-contact-point velocity $\mathbf{v}_A^{**}$ follow from the equations (3.33) and (3.34), and the support angle $\varepsilon^{**}$ from equation (6.10). In the example of Fig. 6.13, the reduction gear has two shafts and therefore reverses the direction of driveshaft and wheel rotation; the vectors of the relative angular velocities $\boldsymbol{\omega}_{Z,K}$ and $\boldsymbol{\omega}_{R,K} = \boldsymbol{\omega}_{Z,K}/i$ (with $i$ = reduction ratio) are opposed.

In analogy to equation (6.9), a 'pole' is derivable from the support angle $\varepsilon^{**}$ but it is now valid for the wheel body and not the wheel carrier. In **Fig. 6.14** the occurrence of such a pole is demonstrated on two 'planar' suspensions - i.e. a wheel driven by a pinion coaxial with a trailing arm's mounting and a train of gears (a), and a wheel carrier with hub reduction gear driven by a transverse shaft (b).

In the example (a), the wheel and the pinion rotate in the same sense and, assuming the pinion to be locked, the wheel will rotate in relation to the trailing arm (or the wheel carrier) with an opposed angular velocity $\Delta\omega_R = \omega_K/i$ if $i$ is the reduction ratio (>0 for same-sense rotation). The absolute wheel velocity is then $\omega_R = \omega_K(i-1)/i$ and the distance of the (virtual) pole $L_A$ of the wheel follows from $\omega_R \rho = \omega_K l$ as

Traction and Braking 121

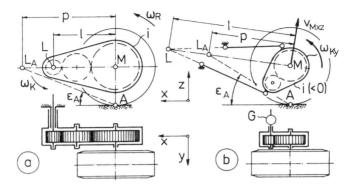

**Fig. 6.14** Reduction gear ratio and virtual pole of the wheel

$$p = li/(i - 1) \tag{6.11}$$

Generally, the distance of the wheel-carrier pole L can be determined from the side-view component $v_{Mxz}$ of the wheel-centre velocity and from the transverse component $\omega_{Ky}$ of the wheel-carrier angular velocity, $l = v_{Mxz}/\omega_{Ky}$, and the distance of the pole $L_A$ follows from equation (6.11) as

$$p = \frac{v_{Mxz}}{\omega_{Ky}} \cdot \frac{i}{i - 1} \tag{6.12}$$

see Fig. 6.14 b.

The lines connecting the poles $L_A$ and the tyre contact points A are inclined to the road surface by the traction-force support angles $\varepsilon_A$.

An arrangement comparable with that of Fig. 6.14a was used in a Scammell heavy truck suspension providing an unusual degree of articulation. Pivot-mounted on the ends of a rigid 'axle beam' (which was connected to the vehicle by conventional leaf springs) were cast-iron housings, each containing a train of gears, the input pinions of which were driven by half-shafts within the beam; each of these gear trains drove two wheels in tandem formation – see Section 2.3.4.

Reversed direction of rotation ($i < 0$) leads to a polar distance $p < l$ and consequently to a high support angle, as on the two-shaft gear of Fig. 6.14 b.

A transverse driveshaft without hub reduction gear represents, so to speak, a reduction ratio $i = 1$. According to equation (6.11), the polar distance is then $p = \infty$, and the support angle $\varepsilon^{**}$ assumes the amount of the wheel travel angle $\varepsilon$, as already shown in Fig. 6.12.

### 6.2.4 Special cases

A transverse driveshaft with a single (cardan) joint at its inner end, lying on the rotational axis of the wheel carrier and often working as a suspension joint too, was utilized for decades in an extremely simple form of swing-axle suspension, **Fig. 6.15**. The traction-force support angle can be calculated by the means given in Section 3.6, assuming for instance that the only driveshaft joint is the 'outer' joint G2 (see also joint 2 in Fig. 3.20) and adding to this a notional driveshaft with an 'inner' joint G1 within the vehicle.

A driveshaft joint that serves also as suspension joint can normally be realized only by a cardan joint, because existing homokinetic joints are not adapted for substantial axial-force transfer.

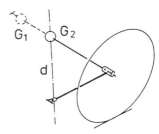

**Fig. 6.15** Driven swing axle (schematically)

Comparable with the trailing gear housing of Fig. 6.14a is the swing-axle system shown in **Fig. 6.16** where the bevel gear of the final drive is fixed to the wheel stub-axle, thus swinging with the wheel about the pinion axis which coincides with the rotational axis d of the pivoting arm and wheel

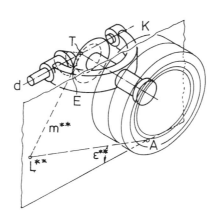

**Fig. 6.16**
Swing axle with swinging bevel gear

Traction and Braking

carrier K. This robust suspension needs no universal joint at all, and the final drive can no longer be regarded as chassis-mounted, since one half of it - i.e. the bevel gear - moves with wheel travel. The support angle can be determined according to Section 3.6, by treating the final drive like a hub reduction gear. The principle of torque transmission is easy to survey: assuming the motor to be locked, the pinion will be locked, too, and the wheel stub-axle must instantaneously rotate about the bevel gear's contact point E as well as about the intersection point T of the wheel and swing-arm axes. Therefore the instantaneous axis $m^{**}$ of the virtual wheel motion runs through E and T, and intersects the longitudinal plane through the tyre contact point A in the 'pole' $L^{**}$. The polar ray A-$L^{**}$ defines the traction-force support angle $\varepsilon^{**}$. The instantaneous axis $m^{**}$ is constant with respect to the wheel carrier, and hence the support angle remains constant with wheel travel. These considerations are exactly valid only for a bevel gear with intersecting axes.

This arrangement can also be analysed using the equations of Section 3.6 while assuming a notional 'outer' universal joint G2 at the intersecting point of the wheel and the swing-arm axes - **Fig. 6.17** - adding a notional drive-shaft in the rotational axis d of the swing arm, and a notional 'inner' joint G1. The bevel-gear (or final-drive) ratio has to be interpreted as a 'hub-reduction gear ratio'. In case of a hypoid gear as indicated in the drawing, the (virtual) output shaft is assumed to be parallel to the wheel axis at a distance of the hypoid displacement HV.

The swing-arm suspension of Fig. 6.16 shows a considerable support angle, while its wheel-travel angle is about zero, as usual on swing-arm layouts. In reality, however, this suspension type is preferably of antimetric design, having one of the two pinions in front of and one behind the wheel axis. This results in support angles of the same value but with opposite signs. The effective support angle of the axle system is then zero, and a roll torque is induced between the vehicle and the wheels, as with the rigid axle of Fig. 6.7a.

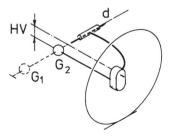

**Fig. 6.17**
Kinematic model of the suspension of Fig. 6.16

On motorcycles, chain drive has been the preferred form until recently, generally combined with a trailing-arm (or trailing-fork) rear suspension, **Fig. 6.18**.

If the brake is anchored to the trailing arm, the latter's pivot joint (or in general the longitudinal pole of the suspension) is the 'braking pole' $L_B$ too, and its polar ray $L_B$-A defines the braking-force support angle $\varepsilon_B$.

Except for the case of a 'drive unit' (see Fig. 6.6a), the engine of a motorcycle is normally installed within the frame. Assuming it to be locked, the engine output sprocket is locked, too. When a traction force is applied, the wheel appears to be guided by a 'linkage' consisting of the trailing arm and the upper strand of the chain, and the 'traction pole' $L_A$ is the intersecting point of both, thus defining the traction-force support angle $\varepsilon_A$. On engine braking, a corresponding pole $L_{MB}$ and a support angle $\varepsilon_{MB}$ follow from the trailing arm and the lower chain strand. A motorcycle suspension with chain drive therefore generally has three support angles.

If a more sophisticated mechanism is chosen for the wheel suspension – as, for example, a double-crank linkage – the poles $L_A$ and $L_{MB}$ are the intersecting points of the respective chain strands and the line connecting the wheel centre M and the longitudinal pole of the suspension.

The motorcycle suspension of Fig. 6.18 shows a certain similarity to the gear-housing suspension of Fig. 6.14a. If the engine-sprocket axis and the trailing-arm mounting were to coincide (which has occasionally been done but – except for avoiding a sprocket-to-wheel distance change with wheel travel – without any advantage to the support angles or to chain and tyre wear), the system would be equivalent to that of Fig. 6.14a, the gear train merely being replaced by the chain drive. As the sprocket radii define the final-drive ratio, the traction pole follows according to the theorem of proportional segments, and in full conformity with equation (6.11).

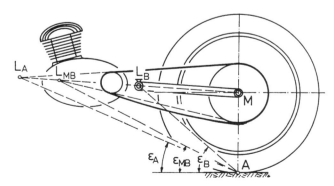

**Fig. 6.18** Support angles on a chain-drive suspension layout

Traction and Braking

## 6.2.5 Effective support angles

The preceding sections gave several examples of how to determine support angles, and here is a general and sufficient method. The support angle is the inclination angle of a virtual velocity vector of the tyre contact point with respect to the vertical line in side view which arises from a wheel displacement, assuming the respective involved torque support to be locked; the support angle is positive if the tyre contact point moves forward for a front wheel with bump, or rearward for a rear wheel.

Traction and braking-force support angles for a wheel-carrier-mounted torque support follow from equation (6.7):

$$\varepsilon_A = \varepsilon^* \qquad \text{and} \qquad \varepsilon_B = \varepsilon^*$$

while for a chassis-mounted torque support with driveshaft transmission and a possible hub reduction gear they follow from equation (6.10):

$$\varepsilon_A = \varepsilon^{**} \qquad \text{and} \qquad \varepsilon_B = \varepsilon^{**}$$

However, these support angles are defined with respect to the vehicle coordinate system. To get the support angles relating to the road surface, the pitch angle $\vartheta$ that is normally caused by horizontal forces must be heeded – **Fig. 6.19** – and, for the vehicle-based support angles and the pitch angle, the effective support angles are

$$\varepsilon_e^* = \varepsilon^* \pm \vartheta \qquad (6.13a)$$
$$\varepsilon_e^{**} = \varepsilon^{**} \pm \vartheta \qquad (6.13b)$$

In each case the upper sign is valid for front wheels.

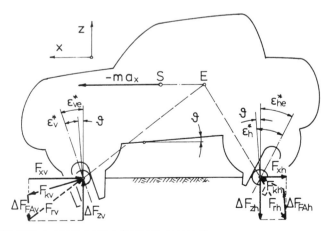

**Fig. 6.19** Effective support angles (braking process)

As already mentioned, the support angles represent the instantaneous paths of the tyre contact points and are realized in Fig. 6.19 in simple form by guiding slots at the vehicle body. Forces $F_k$ acting rectangularly to the slots do not cause any spring deflection. The diagrams of forces visualize the function of (positive) support angles: the spring-force changes $\Delta F_{FA}$ at the front and rear wheels are considerably lower than the 'dynamic' wheel load transfer $\Delta F_z$ and amount to about 70% of the latter, which means 'anti-dive' or 'anti-lift' effects of about 30%. It should be explained here that terms such as 'anti-dive' etc. do not define 'suspension characteristics' since anti-dive depends on the relevant support angle on the one hand and on the diagram of forces and the vehicle dimensions on the other; a suspension with a specific support angle will cause different pitch angles in vehicles with different wheelbases.

The 'reduced' spring force $F_{FA}$ is orientated in the vehicle coordinate system like the support angles. The spring-force changes at the front wheels (upper sign) and the rear wheels follow, in terms of the pitch angle, from

$$\Delta F_{FAv,h} = \frac{\Delta F_z - F_{xv,h} \tan \varepsilon^*_{ev,h}}{\cos \vartheta \pm \sin \vartheta \tan \varepsilon^*_{ev,h}} \qquad (6.14)$$

A realistic determination of the resulting vehicle attitude is, of course, to be carried out by integration, or at least iteration, because the support angles and spring rates normally vary with the pitch angle.

## 6.3 Traction and braking pitching

### 6.3.1 Static and dynamic pitching under traction and braking

For a vehicle with given dimensions, and with linear springs and constant support angles, the diagram on the left of **Fig. 6.20** shows the front and rear wheel travel $s_v$ and $s_h$ and the pitch angle $\vartheta$ versus the deceleration $a_x$ during steady-state braking. Because of the superimposition of the support angles and the pitch angle - see Fig. 6.19 - wheel travel increase is degressive at the front axle and progressive at the rear one, while the pitch-angle increase is linear. In the right-hand diagram, dynamic pitching is plotted against time, assuming a sudden deceleration input of 0.5 g. As equilibrium of wheel loading, braking forces, spring forces and reaction forces ($F_k$ in Fig. 6.19!) is not established immediately, a pitching oscillation arises with the natural pitch frequency, and approximate steady-state values are assumed towards the end of the first cycle.

# Traction and Braking

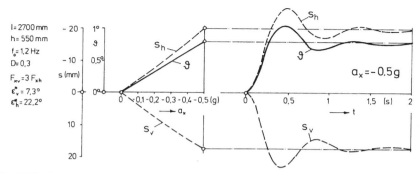

**Fig. 6.20** Static and dynamic pitching under braking

A support angle greater than the 'ideal' one for '100% anti-dive' means over-compensation of the external forces and, for example, lifting of the vehicle's nose with braking. The more effective the anti-pitch measure, the smaller the attitude change of the vehicle even with transient braking or accelerating. If high support angles for the anti-pitch device are provided, good damper functioning is necessary to avoid wheel tramp on an uneven road or with harsh braking or accelerating.

Advantages of an anti-pitch arrangement are:
- avoidance of extreme wheel travel and therefore conservation of the desired spring and damper rates and of an adequate wheel travel even on bad road surfaces;
- reduction of vehicle-attitude variation, giving improved comfort and road-holding, for example on a power variation in cornering;
- possibility of providing a 'softer' spring setting and thus lower damper forces for achieving a desired level of damping.

Anti-pitch measures do **not**, however, cancel the dynamic load-transfer forces between the front and rear axles, as is occasionally presumed.

On the 'down' side of anti-pitch, 'harsher' spring response during braking or accelerating is sometimes reported, which might be caused by the quicker response to longitudinal forces rendered possible by reduced transient vibration, while a possibly negative wheel-travel angle can hardly be blamed for this, considering the longitudinally compliant mountings of modern suspensions. With many wheel suspensions, however, there is an interaction between longitudinal forces and the springing system – see later in Section 6.3.5.

On vehicles with variable braking-force distribution (as for instance motorcycles with independently operated front and rear brakes) it is better to provide support angles that do not exceed the 'ideal' angles for the worst case – i.e. braking of one wheel or one axle only – in order to avoid over-compensation for pitching.

Efficient pitch compensation during braking or accelerating should be achieved only by the measures described above. Influencing the dampers during the braking process merely slows down the pitching movement, without any reduction of the final steady-state pitch angle or wheel deflection, and will cause the risk of over- or under-damping and thus wheel hop on bad road surface. Variations to the springing system, such as readjustment of gas springs, may lead to spring-rate changes (and with this again the danger of over- or under-damping) and require a response time that is not possible on normal self-levelling systems except those of 'active suspensions'.

A dynamic transient vibration as shown in Fig. 6.20 can be avoided only by 100% anti-pitch design at both front and rear axles.

As kinematic braking or accelerating anti-pitch measures need to work in conjunction with the braking or accelerating forces at the suspension, pitching is normally unavoidable on vehicles with single-axle drive or braking. However, there is an exception: the rather theoretical possibility of a rotatably mounted torque support for the braked or driven suspension and the connection of this by a linkage with the suspension of the non-braked or non-driven axle.

A brake-power control device that is influenced by wheel loading – e.g. by monitoring the deflection of a mechanical spring or the pressure of a gas one – cannot perceive a dynamic load change if the relevant suspension is designed for 100% anti-pitch geometry; in such a case, brake-power control by monitoring longitudinal accelerations will be a preferable solution.

### 6.3.2 Single-axle traction and braking

The handbrake system acts in general on one axle only, and that, for installational simplicity, preferably the non-steerable rear axle. On modern vehicles, single-axle braking is possible only via the handbrake. There are two typical cases of single-axle braking, first emergency braking because of failure of the service braking system (improbable today in view of the prevalence of dual-circuit systems), and second the attempt to start off from rest with the handbrake still applied.

As mentioned above, effective anti-lift on the rear axle of a passenger car requires a braking-force support angle of about 20-35°. If the handbrake operates at the same place as the service brake (e.g. at the wheel carrier), the installed support angle may be far greater than the 'ideal' angle for single-axle braking – **Fig. 6.21**, see also Fig. 6.5 – and, with

Traction and Braking

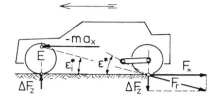

**Fig. 6.21**
Handbrake application on a rear axle with '100% anti-lift' properties

the handbrake applied alone, the vehicle will dive not only at the front axle (as expected anyhow) but at the rear axle too, the rear-wheel travel being possibly greater than that at the front. Similar rear-wheel travel may be noticeable if a front-drive vehicle is started off with its rear-axle handbrake still applied.

This has nothing to do with the considerable vehicle motion that is to be seen on rear-axle-driven vehicles with trailing-link rear suspension, or any suspension with a longitudinal pole near to the rear wheel, if an attempt is made to start off with the handbrake applied. In this case, the driving torque $M_A$ transferred to the wheel via the driveshaft is offset by the handbrake moment $M_B^*$, **Fig. 6.22**, and acts on the wheel carrier. The suspension generates an internal moment consisting of a spring force change $\Delta F_{FA}^*$ and a vertical force $F_z^*$ at the longitudinal pole P, and no traction force $F_x$ is exerted at the tyre contact point. A similar effect may be caused by traction-slip-control systems that use the brakes to influence the angular acceleration of the wheels. The medial or high-frequency braking torque fluctuation excites noticeable bounce, pitch and roll movements of the vehicle if the suspension has a short polar distance. At least, pitching can theoretically be avoided if in side view the pole coincides with the centre of gravity.

As already mentioned, anti-lift or anti-squat measures are possible only on braked or driven axles. **Fig. 6.23** shows the wheel displacement of a rear-wheel-driven vehicle against acceleration, the rear suspension being provided in the first place with '55% anti-squat geometry' (a) and then again with the 'ideal' traction-force support angle (b). Obviously the front-

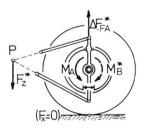

**Fig. 6.22** Starting off with applied handbrake

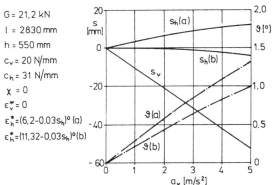

**Fig. 6.23**
Rear-axle drive with 55% (a) and 100% (b) anti-squat design

wheel travel $s_v$ is not influenced by the measures taken at the rear wheels, and, despite nearly suppressed rear-wheel travel $s_h$ in case (b), the pitch angle $\vartheta$ is about 75% of that in case (a). Kinematic measures at one axle contribute benefits to the vehicle's behaviour according only to the axle's share of the forces that are acting.

To avoid any pitching angle on single-axle-driven vehicles, it has been proposed to force parallel vehicle motion on both axles by causing the support angle to be greater than the ideal angle - i.e. to achieve parallel vehicle lifting with rear-axle drive and parallel squat with front-wheel drive (**19**). It should be pointed out here, though, that the effects of such measures on other vehicle-dynamic processes must be appreciated too - as for instance in engine braking on the move (e.g. in abrupt downshifting to a lower gear) with the risk of wheel locking.

### 6.3.3 Torque transmission via driveshafts

Deflection of a driveshaft affects the field of velocities at the wheel suspension, as explained in Section 3.6. There may well already be a deflection angle in the vehicle's 'static load' position, and there certainly will be with wheel travel. The influence of a universal-joint deflection angle on the traction-force support angle is investigated in **Fig. 6.24** (**15**). For easy visualization, a semi-trailing-link rear suspension is chosen as the exemplar, but the results are valid generally.

For an angle of incidence 0° in the normal position (a), the traction-force support angle $\varepsilon_A$ shows up as equal to the wheel-travel angle of the suspension along the entire path of the wheel, which is not surprising: since the driveshaft (on the assumption that the engine or the final drive is locked) prevents any rotation of the wheel in side view, it is clear that the tyre

Traction and Braking 131

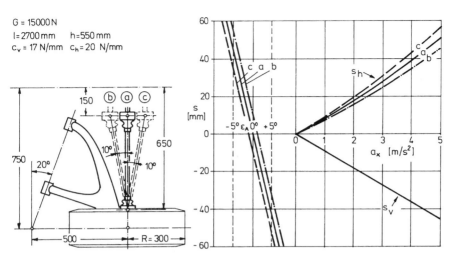

**Fig. 6.24** Influence of a driveshaft deflection angle on the support angle

contact point repeats the path of the wheel centre – see also Fig. 6.12 – and that therefore the support angle must be equal to the wheel-travel angle. The semi-trailing link in question may be horizontal in the static-load position; the wheel-travel angle, and with it the support angle, will therefore be 0° in that position, and both will decrease with wheel bump travel according to the increasing inclination of the link.

If the driveshaft runs obliquely towards the semi-trailing-link axis (b), a small positive support angle occurs in the normal position but follows the decreasing tendency of example (a), albeit with a constant difference. With inverted orientation of the drive shaft, of course, the support angle becomes negative in the normal position (c).

The angle of incidence of the driveshaft obviously has almost negligible influence on the support angle, especially as the (steady) angles of ± 10° assumed here would create serious durability problems for the universal joints.

Analysis of a pure trailing-arm suspension (with a wheel carrier rotating about an exactly transverse axis) under the same conditions as depicted in Fig. 6.24 shows no influence at all on the support angle.

The kinematic properties of a 'planar' semi-trailing-link suspension are easy to survey and therefore allow a simple explanation of this behaviour, **Fig. 6.25**. Considering the statements in Section 3.6 about the relative velocities of the wheel and the wheel carrier influenced by a driveshaft, it becomes obvious that, viewed from above, the vectors of the angular driveshaft velocity $\omega_W$ with respect to the (locked) inner shaft end, as well

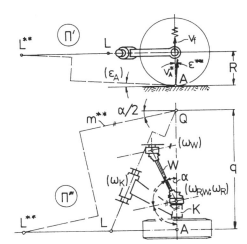

**Fig. 6.25**
Wheel motion analysis of a semi-trailing-link suspension with inclined driveshaft and locked final-drive gear

as the relative velocity $\omega_{R,W}$ of the wheel with respect to the driveshaft, are inclined to the straight-ahead direction by half the shaft deflection angle $\alpha/2$ and run parallel to each other (the shaft is in the 'Z-position', see Fig. 3.16 in Chapter 3). As $\omega_R = \omega_W + \omega_{R,W}$, the absolute angular velocity $\omega_R$ of the wheel must also be inclined by half the shaft deflection angle, and as the intersection point Q of the rotational axis of the wheel carrier and the wheel axis, the 'transverse pole', is instantaneously a fixed point at the vehicle, the instantaneous axis $m^{**}$ of virtual wheel travel (assuming a locked final drive) must run through Q and parallel to the vector $\omega_R$. The distance from the longitudinal pole $L^{**}$ to the wheel axis is then $q/\tan(\alpha/2)$ and its height above the road is, in the normal position, equal to the tyre radius R. Accordingly, the driveshaft contributes a support-angle share

$$\Delta \varepsilon_A \approx (R/q) \tan(\alpha/2) \qquad (6.15)$$

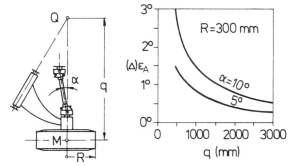

**Fig. 6.26**
Influence of driveshaft deflection on the support angle

and the effective support angle is the sum of the wheel-travel angle $\varepsilon$ and the support-angle contribution $\Delta\varepsilon_A$.

For a pure trailing-arm suspension, the pole distance is $q = \infty$ and, according to equation (6.15), the driveshaft effect vanishes.

**Fig. 6.26** shows the support-angle contribution depending on the pole distance q and the shaft deflection angle $\alpha$ for a given tyre radius R.

### 6.3.4 Hub reduction gear

A hub reduction gear transforms the angular velocity of the wheel with respect to the driveshaft velocity, and may invert its sign too. This will have a considerable effect even on a pure trailing-arm suspension, since the reduction gear is actuated by the relative rotation of the shaft and the trailing arm through wheel travel.

**Fig. 6.27** shows the support angle $\varepsilon_A$ plotted against wheel travel s, and the wheel travel against acceleration $a_x$ for the rear suspension of Fig. 6.24, assuming a transverse driveshaft and different reduction ratios.

With a reduction ratio i = 1 (i.e. no reduction), the result is of course the same as in Fig. 6.24 version (a).

A ratio i = 2 (where i > 0 for same direction of rotation of the reduction gear input and output shafts) generates the considerable support angle of 16° in the normal position; this exceeds the 'ideal' angle for single-axle

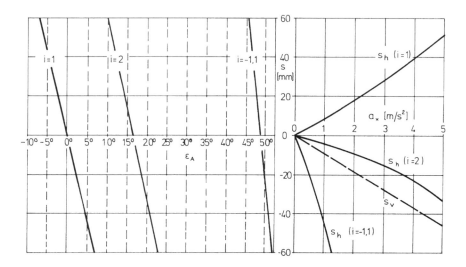

**Fig. 6.27** Support angle and traction pitching of the suspension of Fig. 6.24 but with a hub reduction gear

drive on a passenger car, being at once felt by rebound motion of the rear wheel, causing lifting of the vehicle.

Reversed rotation of the input and output shafts, here realized by a ratio $i = -1.1$, leads to a support angle of nearly 50° and to unstable pitching with increasing vehicle acceleration. Even with a (steady-state!) acceleration of about $2\,m/s^2$, the rear end of the vehicle is pushed upwards by around 100 mm, thus using up the available rebound wheel travel on normal passenger cars (it should be noticed here, however, that such driveline design is not usual on passenger cars, being preferably applied to off-road vehicles or trucks with higher centres of gravity and stiffer springs).

The results shown in Fig. 6.27 will become clearer by determining the longitudinal pole $L^{**}$ of a trailing-arm suspension according to equation (6.11) for different reduction gear ratios i – **Fig. 6.28**. These poles are 'fixed' to the wheel carrier (the trailing arm) on the line connecting the wheel centre M and the pole L of the suspension (here the trailing-arm mounting). Reversal of the rotational direction within the gear (i<0) leads to pole positions $L^{**}$ between the suspension pole L and the wheel centre M, and thus to very large support angles. With a reduction ratio $i = \infty$, the wheel can no longer rotate, regardless of the number of driveshaft revolutions; the wheel would then appear to be fixed to the wheel carrier and the pole L of the suspension to be the traction pole $L^{**}$.

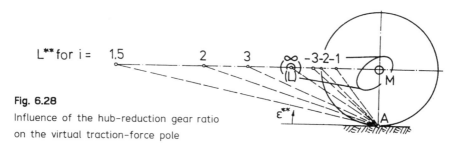

**Fig. 6.28**
Influence of the hub-reduction gear ratio on the virtual traction-force pole

### 6.3.5 Effect of longitudinal forces on the spring rate

As already mentioned in Section 6.3.1, longitudinal forces can influence the effective spring rate. **Fig. 6.29** shows schematically a rear suspension in side view, its type being unimportant. If its wheel travel angle $\varepsilon$ and its support angles $\varepsilon^*$ and $\varepsilon^{**}$ are known, the real suspension can be represented by guiding slots at the wheel centre M and the tyre contact point A, inclined by the angles $\varepsilon$ and $\varepsilon^*$ or $\varepsilon^{**}$. The slots are drawn curved as a reminder that the wheel-travel angle and the support angles may vary with wheel travel.

Traction and Braking

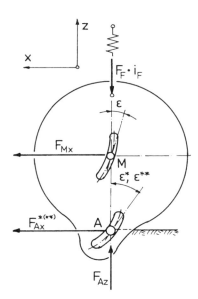

**Fig. 6.29**
Effect of longitudinal forces
on the 'kinematic' spring rate

The wheel receives a dynamic wheel-load transfer $F_{Az}$ caused by the longitudinal forces as explained in Sections 6.1 and 6.2 - a spring force $F_F i_F$ 'reduced' to the tyre contact point, and longitudinal forces $F_{Ax}^*$ or $F_{Ax}^{**}$ at the tyre contact point and $F_{Mx}$ at the wheel centre; the latter force might for example be a rolling resistance or an aquaplaning force.

Except where the wheel-travel angle $\varepsilon$ or the support angle $\varepsilon^*$ or $\varepsilon^{**}$ is zero, longitudinal forces generate vertical force components at the suspension (or, as here, the slots), and the balance of vertical forces results in a wheel-load component

$$F_{Az} = F_F i_F + F_{Mx}\tan\varepsilon \mp F_{Ax}^*\tan\varepsilon^* \mp F_{Ax}^{**}\tan\varepsilon^{**}$$

(upper sign valid for front wheels). The 'reduced' spring rate at the tyre contact point A, derived by analogy from Section 5.4, therefore becomes

$$\begin{aligned}c_{FA} = & \; c_F i_F^2 + F_F(di_F/dz) \\ & + F_{Mx}(1 + \tan^2\varepsilon)(d\varepsilon/dz) \\ & \mp F_{Ax}^*(1 + \tan^2\varepsilon^*)(d\varepsilon^*/dz) \\ & \mp F_{Ax}^{**}(1 + \tan^2\varepsilon^{**})(d\varepsilon^{**}/dz)\end{aligned} \quad (6.16)$$

(upper sign again valid for front wheels).

Equation (6.16) states that the temporary spring rate $c_{FA}$ during the action of a longitudinal force may be increased or decreased with respect to the static rate by 'kinematic' components, if the wheel-travel angle or the support angles vary with wheel travel.

Many rear suspensions show a decrease of the wheel-travel angle $\varepsilon$ and the support angle $\varepsilon^*$ or $\varepsilon^{**}$ with bump, and consequently the derivatives $d\varepsilon/dz$, $d\varepsilon^*/dz$ and $d\varepsilon^{**}/dz$ are negative. With a positive longitudinal force (e. g. an accelerating force) the spring rate will decrease, and with a negative force (e. g. a braking force) it will increase. Since the proportion of rear-wheel braking forces on the total braking force is rather low for passenger cars, the influence of a traction force, especially in the lower gears, is of greater importance than that of a braking force.

Variations of wheel-travel angle and support angle with bump and rebound are naturally the more marked the nearer the longitudinal pole (see Fig. 6.28) is to the wheel - e. g. on trailing-arm rear suspensions.

Front wheel suspensions, which contribute a relatively greater share of the braking force, have a considerable polar distance in order to avoid steering-geometry problems through castor variation with wheel travel etc. - see Section 6.2 and Fig. 6.9. The support-angle change with wheel travel is therefore kept moderate in front suspensions.

The preceding considerations will not be necessary in determining the vehicle attitude during a steady-state acceleration or braking process if done by integration or iteration methods which need in any case to be applied considering any non-linear spring characteristics and the varying support angles with wheel travel. The mentioned effects will then already be integrated into the results of the calculation.

The temporary spring rate according to equation (6.16) is of some importance, however, for transient events, as for example in road-irregularity impacts that occur during braking or accelerating. And, for such situations only, the impression of a 'harsh' spring action reported in Section 6.3.1 may perhaps be explained by a small polar distance or by support-angle variation with wheel travel.

### 6.3.6 Non-symmetrical vehicle attitude

The preceding analysis refers to the vehicle during straight-ahead accelerating and braking. Consequently the traction and braking-force support angles were derived from parallel travel of the wheels. The support angles are, of course, significant in other driving situations, too.

**Fig. 6.30** shows a trailing-arm rear suspension in cornering. Due to the centrifugal force, the vehicle has assumed a roll angle $\varphi$. Since with such a suspension the support angle $\varepsilon^*$ (and the wheel-travel angle $\varepsilon$, not marked) varies with wheel travel, different angles appear at the inner (i) and outer

Traction and Braking

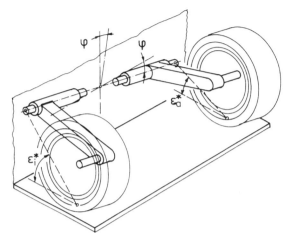

**Fig. 6.30**
Trailing-arm suspension in cornering

(a) wheel, the support angle $\varepsilon_a^*$ of the outer wheel being considerably smaller than that of the inner wheel $\varepsilon_i^*$. When subjected to longitudinal forces, both rear suspensions will therefore react differently: under braking, the inner suspension shows a higher degree of 'anti-lift' than the outer, and consequently offers more resistance to lifting of the vehicle body. Hence, the vehicle is caused to adopt a smaller roll angle; this is equivalent to a 'stabilizer effect' at the rear axle and favours oversteering tendencies.

The wheel-travel angles $\varepsilon$ (not drawn) show approximately the same order of difference. If the axle is driven by transverse shafts with universal joints and without hub-reduction gears, the wheel-travel angles represent (as explained in Section 6.3.3) the traction-force support angles too. In contrast with the braking forces, forward traction forces will then tend to increase the roll angle and thus favour understeer, while engine-braking forces again favour oversteer.

In practice, the effects on driving behaviour are of minor importance in comparison with dominant vehicle properties such as springing-system harmonization and kinematic or elasto-kinematic self-steering effects.

Another feature relating to cornering will be treated in Chapter 7 - namely the effect of the longitudinal component of the sideslip force of the tyre on a steered wheel.

### 6.3.7 Influence of unsprung and rotating masses

In the foregoing, unsprung and rotating masses have not been taken into account, although they amount to perhaps 6-10% of the total mass of the vehicle.

The centre of gravity of the unsprung masses may be assumed to be at the height of the wheel centre and thus somewhat lower than the vehicle's overall centre of gravity. Errors arising through calculation of the dynamic wheel-load transfer are negligible as are the moments of the inertial forces exerted at the longitudinal suspension poles, the latter being approximately at the same height.

The rotating masses – i.e. in the higher gears, the wheels and the brake discs – impose pitching moments on the vehicle with accelerating or braking. On a passenger car, the sum of the moments of inertia of the rotating masses (which represents about 1.5-2 per thousand of the vehicle's total moment of inertia) and the rotational acceleration or deceleration $a_x/R$ of the wheels contribute a pitching moment that amounts to no more than 2-3% of the moment caused by the vehicle's inertia force which acts through the centre of gravity.

Taking very precise account of the unsprung and rotating masses would result in loss of simplicity and clarity of presentation without significant practical benefits.

### 6.3.8 Influence of flexible mountings

Apart from racing cars and motorcycles, the wheel-suspension systems of today are provided with flexible mountings, mainly for noise isolation, but also for elasto-kinematic reasons. These mountings allow displacements of the suspension links under external forces.

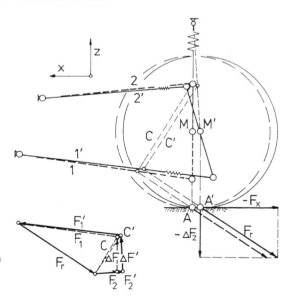

**Fig. 6.31**
Compliant wheel suspension under braking force

**Fig. 6.31** shows a simple wheel suspension which deals with longitudinal forces and moments by two longitudinal links one above the other. The displacements caused by a braking force $F_x$ are exaggerated but, in spite of this, there is negligible difference between the diagrams of forces for the rigid and compliant mechanism. Elastic deflections of a statically constrained mechanism are merely equivalent to a minute displacement of its coordinate system.

However, flexible joints in a suspension system make it difficult to determine braking or traction-force support angles by measuring the path of the tyre contact point with locked brake or engine. This is because very small elastic displacements of the wheel, especially in the longitudinal direction (as may occur even with vertical external loading) will considerably falsify the result. An elastic longitudinal deviation of, say, 5 mm per 100 mm of wheel travel simulates a support-angle contribution of 3°, meaning about 15-20% anti-dive at a front axle of a passenger car when braking, or about 25% anti-squat or anti-lift on a vehicle with single-axle drive.

For the same reason, attempts to measure steady-state braking pitching by pulling a stationary vehicle with locked brakes are also problematic, because achievement of the vehicle's designed braking-force distribution may not be ensured in respect of the compliance of suspensions and tyres. The same is valid for traction pitching on all-wheel-driven vehicles.

The general definition of a support angle (and of all kinematic suspension characteristics such as the roll centre or the scrub radius) must therefore be improved, as follows: the support angle is the deflection angle of the path of the tyre contact point against the vertical direction with wheel bump, assuming a locked torque support and a suspension mechanism with rigid links and stiff joints. This means that suspension characteristics should be determined only by calculation.

There is no point, either, in defining suspension characteristics that are valid for special cases of external forces including elastic properties. As wheel suspensions generally represent 'statically constrained' mechanisms, the influence of compliance on suspension characteristics is negligible, and elastic displacements cause only minute changes to the lines of action of forces and, therefore, minute displacements of 'poles'.

Moreover, it is not advisable to try to 'anticipate' elastic displacements when defining the coordinates of the suspension joints – perhaps with the intention of simulating the behaviour of the flexible suspension mechanism by falsifying the dimensions of the theoretical (rigid) system in the detail drawings or the computer model. Such a process would certainly lead to chaos in the documentation of the design: first, every different state of wheel loading or spring or rubber joint harmonization would necessitate a new 'kinematic model'; second, this model would be valid for one wheel position only, and deviations with wheel travel of the 'kinematic' from the real wheel attitude are unavoidable, because the joint deflections are variable with wheel-load change. The only efficient method is to document the theoretical joint positions of the rigid suspension system as a basis for all investigations, and, either by calculation or by measuring, to superimpose in each case the respective elastic displacements of special loading arrangements. This procedure secures reproducible relationship of the dimensions of the suspension parts (as they are fixed on the drawings) and of the documented kinematic suspension-joint coordinates.

## 6.4 Tandem axles

Heavy trucks frequently have tandem axles to achieve higher load capacity, and each of the two axles may have its own suspension and springing system. Often, though, compound springs are provided (on the scale-beam principle), but kinematic compounding is rare. The dampers are always mounted directly between the axles and the chassis to avoid axle-tramp oscillations.

The simplest possible tandem-axle suspension is shown in **Fig. 6.32**a, effected by leaf springs only (one spring each side) which are fixed to the axles and react braking and traction moments. The springs are rotatably mounted to the chassis and act like scale beams (compound springing). Longitudinal forces are transferred from the axles to the vehicle by the central spring mounting, as demonstrated by a rear-axle braking process: the 'dynamic' axle load-transfer force $\Delta F_{zA}$, caused by the vehicle's centre of gravity height and its (medial) wheelbase, is equally distributed to both axles according to the scale-beam principle, and the total braking force $F_{xA}$ acts on the tandem arrangement by the lever h'. With the tandem-axle wheelbase l', the axle loads result in

$$\Delta F_{z1} = - \Delta F_{zA}/2 + F_{xA}h'/l' \quad \text{and} \quad \Delta F_{z2} = - \Delta F_{zA} - F_{xA}h'/l'$$

and are obviously different. Compared with the static loads, the leading-axle load may be increased or decreased according to the dimensions of the vehicle and the braking-force distribution, while the trailing axle is considerably unloaded. If both axles have equivalent brakes, adhesion at the road surface is therefore disproportionately utilized, and with increasing braking deceleration the second axle will lock prematurely. Hence, the braking-force distribution has to be adapted to the dynamic axle loading, and the leading axle has to contribute by far the greater share.

Traction and Braking                                                                 141

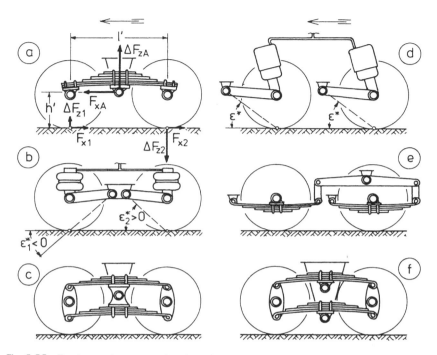

**Fig. 6.32** Tandem-axle suspension layouts

Similar behaviour is shown by the tandem arrangement (b) consisting of 'leading' and 'trailing' longitudinal-link independent suspensions or thrust ball axles. Due to its opposing support angles $\varepsilon_1^*$ and $\varepsilon_2^*$, the leading axle will push the vehicle upwards while the other pulls it downwards, and consequently the wheels of the first axle are pressed against the road whereas those of the second are unloaded. Reference back to Fig. 5.26 in Chapter 5 makes it clear that the leaf-spring halves of example (a) correspond to the leading and trailing suspensions of example (b), thus proving the equivalence of the two systems.

It is indeed helpful to provide equal support angles on both axles. In Fig. 6.32c this has been achieved, of course, at the axle arrangement itself, by controlling both axles via leaf springs one above the other. However, by attaching both leaf springs together to a rotatably mounted pivot bracket, the system shows properties similar to that of Fig. 6.32a: the tandem axle aggregate forms, so to speak, a 'sub-vehicle' that is subject to axle-load transfer caused by the longitudinal force $F_{xA}$ acting on the bracket mounting.

No problems will arise if two separate axle suspensions show equal support angles, as with the trailing-arm or thrust-ball axles illustrated in Fig. 6.32d. Even if this design is provided with compound springs, acceptable braking behaviour is to be expected.

Other familiar good solutions are shown in the examples (e) and (f), the latter with leaf springs similar to those of example (c) but with each spring mounted to its own bracket. The (approximately equal) support angles are here about zero, of course.

The considerations made above are valid also for traction forces, except for the sign of the force vectors.

The suspension types illustrated in Fig. 6.32 are valid for rear wheels. Front-wheel tandem axles normally feature identical axle suspensions (and springing systems) independent of each other.

# Chapter 7

# Cornering

## 7.1 Camber and steering angle with wheel displacement

In Chapter 4 it was explained that the lateral force at a wheel depends on the tyre slip angle and the wheel camber. The tyre slip angle is influenced not only by the lateral drift of the vehicle arising from lateral acceleration but also by possible kinematic or elasto-kinematic steering effects at the suspension. Moreover, toe-in change or, generally, steering-angle change with wheel travel or roll angle may cause a resulting steering angle of the 'axle'. As in the preceding analyses, camber and steering angle are determined in a vehicle-fixed coordinate system.

Wheel camber $\gamma$ is defined as positive if the wheel axis rises towards the medial plane of the vehicle, and steering angle $\delta$ is defined as positive if the wheel axis inclines forward towards the vehicle's longitudinal axis.

This means that a positive 'toe-in angle' $\delta_V$ according to customary definition – i.e. with the wheel planes intersecting ahead of the axle – has on a left wheel to be treated like a negative steering angle in calculations and equations.

Swivelling of the wheel carrier about its instantaneous (or perhaps steady) axis displaces the wheel axis in space. It is irrelevant here whether the instantaneous axis is a screw axis or not, because superimposed axial shift does not influence the relationship of angular velocities or of angles. The wheel axis may be displaced from the instantaneous axis, i.e. not intersecting with it; this is true on steered suspensions with a 'castor offset at the wheel centre', also on a suspension with a 'virtual' kingpin axis (see Chapter 8). A displaced wheel axis generates a hyperbolic surface in space when swivelling about the instantaneous axis, while a wheel axis intersecting the instantaneous axis generates a cone which is easier to appreciate. To simplify the evaluation, any wheel axis can be replaced (see **Fig. 7.1**) by a parallel line 'a' fixed to the wheel carrier K and, at least momentarily, intersecting with the instantaneous axis m at a point $P_0$. If the wheel carrier moves in space, the line a will always show the same projection angles in any view – e.g. the same camber or steering angles – as the true wheel axis.

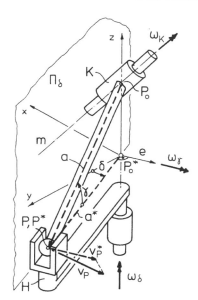

**Fig. 7.1**
Camber and steering-angle change with wheel motion

The state of motion of a wheel suspension (bump and rebound wheel travel, or steering) is characterized by the angular velocity vector $\boldsymbol{\omega}_K$ of the wheel carrier K and the wheel-centre velocity vector $\mathbf{v}_M$.

The instantaneous camber change appears in a vertical plane of projection $\Pi_\delta$ through the wheel axis a (or through its substitute axis). The normal vector $\mathbf{e}$ of this plane has the x- and y-components $-\cos\delta$ and $-\sin\delta$, and no z-component. As the wheel axis is fixed to the wheel carrier, the angular camber-change velocity $\omega_\gamma$ is the component of the angular velocity $\boldsymbol{\omega}_K$ of the wheel carrier in the direction of $\mathbf{e}$, $\omega_\gamma = \boldsymbol{\omega}_K \cdot \mathbf{e}$ or

$$\omega_\gamma = -\omega_{Kx}\cos\delta - \omega_{Ky}\sin\delta \tag{7.1}$$

If $v_{Az} = ds/dt$ is the vertical component of the velocity of the tyre contact point (where s is the vertical wheel travel and t the time), camber change with wheel travel can be expressed by $d\gamma/ds = (d\gamma/dt)(dt/ds) = \omega_\gamma/v_{Az}$. As already explained in Chapters 3 and 6, the determination of suspension characteristics requires preferably a virtual velocity vector $\mathbf{v}_A^*$ of the tyre contact point, assuming it to be fixed to the wheel carrier. The vertical component $v_{Az}^*$ is equal to the component $v_{Az}$ of the velocity vector of the tyre contact point on a freely rotatable wheel, so may be used here instead of $v_{Az}$:

$$d\gamma/ds = \omega_\gamma/v_{Az} = \omega_\gamma/v_{Az}^* \tag{7.2}$$

Cornering                                                                                           145

The steering velocity $\omega_\delta$ is the angular velocity of the projection $a^*$ of the wheel axis a (or its substitute axis) in plan view; $a^*$ is defined by the projection $P_0^*$ of the intersecting point $P_0$ of the wheel axis and the instantaneous axis and by the projection $P^*$ of any point P on the wheel axis. If the wheel carrier K rotates about its instantaneous axis m with an angular velocity $\omega_K$, the point P on the wheel axis assumes a velocity $\mathbf{v}_P = \boldsymbol{\omega}_K \times \mathbf{a}$ which is skew to the plane $\Pi_\delta$ as long as the vector $\boldsymbol{\omega}_K$ does not lie in that plane. Rotation of the projection $P^*\text{-}P_0^* = a^*$ of the wheel axis in plan view is, of course, identical with the steering velocity $\omega_\delta$ whose vector has only a vertical component, and the projection point $P^*$ has the velocity $\mathbf{v}_P^* = \boldsymbol{\omega}_\delta \times \mathbf{a}^*$ which is rectangular to the plane $\Pi_\delta$. The velocity components of P and $P^*$ rectangular to $\Pi_\delta$ must be equal (as depicted by a lever H which is actuated by the point P on the wheel axis a), and consequently their scalar products with the unit normal vector $\mathbf{e}$ must be equal, too:

$$\mathbf{v}_P^* \cdot \mathbf{e} = \mathbf{v}_P \cdot \mathbf{e}$$

The wheel-axis vector $\mathbf{a}$ has the components $a_x = -a\cos\gamma\sin\delta$, $a_y = a\cos\gamma\cos\delta$ and $a_z = -a\sin\gamma$, while the vector $\mathbf{a}^*$ has the same x- and y-components but no z-component. The scalar product equation given above therefore results in

$$\omega_\delta = -\omega_{Kx}\tan\gamma\sin\delta + \omega_{Ky}\tan\gamma\cos\delta + \omega_{Kz} \quad (7.3)$$

and, in analogy to equation (7.2), the steering-angle change versus wheel travel is

$$d\delta/ds = \omega_\delta/v_{Az} = \omega_\delta/v_{Az}^* \quad (7.4)$$

## 7.2 Forces and moments under lateral acceleration

In steady-state cornering, the lateral acceleration $a_y$ acting on the vehicle mass m generates an inertial (or 'centrifugal') force $ma_y$ at the centre of gravity SP, and with it a roll moment that increases the outer-wheel loadings and decreases the inner. The inertial force is balanced by the lateral forces at the tyre contact points, these forces being transmitted to the vehicle body via the wheel suspensions.

**Fig. 7.2** shows a simplified model of a vehicle with rigid front and rear axles that allow the body to move up and down and to roll (the springs are not represented). The axles with 'slot guidances' as shown here are relevant for any suspensions. The lateral forces $F_{yv}$ and $F_{yh}$ of the front and rear axles respectively are transmitted to the vehicle body via the contact points $RZ_v$ and $RZ_h$. As these points are nearer to the road surface than the

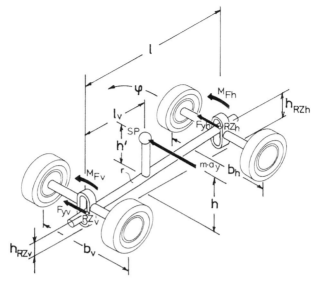

**Fig. 7.2** Forces and moments in cornering

centre of gravity, a moment about the line r connecting the two contact points is generated; this moment must be balanced by restoring moments $M_{Fv}$ and $M_{Fh}$ at the front and rear springing systems, whereby the vehicle body inclines or 'rolls' about the line r by a roll angle $\varphi$. The line r is therefore called the **'roll axis'**, and the force-transmitting points $RZ_v$ and $RZ_h$ are the **'roll centres'**.

The roll centre is the point on the cross-sectional plane through the tyre contact points where the resulting lateral force on an axle is transmitted to the vehicle body, and the point about which the vehicle body begins to roll.

The lever arm of the 'sprung' mass of the vehicle about the roll axis is

$$h' = [h - h_{RZv}(1 - l_v/l) - h_{RZh} l_v/l] \qquad (7.5)$$

The 'unsprung' masses $m_u$ also may generate a roll moment at the vehicle body. If $\omega_\varphi$ is the angular roll velocity of the vehicle body with respect to the road – **Fig. 7.3** – the two wheels of the axle show opposed wheel travel relative to the vehicle body, and camber change according to their camber gradients $d\gamma/ds$ (which are mostly negative in practice); it should be mentioned here that the camber gradients in question must arise from 'antimetric' wheel travel (which, however, affects only rigid-axle and compound suspensions, since independent suspensions show identical camber change with parallel and antimetric motion).

Cornering

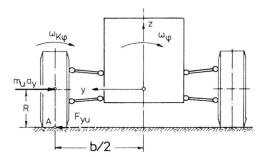

**Fig. 7.3**
Roll moment of the unsprung masses

Assuming positive camber gradients, both the inner (left in Fig. 7.3) and the outer wheel will incline with respect to the vehicle body in the same sense as the roll velocity $\omega_\varphi$ of the body – that is, by a velocity $\omega_{rel} = (b/2)\omega_\varphi(d\gamma/ds)$; hence, the resulting absolute camber velocity of the wheel carriers is

$$\omega_{K\varphi} = \omega_\varphi + \omega_{rel} = \omega_\varphi[1 + (b/2)(d\gamma/ds)]$$

and the roll moment $M_{uF}$ of the unsprung mass $m_u$ of a wheel with its wheel carrier (the centre of gravity of which may be assumed to be approximately at the height of the wheel centre) follows from the balance of virtual work $M_{uF}\omega_\varphi = m_u a_y R \omega_{K\varphi}$ as

$$M_{uFv,h} = m_{uv,h} a_y R[1 + (b_{v,h}/2)(d\gamma/ds)_{v,h}] \tag{7.6}$$

where v and h refer respectively to the front and rear wheels. The moment $M_{uF}$ acts on the vehicle body and so has to be reacted by the vehicle's springing system.

In practice, the wheel's camber gradient $d\gamma/ds$ with respect to the vehicle body is, of course, negative in most cases.

With a rigid axle the camber gradient is $d\gamma/ds = -2/b$ (as is easy to derive). The moment $M_{uF}$ is therefore zero on such an axle; the roll moment of the unsprung mass of a rigid axle does not load the vehicle body and the springs but is supported directly by the road surface.

With the roll angle $\varphi$, the centre of gravity SP of the vehicle body is laterally displaced about the roll axis r by an amount $h'\varphi$. The roll moment of the vehicle body about the roll axis therefore results from the centrifugal-force moment (acting via the lever h') and the weight moment (acting on the lever $h'\varphi$) as $ma_y h' + mgh'\varphi$, and has to be balanced by the restoring moment $M_{Fv,h}$ of the front and rear springing systems. To simplify the calculation, the resulting 'roll spring rates' $c_{\varphi v}$ and $c_{\varphi h}$ of the front and rear axles may be assumed to be constant. The roll spring rate of an axle is composed of the rotational rate of the bump springs – see Chapter 5, equation (5.24) and Fig. 5.12 - and the spring rate of a stabilizer (anti-roll bar), if fitted.

With front-wheel suspensions having a significant support angle $\varepsilon^*$, a further effect may show up with increasing steering angle (**14**). The lateral force $F_S$ at the tyre caused by sideslip and by camber no longer acts in the y-direction of the vehicle system but is inclined by the steering angle $\delta$, **Fig. 7.4**. Its longitudinal component $F_{sx} = F_s \sin\delta$ is therefore transmitted to the suspension like a 'braking force' (this is not true, though, with the Dubonnet suspensions and motorcycle forks, which are swivelled fully with the steering angle) and generates a vertical force $\Delta F^*_{S\varepsilon} = F_{sx}\tan\varepsilon^*$ which helps the vehicle springs to support the (sprung) load. The support angles of the inner and outer wheels may be different according to the different wheel travels, while the outer and inner lateral forces $F_{sa}$ and $F_{si}$ are, of course, different in tight curves or with high lateral acceleration. An additional restoring moment is, therefore, generated at the front axle:

$$M^*_\varepsilon = (F_{sa}\sin\delta_a \tan\varepsilon^*_a - F_{si}\sin\delta_i \tan\varepsilon^*_i)b_V/2 \qquad (7.7)$$

(where $b_V$ is the track of the front axle); if positive, this moment acts like an anti-roll bar. If the support angles change progressively with wheel bump travel, this effect will be reinforced.

The 'stabilizing' moment $M^*_\varepsilon$ favours understeer of the vehicle, resulting, however, in greater tyre slip angles which require greater steering angles and again increase the share of the longitudinal component $F_{sx}$ of the lateral force $F_S$. In this way, self-reinforcing understeer may be produced;

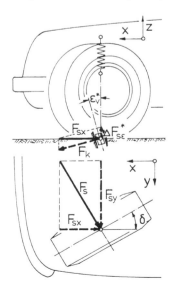

**Fig. 7.4**
Lateral force and support angle for a steered wheel

# Cornering

this could compel a reduction from the desired level of 'anti-dive' harmonization or at least a reduction in the progressive support-angle gradients with wheel travel.

With the roll moments of the 'sprung' mass m of the vehicle body and the 'unsprung' wheel masses $m_u$ and the resulting roll spring rates $c_{\varphi v}$ and $c_{\varphi h}$ of the front and rear axles, the balance of moments

$$m a_y h' + m g h' \varphi$$
$$+ 2 m_{uv} a_y R[1 + (b_v/2)(d\gamma/ds)_v]$$
$$+ 2 m_{uh} a_y R[1 + (b_h/2)(d\gamma/ds)_h]$$
$$= \varphi(c_{\varphi v} + c_{\varphi h}) + M_\varepsilon^*$$

leads to the roll angle

$$\varphi = a_y \frac{mh' + R\{2m_{uv}[1+(b_v/2)(d\gamma/ds)_v] + 2m_{uh}[1+(b_h/2)(d\gamma/ds)_h]\} - M_\varepsilon^*/a_y}{c_{\varphi v} + c_{\varphi h} - mgh'} \quad (7.8)$$

A vehicle's total roll moment is, of course, determined by the centre-of-gravity height h above the road. The springing moment $\varphi(c_{\varphi v} + c_{\varphi h})$ is only one part of this, the rest being transmitted to the road via the roll centres and the suspension mechanism. Of the centrifugal force acting on the vehicle, the front axle's share, $m a_y (l - l_v)/l$, acts at the front roll-centre height $h_{RZv}$, and the rear axle's share, $m a_y l_v/l$, at the rear roll-centre height $h_{RZh}$. The share of the unsprung masses' roll moment that is not transmitted to the vehicle body according to equation (7.6), but acts immediately on the road, is $M_{uR} = m_u a_y R - M_{uF}$ or

$$M_{uRv,h} = - m_{uv,h} a_y R (b_{v,h}/2)(d\gamma/ds)_{v,h}$$

The resulting roll spring moment $\varphi(c_{\varphi v} + c_{\varphi h})$ is divided between the axles according to their roll spring rates - $M_{\varphi v} = c_{\varphi v} \varphi$ and $M_{\varphi h} = c_{\varphi h} \varphi$. With these moments and the roll-centre moments $M_{uR}$, the resulting roll moments at the front (v) and the rear axle (h) are

$$M_v = m a_y h_{RZv}(l - l_v)/l - 2m_{uv} a_y R(b_v/2)(d\gamma/ds)_v + c_{\varphi v} \varphi + M_\varepsilon^* \quad (7.9a)$$
$$M_h = m a_y h_{RZh} l_v/l \quad - 2m_{uh} a_y R(b_h/2)(d\gamma/ds)_h + c_{\varphi h} \varphi \quad (7.9b)$$

and the wheel-load transfer at the front (v) and the rear axle (h) is

$$\Delta F_{zv,h} = M_{v,h}/b_{v,h} \quad (7.10a,b)$$

From the static wheel loads

$$F_{zov} = [(m/2)(l - l_v)/l + m_{uv}]g \quad (7.11a)$$
and $$F_{zoh} = [(m/2)l_v/l + m_{uh}]g \quad (7.11b)$$

(where $m_u$ is half the mass of a rigid axle), we get the resulting wheel-load forces of a four-wheel vehicle in cornering:

$$F_{zva,i} = F_{zov} \pm \Delta F_{zv} \qquad (7.12a)$$
and $$F_{zha,i} = F_{zoh} \pm \Delta F_{zh} \qquad (7.12b)$$

(the upper sign is valid for the outer wheel, subscript 'a'; the lower one for the inner wheel, subscript 'i').

Assuming small steering angles (and hence a large cornering radius) and thus centrifugal forces acting in the transverse or y-direction, the resulting lateral forces at the front and the rear axle are proportional to the static axle loads:

$$F_{yva} + F_{yvi} = a_y[m(l - l_v)/l + 2m_{uv}] \qquad (7.13a)$$
$$F_{yha} + F_{yhi} = a_y[ml_v/l + 2m_{uh}] \qquad (7.13b)$$

With equal loading on both wheels, an axle would be able to transmit at each wheel a lateral force $F_{ym} = (F_{ya} + F_{yi})/2$ by a tyre slip angle $\alpha_0$, which follows from the lateral force characteristic for static load $F_{zo}$ – **Fig. 7.5**. Due to the wheel load transfer $\Delta F_z$, however, the outer and inner lateral forces are to be derived from characteristics valid for wheel loads $F_{zo} + \Delta F_z$ and $F_{zo} - \Delta F_z$ respectively. As already explained in Chapter 4, unequal wheel loadings on an axle cause greater tyre slip angles. To balance the axle's lateral force $2F_{ym}$, the tyre slip angle must increase to a value $\alpha_1$, and the respective lateral forces on the outer and inner wheels are $F_{ym} + \Delta F_{y1}$ and $F_{ym} - \Delta F_{y1}$. Obviously the inner tyre nearly reaches its adhesion limits, while the outer one is still in a stable region.

A toe-in angle $\delta_v$ leads to different outer and inner tyre slip angles and consequently to a greater lateral-force difference $\Delta F_{y2}$. The two wheels are therefore loaded according to their abilities and, at the same time, the (median) tyre slip angle of the axle is reduced to a value $\alpha_2$. In general, toe-in gives an axle better lateral stability.

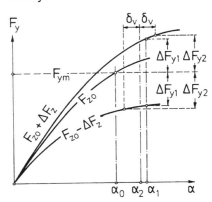

**Fig. 7.5**
Tyre slip angles for an axle with different wheel loads

Cornering                                                                                              151

Due to the non-linear interdependence of numerous variables, calculation of cornering behaviour can be done only by iteration.

Wheel load differences and tyre slip angles are influenced mainly by the following factors:

a) the distribution of roll spring rates between front and rear axles – e.g. by variation of springs, incorporation of anti-roll bars or compound springs; an anti-roll bar on the front axle increases the total roll rate of the vehicle and reduces the roll angle, hence the wheel-load difference at the rear axle at the expense of that at the front axle (on a vehicle with a torsionally flexible frame, as in the following Fig. 7.6, this factor will, of course, have a limited effect!);

b) the height of the front and rear roll centres above the road surface;

c) the front and rear tracks.

With a given wheel-load difference, the tyre slip angles depend on:

d) the lateral force characteristic of the tyre, which can be influenced by, for instance, interior air pressure or tyre dimensions (and, not to be forgotten, by wear);

e) last but not least, the vehicle's static axle-load distribution or the position of the centre of gravity.

Further possible influences on tyre slip angles are superimposed steering angles such as toe-in and kinematic and elasto-kinematic self-steering angles.

In the vehicle model of Fig. 7.2, the vehicle body was assumed to be rigid, which is relatively true for passenger cars. However, the frames of trucks normally have some torsional flexibility, for simpler and cheaper production from 'open' metal sections, or for better off-road capability. The torsional rate $c_{\varphi R}$ of the frame, **Fig. 7.6**, may throughout be within the range of the roll rates of the front and rear axles, or even lower. This

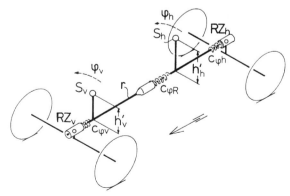

**Fig. 7.6** Vehicle with torsionally flexible frame

requires a refined vehicle model with partial centres of gravity $S_v$ and $S_h$ for the front and rear portions of the vehicle. The roll angles $\varphi_v$ and $\varphi_h$ depend on the respective masses and centre-of-gravity positions, the roll spring rates $c_{\varphi v}$ and $c_{\varphi h}$ of the axles and the torsional rate $c_{\varphi R}$ of the frame. Neglecting the unsprung masses, the conditions of balance are

$$m_v a_y h_v' + m_v g h_v' \varphi_v = c_{\varphi v} \varphi_v + c_{\varphi R}(\varphi_v - \varphi_h) \qquad (7.14a)$$
$$m_h a_y h_h' + m_h g h_h' \varphi_h = c_{\varphi h} \varphi_h - c_{\varphi R}(\varphi_v - \varphi_h) \qquad (7.14b)$$

Rotating wheels form gyroscopes whose angular momentum is forced to change its direction in cornering. The resulting gyrostatic moment acts on the vehicle as an additive roll moment which, however, is of real importance only on solo motorcycles (see Section 7.7). Among the two-track vehicles, agricultural tractors are a special case, with their large-diameter driven wheels which may have the tyres filled with water for better traction.

On front-drive vehicles with transverse engines, opposite rotations of the crankshaft and the wheels is of course advantageous in gyrostatic terms.

## 7.3 The roll centre

### 7.3.1 The vehicle at extremely low lateral accelerations

In Section 7.2, lateral forces were assumed to be transmitted from the wheel suspension to the vehicle body by a joint, the 'roll centre', about which the vehicle body inclines by a roll angle $\varphi$. This is easily imagined with rigid axles but less obvious for independent suspension; it should be appreciated here, too, that the centrifugal force can exert a lifting effect on the suspension relative to the body – something that was not taken into consideration with the simple model of Fig. 7.2. The roll centre is, of course, situated in the vertical cross-sectional plane through the tyre contact points, since external forces caused by lateral acceleration act in this plane. In what follows, the events that occur on a wheel suspension loaded with lateral forces will be analysed in a more detailed manner.

The system of forces is 'over-constrained' even on a single axle, since two points in one plane – the tyre contact points – are loaded with lateral forces with coinciding lines of action, the amounts and distribution of which between the two tyres depend on non-linear laws, namely the interdependence of tyre slip force, wheel load, a possible toe-in angle and wheel camber.

# Cornering

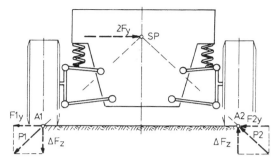

**Fig. 7.7** Diagram of forces on a vehicle with extremely low lateral acceleration

As long as the lateral acceleration is still tiny, a linear relationship of tyre slip force to tyre slip angle may be assumed. If the interaction of a vehicle's front and rear axles is neglected for provisional analysis (or better, if the vehicle is assumed to have identical front and rear axle loads, suspensions and springing systems, and tyres) its front and rear halves may each be treated on their own, similar to a single-axle trailer with a drawbar of infinite length. And if toe-in and camber angles are neglected, too, equal lateral forces $F_y$ and the wheel-load changes $\Delta F_z$ (in opposite directions) will act on both tyre contact points $A_1$ and $A_2$, **Fig. 7.7**, their resulting forces $P_1$ and $P_2$ intersecting with the horizontal inertia force at the vehicle's centre of gravity SP.

As the roll centre is in general nearer to the road surface than the centre of gravity, the inertia force causes a roll moment and a roll angle. As already mentioned, the roll centre is defined as the one point in the vertical cross-section plane of an axle where a lateral force can be transmitted from the road to the vehicle body via the suspension without the aid of the springing system. If loaded with a lateral force at the roll centre, the suspension system must consequently be in a state of equilibrium without requiring additional spring forces.

When acting on the roll centre RZ, the lateral force $2F_y$ at the centre of gravity generates a rolling moment at the suspension about the (still undetermined) roll centre with the lever $h_{RZ}$ of the roll-centre height, **Fig. 7.8**. The initial rolling movement causes antimetric wheel-travel velocities $v_{Aw}$ (subscript 'w' for rolling) at the tyre contact points. According to the theorem of virtual work, the 'power' of the lateral forces $F_y$ and the velocity components $v_{Awy}$ must be equal to the 'power' of the rolling moment $2F_y h_{RZ}$ and the angular roll velocity $\omega_\varphi$: $2F_y h_{RZ} \omega_\varphi = 2F_y v_{Awy}$; the vertical components $v_{Awz}$ (referring to the vehicle coordinate system) of the velocities $v_{Aw}$ follow from the roll velocity $\omega_\varphi$ and the coordinates $y_A$ of the tyre

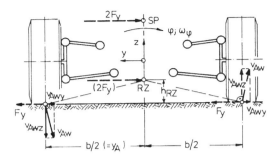

**Fig. 7.8** The roll centre of an independent wheel suspension

contact points, $v_{Awz} = \omega_\varphi y_A$. Hence, the roll-centre height above the road surface is

$$h_{RZ} = y_A (v_{Awy}/v_{Awz}) \qquad (7.15)$$

and since the longitudinal component $v_{Awx}$ has no effect on the roll centre, the vector $\mathbf{v}^*_{A(w)}$ of a tyre contact point that is assumed to be 'fixed' to the wheel carrier may also be applied in equation (7.15).

Accordingly, at the beginning of a roll motion of the vehicle body the roll centre is the intersecting point of the lines perpendicular to the antimetric velocity vectors of the tyre contact points and it lies in the central plane of the vehicle. This definition is valid for any suspension type, the only problem being to find the antimetric velocity vectors of the suspension system.

For the independent suspension of Fig. 7.8, the paths of the tyre contact points are unequivocally defined, and the same for roll and for bump motion, since the mechanism of such a suspension has only one degree of freedom.

If an independent suspension is of planar or spherical type, the intersection point of the instantaneous axis and the cross-section plane can always be defined as the 'pole' of the wheel carrier in cross-section. The antimetric velocity vector $v_{Aw}$ is then rectangular to the polar ray, and the roll centre is the intersection point of the polar rays of both wheels, in the vehicle's medial plane. However, this popular definition cannot be readily applied to other suspension types. With spatial suspensions, such a 'pole' must be reconstructed from the instantaneous state of motion – something that would be a nuisance. For this reason it is advisable to use the generalized definition given by equation (7.15), especially since it offers the simplest application of computer programs for the kinematic analysis of wheel suspensions.

A compound suspension as the general basic mechanism to control two wheels, with two degrees of freedom overall, shows different trajectories

Cornering                                                                                             155

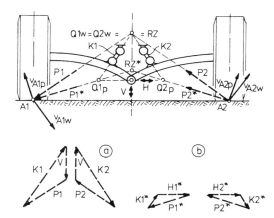

**Fig. 7.9**
The roll centre for a compound suspension

of the tyre contact points in the cases of parallel and antimetric wheel motion. The simple 'planar' model of a compound suspension shown in **Fig. 7.9** enables us to visualize the different velocity vectors of bump and roll motion: each wheel carrier is connected to the vehicle body by a pendulum (or rod link); the axis of this pendulum is, of course, one geometrical locus for the poles of parallel and of antimetric wheel motion. Moreover, both wheel carriers are coupled by a joint in the medial plane of the vehicle. With parallel wheel travel, the medial joint can move only in the vertical direction for reasons of symmetry; its 'polar ray' is therefore horizontal and represents the second geometrical locus of the poles $Q_{1p}$ and $Q_{2p}$ of parallel wheel travel. In contrast with this, with antimetric wheel travel on a rolling vehicle, the medial joint must be displaced in a horizontal direction (because a vertical displacement would be a 'symmetrical' component), and its polar ray is the centre line of the vehicle. The poles $Q_{1w}$ and $Q_{2w}$ of antimetric wheel travel are the intersection points of the pendula with the vehicle's centre line and they coincide; with the vehicle rolling, the simple suspension model chosen here for example obviously acts like a swing-axle layout.

The antimetric velocity vectors $v_{A1w}$ and $v_{A2w}$ at the tyre contact points $A_1$ and $A_2$ are rectangular to the polar rays $A_1Q_{1w}$ and $A_2Q_{2w}$ which intersect in the centre plane of the vehicle at the roll centre RZ (which coincides here with the antimetric poles).

A horizontal force acting on the roll centre must be balanced by reaction forces at the tyre contact points without the aid of spring forces. These antimetric reaction forces $P_1$ and $P_2$ intersect in the roll centre RZ. The reaction force $P_1$ at the tyre contact point $A_1$ on the left side generates a force $K_1$ at the left pendulum and a vertical force V at the medial joint,

– see the diagram of forces in Fig. 7.9a. The force V acts on the wheel carrier of the right side and is balanced by the pendulum force $K_2$ and the reaction force $P_2$. When combining both diagrams of forces, the vertical forces V cancel each other out, and the forces $P_1$ and $P_2$ act on the roll centre by their resulting force.

With parallel wheel travel, symmetrical velocity vectors $v_{A1p}$ and $v_{A2p}$ appear to be as with any independent suspension. However, it would be nonsense to use these vectors for defining a 'roll centre' $RZ^*$ for parallel wheel motion. Such a 'roll centre' would never be valid for an antimetric balance of forces, as shown in Fig. 7.9b: the horizontal forces $H_1^*$ and $H_2^*$ at the medial joint have the same direction and so cannot compensate each other by combining the diagrams of forces. This is, on the contrary, at once possible for a symmetrical diagram of forces – i.e. with opposed lateral forces at the wheels as arise, for instance, from a toe-in angle. These forces will try to lift the vehicle body. The same is valid for independent suspensions with roll centres above or below the road surface (because on independent suspensions the 'true' roll centre RZ and the 'wrong' roll centre $RZ^*$ are congruent).

As already shown, the roll centre has to be determined by analysis of the **antimetric** wheel motion. This will quickly become clear, especially for rigid-axle suspensions: the simple rigid axle in **Fig. 7.10** is laterally controlled by a slot guidance in the centre plane of the vehicle, the fulcrum of which is the point of force transfer and hence the roll centre RZ. With antimetric movement, the axle swings about the fulcrum, and the antimetric velocity vectors $v_{Aw}$ at the tyre contact points are rectangular to the polar rays through the roll centre. Parallel wheel travel, however, generates vertical velocity vectors only; according to this, a 'roll centre' $RZ^*$ defined by them would always be at the road surface (this of course may explain why toe-in forces on rigid axles cannot exert lifting forces at the vehicle body).

Obviously a roll-centre height other than zero on an independent suspension necessitates track change with wheel travel. The tyre slip angles and

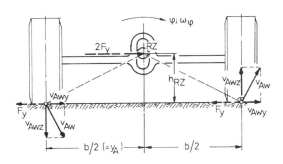

**Fig. 7.10**
Roll centre on a rigid axle

Cornering

consequently lateral forces arising from this are negligible at normal driving speeds. At extremely low velocity or with the vehicle at rest, though, the lateral tyre deflection increases the spring rate.

In the foregoing, compliance of suspension joints and lateral spring rates of the tyres have been neglected. As already set out in Section 6.3.8, compliance of a statically constrained mechanism causes only minute displacements of force lines of action, and takes effect in the cross-section of the vehicle rather like a parallel shift of the vehicle coordinate system with respect to the road. Influence on the diagrams of forces and the rolling behaviour is insignificant.

Like the support angle, the roll centre is a wheel suspension 'characteristic', and it has to be determined on a 'rigid' suspension mechanism. According to the considerations made in this section, it may be defined through the geometry of motion of an 'axle' as follows.

The roll centre is the instantaneous pole of the vehicle body with respect to the road surface in the cross-sectional plane of an axle at the very beginning of lateral acceleration and so of roll movement; on a rigid suspension mechanism, it represents the intersection point of the lines normal to the displacement vectors of the tyre contact points with antimetric wheel travel.

### 7.3.2 High lateral accelerations

With high lateral acceleration, the simplified vehicle model with an exactly antimetric state of motion and diagram of forces is no longer applicable. The considerable roll angle generally leads to non-symmetrical suspension positions, **Fig. 7.11**. Consequent on the increased wheel loading, the lateral force at the outer wheel is considerably greater than that at the inner wheel, and the resultants of the lateral forces $F_{qa}$ and $F_{qi}$ at the outer and inner wheel respectively and of the opposed wheel-load transfer forces $\Delta F_{FRa}$ and $\Delta F_{FRi}$ (still of the same amount!) intersect with the inertial force $m a_q$ at the height of the centre of gravity SP; however, the intersection point is displaced towards the inner side of the corner. In practice, a non-linear (and generally progressive) spring characteristic is to be taken into consideration, too.

The independent suspension in Fig. 7.11 is designed for a straightened path of the tyre contact point A, and hence for a constant roll-centre height $h_{RZ}$ above the road with parallel wheel travel. With double-wishbone suspension, this demands transverse links of unequal length, the lengths being approximately inversely proportional to the distance from the road (see upper sketch).

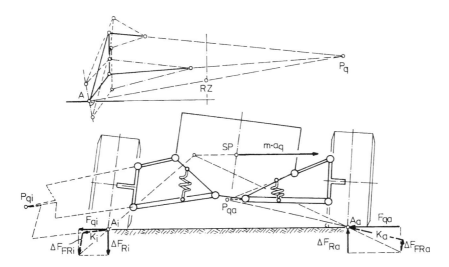

**Fig. 7.11** Independent wheel suspension with high lateral acceleration (constant roll-centre height)

With the simple 'planar' suspension, the poles $P_{qa}$ and $P_{qi}$ can easily be determined in cross-section by intersecting the respective wishbone centre lines. The polar rays $A_a P_{qa}$ and $A_i P_{qi}$ define the directions of forces that may act at the tyre contact points $A_a$ and $A_i$ and that can be transmitted to the vehicle body without loading the springs. On the suspensions, the resultant forces from the lateral forces $F_q$ and the wheel-load transfer forces $\Delta F_R$ at the tyre contact points are therefore split into respective 'kinematic' forces K in the direction of the polar rays and spring-force changes $\Delta F_{FR}$. In Fig. 7.11, the spring-force increase $\Delta F_{FRa}$ at the outer wheel is considerably lower than the corresponding decrease $\Delta F_{FRi}$ at the inner wheel; even when assuming a linear spring characteristic, the vehicle will sink at the outer side by a smaller amount than it will be lifted at the inner side. This means (in addition to the roll angle) a resultant lifting of the vehicle body – it will be 'jacked up' – and the effective centre of rotation is displaced towards the outer side.

In these circumstances the 'roll centre' can no longer be defined by the simple geometrical considerations which were acceptable with small lateral accelerations. The reason is that even on the simple 'one-axle model' of Fig. 7.11, the system represents a 'statically over-constrained' mechanism; the distribution of lateral forces follows from non-linear tyre characteristics and, thus, from laws that can at best be influenced by merely non-essential suspension design means – e. g. toe-in or camber-angle variation.

Cornering 159

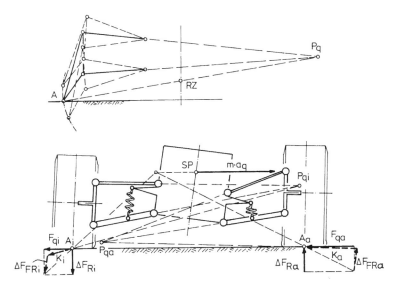

**Fig. 7.12** Double-wishbone suspension with variable roll-centre height

'Jacking up' of the vehicle body can obviously be avoided by changing the directions of the polar rays (and with them the directions of the reaction forces K) in order to get spring-force changes of same amount (presuming the spring characteristics to be linear); this requires a reduced inclination of the outer polar ray $A_aP_{qa}$ with respect to the road surface and an increased inclination of the inner polar ray $A_iP_{qi}$, **Fig. 7.12**. These measures cause the path of the tyre contact point to become curved, however (see the upper sketch), so the roll centre RZ will move downwards with wheel bump travel and will be raised with rebound travel.

In Fig. 7.12, the spring-force variation $\Delta F_{FRa}$ at the outer wheel is slightly greater than the variation $\Delta F_{FRi}$ at the inner wheel; in contrast with Fig. 7.11, the vehicle body will therefore sink a little towards the road with rolling (unless or until a progressive spring characteristic comes into action). The height change of the vehicle body during roll can obviously be influenced by suspension geometry design.

Rigid axles, too, may provoke a jacking-up effect though the distribution of lateral forces at the tyre contact points is of no importance here because the axle beam transmits them 'in sum' to the vehicle body. **Fig. 7.13** shows two variants of simple rigid-axle suspension (see also Fig. 7.10) with lateral control by slot guidance. Variant (a) has a roll centre RZ fixed to the axle, and the slot is provided at the vehicle body. With high lateral acceleration $a_q$, and consequently considerable roll angle, the slot tilts together with

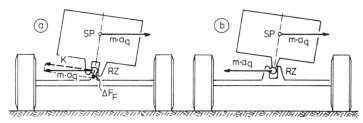

**Fig. 7.13** Rigid-axle suspensions with a) constant and b) variable roll-centre height

the vehicle body, and the inertial force $ma_q$ is split into a force component K applied to the slot and a component $\Delta F$ which tries to pull the axle downward with respect to the vehicle body against the resistance of the springs (not drawn). The vehicle will therefore be jacked up with roll angle. In contrast, the slot in variant (b) is fixed to the axle and so always remains vertical independently of the roll angle (neglecting, of course, tyre deflections caused by wheel-load transfer), and there is no lifting force $\Delta F_F$. Here, though, the roll centre is variable and will sink towards the road with wheel bump movement; the roll-centre height variation with wheel travel is $dh_{RZ}/ds = -1$. As already explained for independent suspensions, jacking-up can obviously be avoided on rigid axles, too, by a suspension layout providing decreasing roll-centre height with wheel bump travel.

The two variants of rigid-axle suspension in Fig. 7.13 represent special cases of roll-centre geometry, namely a roll centre fixed to the axle on the one hand and one fixed to the vehicle body on the other; that means constant roll-centre height with parallel wheel travel in case (a) and roll-centre descent with wheel bump - and by the same amount as wheel travel - in case (b). As will be shown in what follows, quite a free choice of roll-centre geometry is possible (at least theoretically), with rigid-axle suspensions as well as with independent systems.

Slot guidance for the lateral control of rigid axles has occasionally been applied to racing cars, as already mentioned, but not to production cars for obvious reasons (noise, percussion, lubrication). An elegant friction- and maintenance-free substitute is the 'Watt linkage' - **Fig. 7.14** - which steam-engine pioneer James Watt used typically for the straight-line guidance of steam-pressure indicator needles. In its normal form, it is a four-joint chain with opposed links of equal length. In its polar-symmetrical position (a), the pole P of the coupler AB is at infinity, and all displacement vectors of coupler points are rectangular to the links. The centre of curvature $C_0$ of the centre C of the coupler is at infinity, too, as is easily proved by Bobillier's method (see Chapter 3, Fig. 3.2b; the distance e corresponds to the angle $\delta$ of that figure, here decreased to zero). In symmetrical positions of

Cornering 161

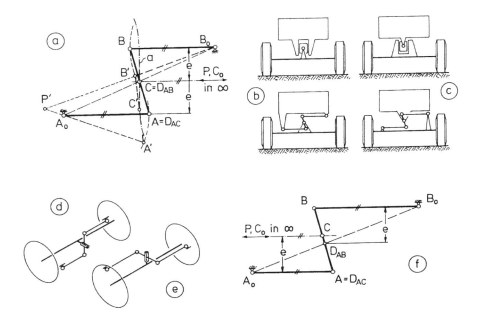

**Fig. 7.14** The Watt linkage

mechanisms, the radii of curvature approximate closely to the trajectories of points at the coupler; point C is, therefore, guided along a straight line over a wide range. Its path (a) is indicated by a chain-dotted line. In a deflected position of the mechanism, the pole returns from infinity (position P'), but the straight-line path of C' is still maintained since the pole P' travels at much the same height as C'.

The analogy between a Watt linkage and a slot guidance is illustrated by Fig. 7.14b and c: the attachment of the links to the vehicle body (the most common application on vehicles) corresponds to the slot on the body, while attachment to the axle corresponds to the slot on the axle.

In addition to the arrangement in a vertical plane - see (d) in Fig. 7.14 - the Watt linkage has also been applied horizontally (e), when of course it no longer remains 'planar' with wheel travel.

With unequal lengths of the links (Fig. 7.14f), the coupler point C that is guided along a straight line is no longer medial, being nearer to the longer link.

On rigid-axle suspensions consisting of several links, roll-centre geometry can be influenced within limits. **Fig. 7.15** shows examples of 'kinematically exact' suspensions with a degree of freedom $F = 2$.

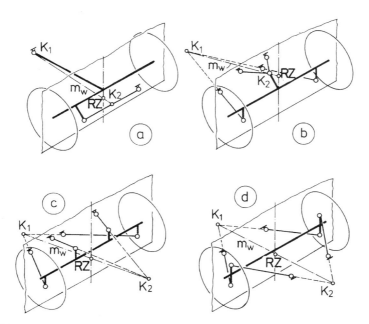

**Fig. 7.15** The roll centre for different rigid-axle suspensions

The thrust-ball axle (a) - here with lateral control by a Panhard rod - can transmit lateral forces by the thrust ball $K_1$ and the Panhard rod only. As long as the rod is horizontal, the connecting line of its intersection point $K_2$ with the centre plane of the vehicle and the thrust ball $K_1$ (which can be loaded with lateral forces without causing reaction forces at the springs) may be regarded as the instantaneous roll axis $m_w$ of the axle with respect to the vehicle body. Consequently its intersection point with the cross-sectional plane is the roll centre RZ. If the Panhard rod is inclined, a lateral force exerts a lifting or diving component at the vehicle body independent of the roll angle; the rod acts in a similar way to the pole of a pole-vaulter! This has nothing to do with the previously mentioned 'jacking-up effect' which may, of course, occur in addition if the roll-centre shift with wheel travel is too small, and will preferably be avoided by positioning the vehicle-side joint in the centre plane, meaning a roll-centre movement equal to the wheel travel - see Fig. 7.13b.

The suspensions of the examples (b) to (d) are symmetrical about the centre plane of the vehicle. The resulting lateral force of the two lower links acts through their intersection point $K_1$ while that of the upper links acts through their intersection point $K_2$ or, in the case of the triangular upper link, on its apex. $K_1K_2$ is the instantaneous roll axis $m_w$ which

Cornering

contains the roll centre RZ. For layout (b), the roll-centre shift with wheel travel will be tiny because the roll centre is very near to the axle-fixed joint $K_2$ of the triangular link, while the roll axis is defined mainly by the inclination of the lower links and will therefore move with wheel travel. The axle (c) shows four links orientated to the same side of the axle, $K_1$ and $K_2$ being on opposite sides. With wheel bump, $K_1$ is shifted downward and $K_2$ upward; the inclination of the instantaneous roll axis $m_W$ varies considerably, but the roll-centre height remains roughly constant. In contrast, for axle (d) with its opposed links, there is little change of the inclination of the axis $m_W$, while the roll centre moves downward with wheel bump.

The inclination of the instantaneous roll axis $m_W$ is of great importance in respect of kinematic self-steering, as will be shown in Section 7.5.

While the cause of the jacking-up effect is easy to explain on independent and rigid-axle layouts, problems arise with compound suspensions – not only because of the wide variety of possible kinematic systems. It is to be expected, of course, that mixed properties will be revealed according to the proximity of the relevant axle to an independent or a rigid-axle suspension. In the following section, a simple method will be described for estimating the cornering behaviour of all suspension types.

Unlike independent suspensions, rigid-axle and compound layouts present some possibility of influencing the jacking-up effect by the geometrical arrangement of the springs. While in independent suspensions a spring characteristic 'reduced' to the tyre contact point can always substitute for the actual spring in any situation, in rigid-axle and compound suspensions the spring action differs as between parallel and antimetric wheel travel (see also Chapter 5, Fig. 5.22); moreover, the attitude change of a spring in relation to the vehicle body, or to the axle, with antimetric wheel travel is of some interest.

**Fig. 7.16** shows an axle suspension where in case (a) the lower and in case (b) the upper spring mountings are positioned at the height of the roll centre RZ. With rolling of the vehicle, the springs in case (a) incline in unison with the vehicle body, thus exerting a horizontal force component which acts in the direction of the centrifugal force, while the vehicle weight

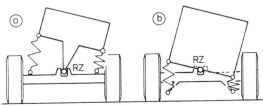

**Fig. 7.16** Attitude of springs on rigid axles with body roll

is supported by a vertical force component that is slightly lower than the spring force. In case (b), however, the springs remain more or less vertical relative to the road.

### 7.3.3 Simple method for suspension-geometry approximation in cornering

As was already clear for the compound suspension of Fig. 7.9, in general an 'axle' shows two typical modes of wheel displacement that may be superimposed in driving practice: first a symmetrical (parallel) travel of the two wheels (with velocity vectors $v_{A1p}$ and $v_{A2p}$ in Fig. 7.9) and second an antimetric travel (with velocity vectors $v_{A1w}$ and $v_{A2w}$) resulting for instance from vehicle roll. With rigid axles, the 'parallel' wheel travel can, of course, occur only in the vertical direction with respect to the vehicle coordinate system, while with independent suspensions the two modes coincide.

In symmetrical or parallel wheel travel, each tyre contact point is displaced along a precisely defined path in the vehicle's cross-section, while the trajectories of antimetric wheel travel depend on the medial displacement of the tyre contact points (or on the 'symmetrical' starting position of the antimetric motion). **Fig. 7.17** makes this clear through examples of two compound suspensions of very different types.

Based on the path $g_p$ of parallel wheel travel of the left tyre contact point, several trajectories $g_w$ of antimetric movement are drawn – in addition

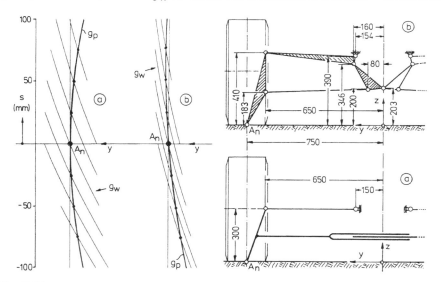

**Fig. 7.17** Trajectories of tyre contact points with symmetrical and antimetric wheel travel for two different compound suspensions

# Cornering

to the movement starting from the 'normal position' of the vehicle (tyre contact point $A_n$), other trajectories each starting from three positions in the bump and rebound directions, e.g. from each of three static-load variations.

With both suspensions, the antimetric trajectories $g_w$ are obviously more inclined relative to the vertical line than the symmetrical trajectories $g_p$; this is, of course, in line with one of the reasons for adopting compound suspensions - the ability to combine a relatively high roll centre with a small track change, with parallel wheel travel.

For a rigid-axle suspension, the trajectories $g_p$ are vertical lines and the trajectories $g_w$ are circles about points on the centre plane of the vehicle. An independent suspension, however, shows the two trajectories $g_p$ and $g_w$ coinciding.

This method of representation encourages the establishment of a generalized geometrical model for comparative investigations on independent, rigid-axle and compound suspensions (15). With the simplifying assumption that tyre slip forces are essentially of more importance than camber forces, wheel camber can be ignored, and the description of the trajectories of the tyre contact points of both wheels will be sufficient for the fundamental analysis of suspension behaviour in cornering.

The basic trajectory $g_p$ of symmetrical wheel travel may be represented by a parabola, which follows from the coordinate $y_0$ of the tyre contact point in the 'normal' position and from the wheel travel s, **Fig. 7.18**:

$$y_p = y_0 + k_1 s + k_2 s^2 \tag{7.16}$$

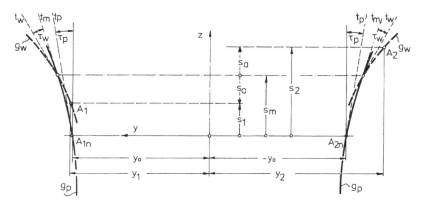

**Fig. 7.18** Symmetrical and antimetric trajectories of tyre contact points

In addition to the 'normal' position ($A_{1n}$ and $A_{2n}$), two tyre contact points $A_1$ and $A_2$ are drawn in different bump positions $s_1$ and $s_2$ which lie on laterally opposed trajectories $g_w$ originating from the medial wheel displacement

$$s_m = (s_1 + s_2)/2 \tag{7.17}$$

The tangent $t_w$ of the trajectory $g_w$ which is superimposed on the basic curve $g_p$ is inclined to the latter's tangent $t_m$ at the medial wheel travel $s_m$ by an angle $\tau_w$. This angle may be variable with $s_m$, and the inclination of $g_w$ relative to $g_p$ is therefore to be described by a first (and constant) coefficient $k_3$ and a second one $k_4$ depending on the medial displacement $s_m$. The curvature of $g_w$ relative to $g_p$ may be characterized by a coefficient $k_5$ which proves, however, to have a negligible influence on rolling behaviour. With the antimetric displacement of the points $A_1$ and $A_2$

$$s_a = (s_2 - s_1)/2 \tag{7.18}$$

the antimetric trajectory $g_w$ superimposed on $g_p$ is defined by

$$y_w = \pm (k_3 + k_4 s_m)s_a \pm k_5 s_a^2 \tag{7.19}$$

The parameters $k_2$ and $k_5$ are positive for centres of curvature outside the trajectories as shown in Fig. 7.18. In practice, though, these trajectories are normally curved towards the middle of the vehicle – see also Fig. 7.17 – and both coefficients are negative.

The equations of the resulting trajectories of the left-hand tyre contact point $A_1$ with lower deflection $s_1$ (so to say, the point on the 'inner' side of the corner) and of the right-hand (or 'outer') point $A_2$ follow from the equations (7.16) and (7.19) as:

$$y_1 = y_0 + k_1 s_1 + k_2 s_1^2 - (k_3 + k_4 s_m)s_a + k_5 s_a^2 \tag{7.20}$$
$$y_2 = -y_0 - k_1 s_2 - k_2 s_2^2 - (k_3 + k_4 s_m)s_a - k_5 s_a^2 \tag{7.21}$$

For an independent suspension, which (as already stated) shows only one trajectory $g_p = g_w$, the coefficients $k_3$, $k_4$ and $k_5$ are zero. A rigid-axle suspension, on the other hand, gives no track change with parallel wheel travel, and that is why $k_1$ and $k_2$ are zero. Only with compound suspensions are all five coefficients normally other than zero.

With this simplified model, the wheel-suspension mechanism is reduced to the trajectories of the two tyre contact points. Hence, it has to be assumed that the spring elements act immediately on the tyre contact points and are parallel to the centre plane of the vehicle body. Special effects of spring attitude change in rigid-axle or compound suspensions, as shown in Fig. 7.16, cannot be taken into consideration, and investigations

Cornering                                                                 167

into these suspension types will therefore reveal more or less typical tendencies in rolling behaviour.

The model allows us to study the essential properties of lateral dynamics on independent, rigid-axle and compound suspensions by a common system of equations. The coefficients $k_1$ to $k_5$ are defined by the kinematic 'characteristics' of the suspensions: the relationship between roll-centre height and (parallel) wheel travel is

$$h_{RZ}(s) \approx y_0(k_1 + 2k_2 s + k_3 + k_4 s)$$

and the roll-centre height change with wheel travel is

$$dh_{RZ}/ds \approx y_0(2k_2 + k_4)$$

The local radius of curvature of the resultant trajectory $g_p + g_w$ is

$$\rho = 1/(2k_2 + 2k_5)$$

For a rigid axle, $\rho$ is approximately equal to half of the track width (or to the y-coordinate of the tyre contact point).

## 7.4 Vehicle attitude in steady-state cornering

In what follows, the method described in Section 7.3.3 will be used to investigate the effects of different design measures on vehicle attitude in cornering. To make the typical properties of the suspensions as clear as possible, any influence of other axles can be neglected or, as already covered in Section 7.3, the vehicle can be assumed to have identical front and rear axles or to be a single-axle trailer with a drawbar of infinite length. Possible toe-in angles are neglected, too, because toe-in is not a typical property for some axle systems. Moreover, the suspension may not be loaded with longitudinal traction or braking forces (which, of course, would influence the tyre slip angles, see Chapter 4). For the following studies, an average passenger car is assumed, with typical (but linear) spring rates and typical tyre properties (**15**).

**Fig. 7.19** shows the change of vehicle attitude and forces plotted against lateral acceleration $a_q$ of a vehicle with independent suspension and with a roll-centre height $h_{RZ} = 150$ mm (constant with wheel travel; $dh_{RZ}/ds = 0$).

Starting from the static wheel load $F_R = 4$ kN with lateral acceleration $a_q = 0$, the outer and inner wheel loadings respectively increase and decrease almost linearly with $a_q$. The capability of lateral grip thus increases at the outer tyre and decreases at the inner one, and the outer lateral force $F_{qa}$

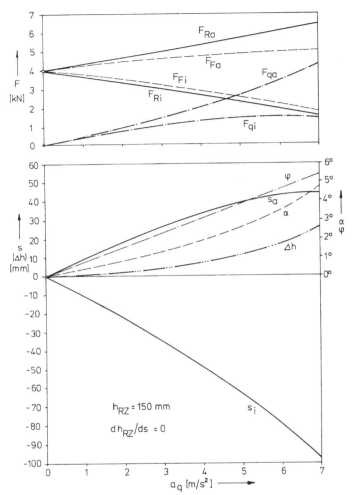

**Fig. 7.19** Vehicle attitude and forces with lateral acceleration (independent suspension)

shows progressive growth while that of the inner force $F_{qi}$ is less than linear until it reaches a maximum and begins to decrease.

The increase of the roll angle $\varphi$ with $a_q$ is linear, while that of the tyre slip angle $\alpha$ is clearly progressive due to the 'degressive' characteristics of the lateral force against the slip angle as well as the wheel load, see Chapter 4, Fig. 4.2b.

The suspension tends to jack-up because the outer wheel travel $s_a$ increases less than proportionally to $a_q$ (although linear springs are here assumed!); this is a consequence of constant roll-centre height, see Fig. 7.11

# Cornering

or 7.13. As already indicated by Fig. 7.12, a decreasing roll-centre height with (parallel) wheel travel may tone down the lifting effect $\Delta h$ on the vehicle's centre of gravity.

For independent suspensions, **Fig. 7.20** shows the influence of the nominal roll-centre height $h_{RZ}$ above the road in the 'normal' position of the vehicle, and the change of this height $dh_{RZ}/ds$ with wheel travel s, on the rise $\Delta h$ of the centre of gravity with a lateral acceleration $a_q = 7\,m/s^2$ (again with linear springs assumed).

As expected considering Figs 7.11 and 7.12, the tendency to jack-up is the stronger the higher the roll centre is in the normal position, and so the greater must be the (negative) roll-centre height change $dh_{RZ}/ds$ to avoid jacking-up.

Independent suspensions with a roll-centre height as much as 250-300 mm are no longer used for passenger cars. Swing axles belong to this group. For the 'single-joint' swing axle with swing-arm joints coinciding in the centre plane of the vehicle, the roll-centre height change is $dh_{RZ}/ds = -1$, while normal (two-joint) swing axles show $dh_{RZ}/ds \approx -1.1$ to $-1.2$. According to Fig. 7.20, swing axles will lift the centre of gravity by about $\Delta h \approx 25-30$ mm at a lateral acceleration of $7\,m/s^2$ even with linear springs! However, the

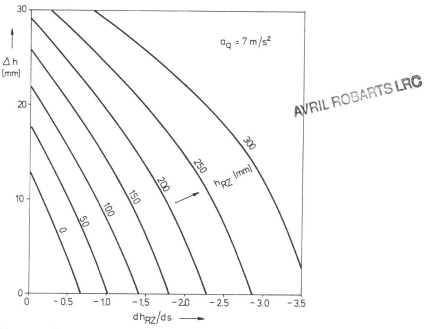

**Fig. 7.20** Jacking-up effect depending on roll-centre height and the change of this for independent wheel suspensions with linear springs

pure trailing-arm suspension, which is frequently applied to the rear wheels of front-drive cars and which shows the parameters $h_{RZ} = 0$ and $dh_{RZ}/ds = 0$, lifts the vehicle body by $\Delta h \approx 13$ mm. With progressive springs, these figures will become even greater. The reason for the widespread condemnation of the swing-axle suspension for jacking-up the vehicle may be that its 'advantageous' negative camber angle at the onset of body roll changes – caused by extreme jacking-up – rather quickly to 'positive' camber towards the adhesion limits. In contrast, with pure trailing-arm layouts, both wheels lean together with the vehicle body, assuming steadily increasing 'positive' wheel camber with respect to the cornering centre. A trailing-arm suspension thus cannot provide the optimum potential of lateral grip, but a steady transition to the stability limits is possible.

Double-wishbone suspensions are today preferably designed with a roll-centre height between zero and about 150 mm; the roll-centre height change with wheel travel $dh_{RZ}/ds$ is usually between -1 and -2 mm/mm.

At least on front wheels, strut suspensions feature relatively short transverse links matching short track rods, especially if rack-and-pinion steering is used. Since in effect the strut is substituting for an 'upper transverse link' of infinite length, roll-centre change with wheel travel is in the range of $dh_{RZ}/ds \approx -2.5$ to $-3.5$. Consequently, strut suspensions tend to cause diving of the vehicle's front in cornering which is usually limited in practice, however, by the progressive auxiliary springs.

Anti-roll bars (stabilizer springs) increase the spring rate of the vehicle in roll without changing the bump rate. Their effect on the vehicle model in steady-state cornering is shown in **Fig. 7.21** by examples of two independent suspension systems, both with a roll-centre height of $h_{RZ} = 150$ mm in 'normal' position, assuming a lateral acceleration of $7$ m/s$^2$.

On an independent suspension with constant roll-centre height ($dh_{RZ}/ds = 0$) – which belongs to the 'jacking-up' types (see Fig. 7.20) – increasing the anti-roll bar rate $c_S$ leads to a reduction in the lifting tendency. The bar reduces the roll angle and hence the disadvantageous change of inclination of the trajectories of the tyre contact points with respect to the road – see Fig. 7.11 – which leads to the jacking-up. Usual anti-roll bar rates $c_S$ for the front wheels of passenger cars amount to 50% (or even more) of the suspension spring rates $c_F$. Simple 'active suspension systems' which control only the roll angle but not the vehicle ride height (see Section 5.7) may be assumed to be represented here by high 'bar' rates beyond $c_S = 10 c_F$ (right side in the diagram); the roll angle having almost vanished, lifting of the vehicle becomes more conspicuous.

According to Fig. 7.20, a suspension with a roll-centre height $h_{RZ} = 150$ mm and a roll-centre height change $dh_{RZ}/ds = -1.8$ should totally

Cornering                                                                                             171

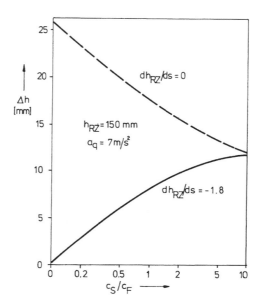

**Fig. 7.21**
Influence of anti-roll bars
on vehicle attitude
(independent suspension)

suppress jacking-up, but that of Fig. 7.21 shows contrary behaviour with increasing anti-roll-bar rate: the higher the rate, the greater the lifting of the vehicle. By reducing the roll angle, the bar reduces the suspension's ability to orientate the trajectories of the tyre contact points and hence their polar rays for optimum support of the lateral forces – see Fig. 7.12.

Since the roll angle is nearly suppressed with extremely high anti-roll-bar rates, the influence of the roll-centre height change $dh_{RZ}/ds$ vanishes, and vehicle lifting $\Delta h$ finally depends merely on the roll-centre height $h_{RZ}$. For this reason, both suspension variants in Fig. 7.21 show similar behaviour with rising rate of the anti-roll bar.

A transverse compound spring (see Chapter 5, Fig. 5.14) helps the main spring to support the vehicle's weight but is ineffective in roll; hence, it represents the opposite of an anti-roll bar. It follows that a vehicle with constant roll-centre height will show increasing body lift with increasing compound spring rate while one with a negative roll-centre change ($dh_{RZ}/ds < 0$) will show decreasing lift or will even lower the body (remember: linear springs are assumed!).

A rigid axle transmits all the lateral forces of the tyres 'in sum' to the vehicle body. Rolling behaviour is therefore influenced only by the roll-centre height (or more precisely, the distance between the roll centre and the centre of gravity), the roll-centre height change and the springing system, but no longer by the tyre properties. **Fig. 7.22** shows the lifting $\Delta h$ of the vehicle for rigid axles with different roll-centre heights $h_{RZ}$ against the

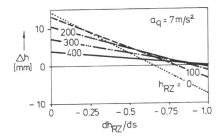

**Fig. 7.22**
Jacking-up effect on rigid-axle suspensions (average passenger car)

roll-centre height change $dh_{RZ}/ds$, assuming a lateral acceleration of $7\,m/s^2$.

For a given centre-of-gravity height with rigid axles, jacking-up is inversely proportional to the roll-centre height, in contrast to independent suspensions. The distance from roll centre to centre of gravity essentially determines the roll angle and thus the inclination of the (perhaps virtual) 'guidance' of the vehicle body at the axle - see Fig. 7.13. With a high roll centre - e.g. $h_{RZ} = 400$ mm - the very small roll angle makes the vehicle insensitive to any inclination change of that guidance. Jacking-up is obviously avoided with rigid axles by having a roll-centre height change $dh_{RZ}/ds \approx -0.75$ to $-1$ (with linear springs). Theoretically, for a roll-centre height change $dh_{RZ}/ds = 0$ the axle suspension corresponds to a (virtual) guiding slot that is fixed to the axle, and this means that the lateral force cannot exert any lifting component on the vehicle body - see Fig. 7.13. The results of Fig. 7.22 were calculated, however, assuming the springs to act parallel to the medial plane of the vehicle body, and not always perpendicularly to the road surface. This is acceptable on independent suspensions but may not always apply to rigid-axle and compound suspensions, see Fig. 7.16. If the springs are assumed to stay always perpendicular to the road surface, jacking-up will vanish with a roll-centre height change $dh_{RZ}/ds = -1$, independently of the roll-centre and centre-of-gravity heights.

With a compound suspension, jacking-up depends on the degree of relationship to either independent or rigid-axle layouts.

For the compound suspension of Fig. 7.17a, which is similar to a rigid-axle suspension, **Fig. 7.23** gives the parameters $k_1$ to $k_5$ for the application of the method described in Section 7.3.3. The roll angle $\varphi$ increases nearly linearly with lateral acceleration $a_q$ and, in spite of the considerable roll-centre height $h_{RZ}$, the suspension causes a slight sinking of the vehicle body with lateral acceleration, which may be explained by the large roll-centre height change $dh_{RZ}/ds = -1.3$.

In contrast, the compound suspension of Fig. 7.17b is rather similar to an independent layout with very long transverse links. Its parameters $k_1$ to

Cornering

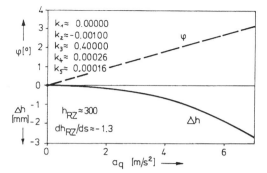

**Fig. 7.23**
Rolling behaviour of the compound suspension shown in Fig. 7.17a

$k_5$ are quoted in **Fig. 7.24**. With the suspension's roll-centre height of 186 mm and a roll-centre height change of −0.8, the vehicle body is lifted slightly with high lateral acceleration, but by a much smaller amount than with a comparable independent suspension, see Fig. 7.20.

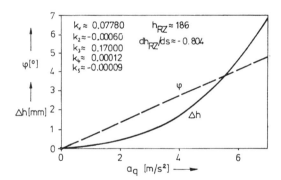

**Fig. 7.24**
Rolling behaviour of the compound suspension shown in Fig. 7.17b

In worldwide use on front-drive vehicles is a compound rear suspension of very simple construction, which enables the designer to obtain compound properties ranging from independent to rigid-axle systems and beyond, **Fig. 7.25**. Two rigid trailing arms carrying the wheel stub-axles are mounted to the vehicle body at their front ends by rubber joints. A turning joint with transverse axis connects the trailing arms. In practice, that joint is preferably replaced by a bending-resistant but torsionally elastic member with an 'open' cross-section; the effective rotational axis then runs through the 'shear centre' of the member.

If the joint connecting the trailing arms is a turning and sliding component, the trailing-arm joints may kinematically be modelled by ball joints. With a turning joint or a transverse beam with open section, though, the distance between the trailing-arm joints will vary with antimetric wheel travel; this is, of course, possible in practice with rubber joints but requires a kinematic

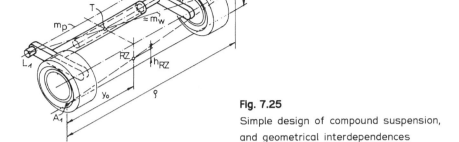

**Fig. 7.25**
Simple design of compound suspension, and geometrical interdependences

model that provides mirror-image (or 'antimetric') lateral motion of the joints - e.g. by a notional Watt linkage connecting both joints with the vehicle body.

The kinematic relationships of this suspension type are easily to visualize. With parallel wheel travel, the whole assembly rotates about the axis $m_p$, and the tyre contact points remain in their respective vertical planes. With antimetric wheel travel or vehicle roll, the point T of the 'turning joint' in the medial plane of the vehicle (or the shear centre of the open member connecting the trailing arms) must stay 'fixed' for reasons of antimetric motion, and the axes $m_w$ of antimetric wheel travel must run through the trailing-arm joints and the point T. The axis $m_w$ of the left wheel intersects the cross-sectional plane through the tyre contact points in the 'transverse pole' $Q_1$. Its connecting line to the left tyre contact point $A_1$ intersects the medial plane of the vehicle at the roll centre RZ. With parallel bump or rebound wheel travel about the axis $m_p$, the pole $Q_1$ remains 'fixed' relative to the axle assembly and thus to the roll centre too. This type of compound suspension therefore shows a constant roll-centre height with wheel travel.

For basic investigations into this suspension type, the distance between the trailing-arm joints at the vehicle body may be assumed to be equal to the track width of the axle, and the transverse member to be in the same horizontal plane as the wheel centres and the mentioned joints (practice may deviate from this with respect to kinematic self-steering). That enables simplified comparative calculations by varying the roll-centre height through longitudinal displacement of the transverse member. The table (next page) gives the parameters $k_3$ and $k_5$ for applying the method described in Section 7.3.3 (the parameters $k_1$, $k_2$ and $k_4$ are zero), assuming a track width of 1500 mm and a tyre radius of 300 mm, and, as in the foregoing, linear springs.

Cornering

| $h_{RZ}$ (mm) | $k_3$ | $k_5$ |
|---|---|---|
| 0 | 0.0000 | 0.00000 |
| 50 | 0.0667 | -0.00011 |
| 100 | 0.1333 | -0.00022 |
| 200 | 0.2667 | -0.00044 |
| 300 | 0.4000 | -0.00067 |
| 400 | 0.5333 | -0.00089 |

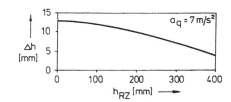

**Fig. 7.26**
Jacking-up effect of compound suspensions according to Fig. 7.25

The results of these calculations are shown in **Fig. 7.26**. Jacking-up effect is very moderate with all variants, as demonstrated by the vehicle height change $\Delta h$ (at a lateral acceleration of $a_q = 7\,m/s^2$) against the nominal roll-centre height $h_{RZ}$.

With zero roll-centre height, the axle system becomes in effect an independent suspension with pure trailing arms and therefore shows the same jacking-up as the latter - see Fig. 7.20. A roll-centre height of 300 mm, equal to the assumed tyre radius, is achieved by positioning the transverse member in line with the wheel axes; this version behaves like a rigid axle with a roll-centre height change of zero - see Fig. 7.22. If the roll-centre height is above 300 mm, the transverse member lies behind the wheel centres. The geometrical properties with vehicle roll then correspond to those of a swing-axle layout with the swing-arms shorter than half the track, and in cornering the wheels incline towards the inside of the corner. Such behaviour (unlike that of a true swing-axle suspension) not being influenced by even a jacking-up effect, this variant of the compound-suspension family generates negative wheel camber as with a motorcycle, and was used on a Mercedes-Benz racing car in the 1930s (see Chapter 14, Fig. 14.4) and on a Honda passenger car in the early 1980s.

Investigations into the jacking-up characteristics of compound suspensions of very different types, using the method given in Section 7.3.3, confirm their kinematic properties in cornering to be a mixture of those of independent and rigid-axle suspensions (**15**).

It should be reiterated that the foregoing investigations were carried out on a 'single-axle' vehicle model, outlining, of course, the typical properties of the respective suspension systems, but neglecting the mutual influencing of front and rear axles (normally of different design and with different springing characteristics) on a real vehicle.

Concerning the various requirements and compromises met with in suspension harmonization, and above all the usual progressive spring characteristics, a certain amount of jacking-up effect can hardly be avoided in practice. The increased rolling moment caused by the lifting of the vehicle body is, however, of less importance than the camber change (i.e. loss of 'negative' camber) that is associated with many suspension types and that leads to a progressive increase of the tyre slip angle when cornering adhesion limits are approached. Since wheel bump and rebound are limited for various reasons (interior space, deflection angles of mountings or universal joints etc.), jacking-up causes premature contact with the rebound stop for the inner wheel, thus unloading the latter or even lifting it off the road, and in any case reducing driving stability, especially on driven wheels by the loss of traction capability.

Most of the recently developed multi-link suspensions reveal, through their kinematic attunement, the designer's intentions of reducing the jacking-up effect.

## 7.5 Kinematic self-steering

Kinematic self-steering of a vehicle depends on the axle's bump-steer properties on the one hand and on the roll angle $\varphi$ and the wheelbase l on the other. In cornering, the vehicle body rotates approximately about the roll axis r - the line connecting the front and rear roll centres $RZ_v$ and $RZ_h$ - while the inner and outer wheels are displaced respectively by rebound and bump wheel travel, and may at the same time assume modified steering angles. This is visualized in a simplified manner in **Fig. 7.27** by means of rigid axles with instantaneous roll axes $m_w$ which are inclined with respect to the vehicle's roll axis r by angles $\varkappa_v$ and $\varkappa_h$. With vehicle roll, and the angles $\varkappa_v$ and $\varkappa_h$ other than zero, the instantaneous axes $m_{wv}$ and $m_{wh}$ assume inclined positions with respect to the projection of the roll axis r in plan view. Being rectangular to the instantaneous axes $m_w$ in reality, the rigid axles must remain rectangular also to the projections of the axes $m_w$ in plan view, and assume steering angles $\delta$ which are proportional to the roll angle $\varphi$ and the angles $\varkappa_v$ or $\varkappa_h$. A typical example of the vehicle model in Fig. 7.27 is the well-known steerable roller skate.

A positive steering angle $\delta$ on a rear axle increases the cornering radius in a left hand corner and consequently promotes 'under-steering' while a positive steering angle on a front wheel induces 'over-steering'.

# Cornering

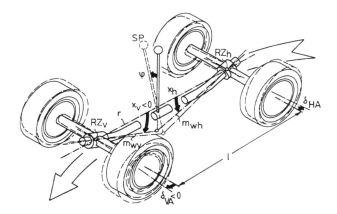

**Fig. 7.27** Kinematic self-steering of a vehicle, shown schematically

The angle ϰ between a vehicle's roll axis r and the instantaneous roll axis $m_w$ of an axle may be defined as positive if the roll axis r rises above the instantaneous axis $m_w$ in the straight-ahead direction. When the roll angle of the vehicle is φ, the kinematic self-steering angle is

$$\delta = \varkappa \varphi \qquad (7.22)$$

For a rigid-axle suspension that is symmetrical to the centre plane of the vehicle, the instantaneous axis $m_w$ of antimetric wheel travel is easily recognized: the four-link suspension of **Fig. 7.28** can transmit lateral forces to the vehicle body only by the intersecting points $P_v$ and $P_h$ of the lower and upper links; the roll centre RZ must therefore be on the line connecting those two points. Moreover, the axle can instantaneously swivel with respect to the vehicle body without constraint only about the line $P_v P_h$, which thus becomes the instantaneous roll axis $m_w$ (see also Fig. 7.15).

For independent and compound suspensions, too, a 'suspension characteristic' ϰ would be feasible to define the kinematic self-steering properties

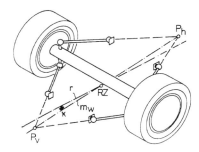

**Fig. 7.28**
Instantaneous axis $m_w$ of antimetric wheel motion and roll centre RZ of a four-link rigid-axle suspension

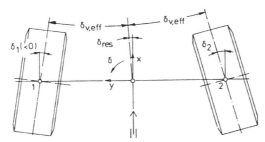

**Fig. 7.29** Toe-in angle $\delta_{v,eff}$ and resulting steering angle $\delta_{res}$

of the systems. It would hardly be worthwhile, however, because such a (virtual) angle x has to be determined anyway by calculating and averaging the steering angles of the wheels of the axle.

With parallel as well as with antimetric wheel travel, kinematic (also elasto-kinematic) self-steering angles can occur at the wheels. The resulting steering angle of an axle with respect to the vehicle coordinate system – see **Fig. 7.29** – is

$$\delta_{res} = (\delta_1 + \delta_2)/2 \tag{7.23}$$

If $\delta_1 \neq \delta_2$, each wheel shows an effective toe-in angle $\delta_{v,eff}$ with respect to the steering angle $\delta_{res}$. According to the usual definitions, the toe-in angle is positive if the wheel planes intersect in front of the wheels. On left-hand wheels, positive toe-in angles are, therefore, equivalent to negative steering angles. The effective toe-in angle per wheel is

$$\delta_{v,eff} = (\delta_2 - \delta_1)/2 \tag{7.24}$$

Normally the toe-in angle changes with wheel travel. **Fig. 7.30** shows the toe-in angle $\delta_v$ against wheel travel s on a true-scale planar double-wishbone suspension, consisting of two triangular links one above the other with horizontal turning joints, and of a rod link behind the axle (or a track rod on a steerable suspension). The inner joint of the rod is fixed in the normal position (0) for toe-in change with wheel travel $d\delta/ds = 0$, or for a vertical tangent of the toe-in against wheel-travel diagram. Towards the end of wheel travel, however, an increasing deviation occurs with positive values at bump and negative values at rebound; the toe-in curve is s-shaped – typical for multi-link suspensions. Except for special cases of symmetry, the circles of curvature of the trajectories of points at the 'coupler' of the suspension mechanism (i.e. the wheel carrier) touch the trajectories at 'three points' infinitesimally close together, and consequently pass over the trajectories (i.e. they appear on *different* sides of the trajectories before and after the contact point). As the outer end of the rod can move

Cornering 179

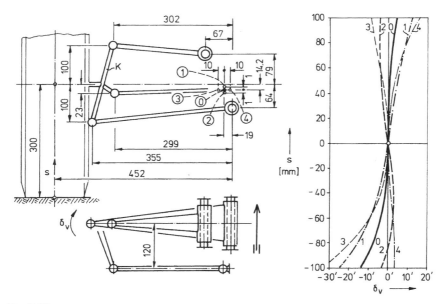

**Fig. 7.30** Toe-in against wheel travel for a planar double-wishbone suspension (rear view and plan view)

only on a circle about the inner joint, deviations from the theoretical path of the outer joint in opposite directions (and, hence, toe-in change) are unavoidable.

Displacements of suspension joints, either by error or even deliberately, 'detune' the suspension mechanism and affect the toe-in curve.

Vertical displacement of the inner joint of the rod link (positions 1 and 2) by a minute amount of 1 mm causes an inclination of the tangent of the toe-in curve and thus a considerable gradient of toe-in against wheel travel. The basic s-shape of the curve remains, however.

Changing of the rod radius, here by ± 10 mm (positions 3 and 4), does not affect the tangent of the toe-in curve in the normal position (which remains vertical) but leads to a changed radius of curvature. In the case of a shortened rod (3), the curve bends towards toe-out, but its shape is very different from a circle, because the s-shape of the basic curve (0) still makes itself felt. Accordingly, a lengthened rod (4) bends the curve towards toe-in.

If the rod link is positioned in front of the suspension, contrary to Fig. 7.30, the tendencies shown in the diagram will be reversed.

What has just been discussed, with the example of the inner joint of the rod link, is valid for **any** joint of the suspension. Vertical displacement of a joint with respect to the 'ideal' position for minimum toe-in change with wheel travel inclines the tangent of the toe-in curve and so changes the self-steering tendency (e.g. understeer → oversteer), while varying the length of a link changes the radius of curvature of the toe-in diagram and hence its shape. In production, therefore, the tolerances of vertical joint coordinates, in particular, must be carefully controlled.

If kinematic toe-in change with wheel travel is unwanted in the 'normal position' (i.e. the straight-ahead position of a steerable wheel), equation (7.3) and the conditions $\delta = 0$ and $\omega_\delta = 0$ give us

$$\omega_{Kz}/\omega_{Ky} = -\tan\gamma$$

and this means that the vector $\omega_K$ characterizing the instantaneous state of motion of the wheel carrier must appear parallel to the wheel axis in the cross-section of the vehicle. Consequently, **the gradient $d\delta/ds$ of the toe-in curve against wheel travel is zero if, in that cross-section, the wheel axis appears parallel to the projection of the instantaneous axis.**

This statement is, of course, of merely theoretical importance for multi-link suspensions, where instantaneous axes can be determined only by computer programs; it could, however, be very useful for the design of spherical or planar suspensions with their easily obtainable instantaneous axes. In particular a semi-trailing-link suspension, with its fixed axis of rotation at the vehicle body, allows us clearly to visualize the relationship between self-steering and wheel attitude in cross-section, **Fig. 7.31**. The axis of the trailing link descends by 1° towards the middle of the vehicle; so, with a negative wheel camber of $\gamma = -1°$, instantaneous wheel-camber

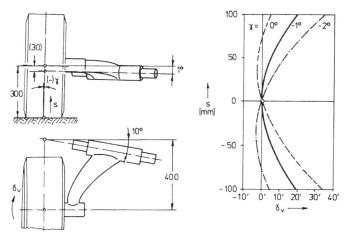

**Fig. 7.31** Toe-in against wheel travel for a semi-trailing-link suspension

change with wheel travel is zero. With that travel the wheel axis moves on a hyperbolic surface (or, if intersecting the link's rotational axis, on a cone) and, since the rotational axis of the link is fixed, there is only one position of the wheel where its axis and that of the trailing-link appear parallel in cross-section. Hence, semi-trailing-link suspension shows a bent toe-in curve with wheel travel. If the wheel camber in the normal position differs from $-1°$, the toe-in curve changes its direction while retaining its basic shape. For a nominal wheel camber of $\gamma = 0°$ in the normal position, minimum toe-in occurs with about 40 mm bump wheel travel and, for $\gamma = -2°$, with about $-40$ mm rebound wheel travel. In these positions, the wheel and trailing-link axes are parallel in cross-section.

For steerable suspensions, the toe-in curve can always be influenced by suitably arranging the positions of the track rod and the suspension links, as already explained for Fig. 7.30.

Special cases of steerable independent wheel suspensions are shown in **Fig. 7.32**. The Dubonnet system (a) is an independent suspension with a 'kingpin' fixed to the vehicle body or the chassis. The suspension itself is placed 'outside' the steering mechanism, in the same way as with motorcycle forks (see Fig. 6.10 in Chapter 6), and is normally realized by a leading arm (here drawn without the usual brake-reaction rod); the whole suspension swivels about the kingpin when steered. Since the steering mechanism is not affected by wheel travel, a single transverse track rod can be applied, as with a rigid axle.

The telescopic-guidance or 'sliding-pillar' layout (b) shows infinite radii of curvature of all wheel-carrier points, and toe-in change with parallel wheel travel can be avoided only by parallel guidances and a single transverse track rod. Because of the finite length of the longitudinal drag-rod connecting one of the wheel carriers to the steering box L, however, resultant steering angles with parallel bump and rebound wheel travel are unavoidable (the vehicle travels in a zigzag fashion). On the other hand, a 'divided'

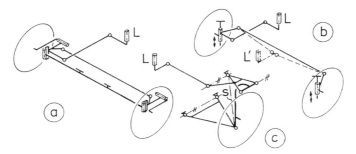

**Fig. 7.32** Special cases of steerable independent wheel suspensions

steering linkage with two transverse track rods (dotted lines) will cause toe-in angles with parallel wheel travel if the track rods are in front of the axle but toe-out angles if they are behind it, while the resultant steering angle is about zero. With antimetric wheel travel, steering angles cannot be avoided at all.

Theoretically, double-wishbone suspensions can be made nearly independent of steering geometry by the inversion of one of the triangular links (c), its turning joint being fixed to the wheel carrier and its ball joint to the chassis.

Rigid axles with the wheel axes in line cannot generate toe-in angles with wheel motion. On non-driven 'dead' axles and on driven 'de Dion' axles (see Chapter 6, Fig. 6.11b), however, it is possible and usual to provide toe-in angles at the axle beam; in this case, toe-in change and (negligible) camber change may occur even with parallel wheel travel if the axle swivels about a transverse instantaneous axis - e.g. about a 'longitudinal pole' in side view that follows from traction or braking anti-pitch measures (see Chapter 6, Figs 6.9, 6.14, 6.22 or 6.28).

With rigid axles, additional steering angles may occur if the axles are laterally controlled by a Panhard rod. Such a rod induces non-symmetrical geometry by the arcuate path of its joint A at the axle which leads to the latter's lateral displacement - **Fig. 7.33**. A thrust-ball axle (a) will therefore show a steering angle with parallel wheel travel, and the 'drawbar' is, of course, best made as long as possible. With longitudinal links (b), the Panhard rod causes lateral (but parallel) displacement of the axle with respect to the vehicle body; this will disturb directional stability, however,

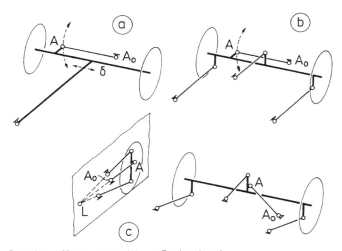

**Fig. 7.33** Steering effects caused by a Panhard rod

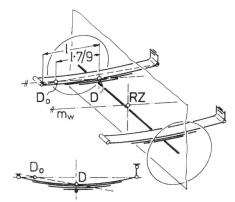

**Fig. 7.34**
Instantaneous axis $m_w$ of antimetric wheel travel for a leaf-sprung rigid axle

due to the finite wheelbase length. That disturbance can be avoided by a diagonal arrangement of the Panhard rod (c), the projection of its joint $A_0$ at the chassis in side view being the centre of curvature of the projection of the path of the joint A at the axle. When loaded with a lateral force, though, the Panhard rod will now generate a longitudinal force component that causes steering angles on a longitudinally compliant suspension, unless the joint A is in the medial plane of the vehicle, as shown in the drawing.

The 'classic' rigid-axle suspension is achieved by leaf springs alone, **Fig. 7.34**. In cross-section of the vehicle, these springs act like vertical slot guidances, and in the lateral view they act like longitudinal links with a 'joint' D at the axle near the upper leaf and a 'joint' $D_0$ at the chassis which can be found on the 'chord' of the spring at a distance of about $7/9$ of the spring's half-length, see Chapter 5, Fig. 5.26. The instantaneous axis $m_w$ of antimetric wheel travel runs roughly parallel to the line $DD_0$ and determines the roll centre RZ. For a steerable axle, the longitudinal drag-rod of the steering linkage should best be parallel to the line $DD_0$ in side view – **Fig. 7.35**a – and its joint at the axle should coincide with the point about which the axle winds up with braking or traction torque – see Chapter 5, Fig. 5.28.

If the axle is laterally controlled by a Panhard rod 1 (Fig. 7.35b), the drag-rod 2 is best arranged parallel to the Panhard rod; at least in the straight-ahead position, this gives unobjectionable steering behaviour with parallel as well as with antimetric wheel travel, because the Panhard rod is usually provided with very stiff joints.

For a four-link axle suspension (c), additional conditions have to be met, since the instantaneous axes $m_p$ and $m_w$ of respectively parallel and antimetric wheel travel do not normally intersect in space. The drag-rod $DD_0$ should intersect $m_p$ as well as $m_w$, and the projection $D'_0$ in side view

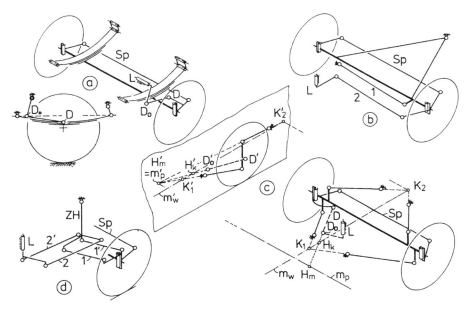

**Fig. 7.35** Drag-rod layouts for rigid-axle suspensions

should be the centre of curvature of the path of the projection D'. The axle will then show a self-steering behaviour that is determined mainly by the instantaneous axis $m_w$ – see also Fig. 7.28 and equation (7.22). If this is not suitable, a steering angle can be superimposed by changing the position of the drag-rod; in this case, however, the line $K_1K_2$ will no longer be correct for the determination of kinematic self-steering, though it remains approximately the geometrical locus of the roll centre RZ.

If a steerable rigid axle (or any wheel suspension) is made longitudinally compliant for better ride comfort, longitudinal drag-rods are difficult. Exact transmission of the steering angle from the steering box L to the wheel carrier can then be achieved by means of a 'floating' intermediate lever ZH, Fig. 7.35d, which is connected to the wheel carrier by a parallelogram consisting of a track-rod 1 and a control rod 1', and to the steering box by a second parallelogram consisting of a drag-rod 2 and a control rod 2'.

Normally, the self-steering properties of compound suspensions can be determined only by computer calculation, except for the simple rear-axle system shown already in Fig. 7.25 (so-called 'H-frame' because of its shape in plan view). The instantaneous axis $m_w$ of antimetric wheel travel, **Fig. 7.36**, runs through the relevant trailing-arm joint L and the intersecting point M of the transverse turning joint (or the shear centre T of the open-profile beam replacing it) with the medial plane of the vehicle. The longitudinal

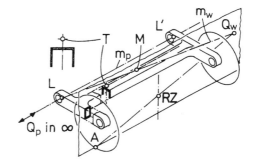

**Fig. 7.36**
Instantaneous axes of the
'H-frame' compound suspension

position of the transverse rotatable interconnection can be chosen without difficulty, as can its vertical coordinate. The coordinate of the point M and, with it, the axis $m_w$ can therefore be fixed within a wide range, and consequently its attitude relative to the wheel axis (which is responsible for the gradient of toe-in against wheel travel), and the roll centre RZ too. The intersection point $Q_w$ of the axis $m_w$ in the cross-sectional plane is the 'transverse' pole and determines the camber change of the relevant wheel with antimetric wheel travel. The instantaneous axis $m_p$ of parallel wheel travel runs through the trailing-arm joints L and determines the wheel travel angle and the support angle, as described in Chapter 6.

Kinematic self-steering of wheel suspensions depends **primarily** on wheel travel and **not** on external forces. Because of this, it is always preferable to achieve the desired steering effects for the cases of braking, traction or cornering by elasto-kinematic measures rather than by kinematic ones, if this is practicable.

## 7.6 Driving stability of two-track vehicles

**Fig. 7.37** shows a vehicle represented in a simplified manner by combining both wheels of each axle into a 'medial' wheel ('single-track model'). Such a model is quite well suited for primary investigations into vehicle dynamics. The effects of wheel load transfer on the real vehicle can be covered, too, by using the medial calculated resultant tyre slip angles of the axles for the slip angle of the single wheel. The front wheel is swivelled by the medial steering angle of the real front wheels.

At extremely low speeds, the intersecting point $M_0$ of the wheel axes is the centre of vehicle motion. With increasing speed and lateral acceleration, and a small steering angle $\delta$, lateral forces $F_v$ and $F_{yh}$ at the front and rear

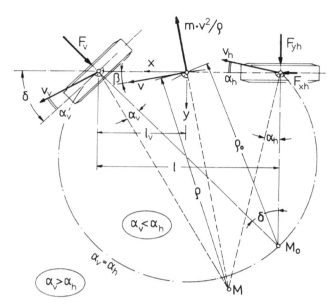

**Fig. 7.37** Steering angle and tyre slip angles on a vehicle in cornering (simplified to a 'single-track model')

wheels respectively are proportional to the static axle loads, and their values are (according to the dimensions of the vehicle) $F_v = m a_y (l - l_v)/l$ and $F_{yh} = m a_y l_v / l$. Assuming linear interdependence of tyre slip angle and lateral force, characterized by parameters $k_v$ and $k_h$, the front and rear tyre slip angles are $\alpha_v = F_v / k_v$ and $\alpha_h = F_{yh}/k_h$. With these angles, the directions of wheel motion depart from the very-low-speed ones, and the new cornering radius is

$$\rho \approx l/(\delta - \alpha_v + \alpha_h)$$

For $\alpha_v = \alpha_h = 0$ (i.e. extremely low speed), the radius is $\rho_0 = l/\delta$. From $\rho_0$ and the lateral acceleration $a_y = v^2/\rho$ we get

$$\rho/\rho_0 \approx 1 + (mv^2/l^2)\{(l - l_v)/k_v - l_v/k_h\} \qquad (7.25)$$

If $\rho > \rho_0$, the steering angle $\delta$ must be increased with increasing lateral acceleration $a_y$ to achieve the desired radius $\rho_0$; in the past, this behaviour was called 'understeering'. For $\rho = \rho_0$, the vehicle is 'neutral', and the steering angle $\delta$ applied by the driver is correct for a constant effective cornering radius independently from lateral acceleration; due to increasing tyre slip angles, however, the centre of cornering M moves forward on a circle whose centre (not drawn) is the apex of an isosceles triangle over the wheelbase, whose sides are spread by $2\delta$. Outside that circle are the centres of cornering of 'understeering' vehicles ($\alpha_v > \alpha_h$) and inside it those of 'oversteering' vehicles ($\alpha_v < \alpha_h$). With increasing speed and lateral acceleration, the 'oversteering' vehicle will spin ($\rho = 0$), and the 'critical speed' follows from equation (7.25) as

$$v_{cr} = 1/\sqrt{m\{l_v/k_h - (l - l_v)/k_v\}} \tag{7.26}$$

Beyond this 'critical speed', an 'oversteering' vehicle requires continuous steering correction by the driver since it is no longer capable of stabilizing itself. Equations (7.25) and (7.26), however, cover only the forces acting between the road and the tyres; aerodynamic forces will generally improve the directional stability and raise the critical speed. An 'understeering' vehicle is self-stabilizing through its tendency to return to straight-ahead motion after any external disturbance.

The equations (7.25) and (7.26) are suited for principal investigations and for understanding the most important laws of steady-state cornering, but the actual conditions are more complex. Vehicle parameters may change with vehicle speed, and the same vehicle may show speed ranges of both under- and oversteer. For this reason, the quotient $\rho/\rho_0$ is today no longer used to specify driving stability. Instead, the derivative of the steering-wheel angle $\delta_H$ over lateral acceleration $a_y$ is compared with the relevant value for a 'neutral' vehicle. For such a vehicle the lateral acceleration is $a_y = v^2/\rho_0 = v^2\delta/l$ and, with the steering ratio $i_S = d\delta_H/d\delta$, we get $d\delta_H/da_y = i_S l/v^2$. The characteristic

$$\lambda = d\delta_H/da_y - i_S l/v^2 \tag{7.27}$$

is > 0 in the state of **understeer**, = 0 in **neutral** state and < 0 for **oversteer**.

In the case of fast vehicles, usually the front wheels are steered, **Fig. 7.38**a. With increasing lateral acceleration, an understeering vehicle will reach the adhesion limits first at the front wheels, and an oversteering vehicle first at the rear wheels, to lose control of the respective tyre slip angles $\alpha_v$ and $\alpha_h$. In so doing, the understeering vehicle stabilizes itself by increasing the turning radius and so reducing lateral acceleration. On the other hand, an oversteering vehicle must be stabilized by the driver reducing the steering angle from a previous value $\delta_1$ to a minor value $\delta_2$ in order to reduce the lateral acceleration; hence, of course, the tyre slip angle of the front wheels is reduced from its previous value $\alpha_1$ to $\alpha_2$, and a considerable correction of the steering angle may be required, even as far as countersteering which requires some experience to do with precision. A neutral vehicle shows no clear pattern of behaviour and may break away

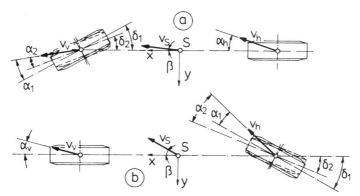

**Fig. 7.38** Directional stability with front (a) and rear wheel steering (single-track model)

at either the front or the rear wheels. It follows that moderate understeering behaviour is preferable for an ordinary road car.

Steering a vehicle via the rear wheels would be difficult even at low speeds since the driver cannot watch the rear end of the vehicle and must steer very cautiously to avoid damage in traffic. At higher speeds additional stability problems arise, as will be clear from the considerations of Fig. 7.38b in which a vehicle with rear-wheel steering is represented schematically. First, the vehicle has a considerable attitude angle $\beta$ relative to the direction of the forward velocity $v_S$. With an understeering vehicle, first the tyre slip angle $\alpha_v$ of the front wheels will get out of control when approaching adhesion limits, and the turning radius will of course increase and the lateral acceleration will decrease, but the attitude angle $\beta$ will increase. If the driver reacts by reducing the steering angle from $\delta_1$ to $\delta_2$, the tyre slip angle will increase by the value $\delta_1 - \delta_2$ from $\alpha_1$ to $\alpha_2$ and thus reduce the reserve of adhesion capacity, or perhaps use it up. With an oversteering vehicle, however, adhesion limits are first approached at the rear wheels, and reducing the steering angle would be useless because of increasing the tyre slip angle even more. On the other hand, increasing the steering angle would merely effect a temporary reduction of the tyre slip angle – but of the turning radius, too – with a consequent further increase of lateral acceleration. For this reason, off-road vehicles equipped with all-wheel steering, for better manoeuvrability, normally have the rear-wheel steering mechanism locked when they have to be used in road traffic at any sort of speed.

Among transient events in cornering, 'power change' and steering input are of major importance.

# Cornering

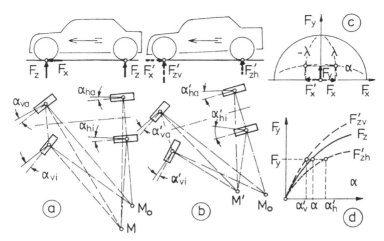

**Fig. 7.39** The power-off effect

'Power change' is transition from a forward tractive force $F_x$ to an engine braking force $F_x < 0$. **Fig. 7.39a** shows an understeering vehicle with front drive in cornering. After release of the accelerator pedal, the traction force $F_x$ is reversed (b). As long as the adhesion reserve is sufficient, this reversal of the force alone will hardly cause a change of tyre slip angles (Fig. 7.39c, see also Fig. 4.4 in Chapter 4). However, the simultaneous 'dynamic' increase of front-wheel loading and decrease of rear-wheel loading leads to reduced tyre slip angles $\alpha_v'$ at the front wheels and increased ones $\alpha_h'$ at the rear wheels – Fig. 7.39d. The degree of understeer is thereby reduced, and with it the turning radius too. The driver may balance this effect by reducing the steering angle. Within narrow limits, the power-off effect can be influenced by kinematic or elasto-kinematic self-steering measures using the spring deflections or the traction-force changes that occur during the power change; such steering angles, though, must not amount to more than a few angular minutes if straight-ahead stability is to be maintained. In practice, no significant behavioural difference is noticed between front- and rear-driven vehicles; the latter tend at best towards reduced understeer, while the former are more strongly resistant to steering corrections.

Cornering begins with a steering input. Maximum angular velocities $d\delta_H/dt$ achieved at the steering wheel are about 500°/s, and with steering ratios $i_S = d\delta_H/d\delta$ usually about 20:1 this is equivalent to a steering-angle gradient $d\delta/dt$ of about 25°/s.

**Fig. 7.40** shows the steering angle $\delta$, the lateral forces $F_{yv}$ and $F_{yh}$ at the front and rear axles, the lateral acceleration $a_y$ and the yaw velocity $\dot\psi$

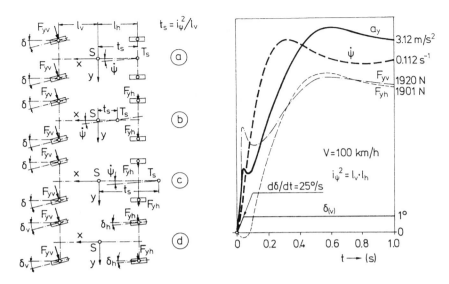

**Fig. 7.40** Steering input on an average passenger car

against the time t for a medium-class passenger car with an understeering set-up if a steering angle $\delta = 1°$ is achieved within 0.04 s at a forward velocity of 100 km/h.

As initially the steering angle is immediately transformed into a tyre slip angle at the front wheels, their lateral force $F_{yv}$ increases approximately proportionally to the steering angle $\delta$, and with it the lateral acceleration $a_y$. The final steering angle of 1° is achieved after 0.04 s; the tyre slip angle at the front wheels therefore increases no farther and is, on the contrary, slightly reduced by the increasing yaw velocity $\dot{\psi}$. This causes a temporary reduction of the lateral force at the front axle and of the lateral acceleration. The rear axle cannot establish a tyre slip angle until the vehicle acquires a yaw angle; in what follows, the increase of yaw velocity fades away, thus leading to steady-state cornering.

Noteworthy is the negative sign of the lateral force $F_{yh}$ at the rear wheels during the first hundredths of a second, thus additionally delaying the establishment of lateral forces. The schematic sketches on the left of Fig. 7.40 may help the visualization of this phenomenon (1).

Fig. 7.40a shows in plan view a passenger car with an average distribution of masses. From the mass m and the yaw moment of inertia $\Theta_\psi$ we get the radius of gyration $i_\psi = \sqrt{\Theta_\psi/m}$ and, with the distance $l_v$ of the centre of gravity from the front axle, the distance $t_s$ of the 'percussion point' $T_s$ corresponding to the front axle. From equation (5.23) in Section 5.2.3

follows $t_s = i_\psi^2/l_v$. In the given example (a), $t_s$ is equal to the distance $l_h$ between the rear axle and the centre of gravity, and consequently the percussion point is on that axle. If the quick establishment of the lateral force at the front wheels is (roughly simplifying) regarded as being a 'percussion', the rear axle will initially 'perceive' nothing, as the vehicle begins to rotate about the percussion point $T_s$ on the rear axle.

In the example (b), a very small yaw moment of inertia is assumed, and the percussion point $T_s$ is therefore positioned in front of the rear axle (this is true, for instance, on empty trucks or on front-drive passenger cars occupied by the driver alone). The vehicle begins to rotate about $T_s$ and, with this, displaces the rear wheels towards the outside of the corner, thus immediately establishing lateral forces at the rear wheels.

In contrast, vehicle (c) with a very high yaw moment of inertia (e.g. a fully laden one) shows a percussion point $T_s$ behind the rear axle. At the beginning of yaw motion, the rear axle is displaced towards the inside of the corner, thus provoking a reversed lateral force $F_{yh} < 0$.

One possible remedial measure could obviously be an increased wheelbase, but this would cause problems with vehicle weight and, especially for trucks, with manoeuvrability. Another way of achieving the quicker establishment of lateral forces at the rear wheels is to steer them simultaneously with the front wheels by an (of course smaller) angle $\delta_h$ – Fig. 7.40d. This version of all-wheel steering may improve driving stability (unlike the variety with opposed steering angles for a reduced turning circle), but naturally leads to a larger turning circle. For this reason, the steering angle $\delta_h$ is usually limited to some few degrees and is better cut out with increasing steering angle at the front wheels.

## 7.7 Cornering of motorcycles

Motorcycles react to the moment caused by the centrifugal force by inclination towards the inside of the corner – **Fig. 7.41**. The moment of the centrifugal force is $M_a = m a_y h \cos\varphi$, and that of the weight is $M_G = (-) m g h \sin\varphi$. The (negative) wheel camber is equal to the 'roll angle' $\varphi$, for which reason motorcycle tyres have a rounded tread cross-section.

The masses and moments of inertia of the wheels and the engine being high in relation to the total weight, gyroscopic moments are no longer negligible.

Fig. 7.41a depicts the effects of a gyroscopic moment. If a gyroscope rotating about the y-axis with an angular velocity $\omega$ is forced to swivel about the z-axis with an angular velocity $d\delta/dt$, its mass elements which

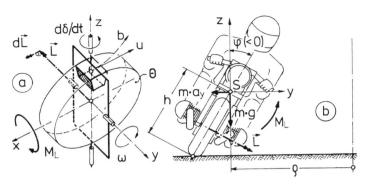

**Fig. 7.41** Motorcycle in steady-state cornering
a) gyroscopic moment   b) diagram of forces

run with a circumferential velocity u are accelerated in the y-direction, this referring especially to the elements coinciding with the z-axis (trajectory b), thereby causing forces that act in y-direction at the upper mounting and, in the opposite sense, at the lower mounting. These are the 'Coriolis' forces which are, for instance, responsible for the easterly deviation of northward-flowing rivers and for the clockwise rotation of high-pressure areas in the northern hemisphere of the globe. The gyroscope acts on its mountings by a moment $M_L$ about the x-axis following from its angular momentum

$$L = \Theta \boldsymbol{\omega} \tag{7.28}$$

opposed to the momentum change:

$$M_L = - dL/dt \tag{7.29}$$

In Fig. 7.41, the vector **L** of the angular momentum rotates about the z-axis with the angular velocity $\omega_\delta = d\delta/dt$, and consequently the vector d**L** is rectangular to **L** in the positive x-direction, $d\mathbf{L}/dt = \boldsymbol{\omega}_\delta \times \mathbf{L}$.

In plan view, the motorcycle rotates with an angular velocity $\omega_z = v/\rho$, and the angular momentum **L** of the wheels has the components $L_y = L\cos\varphi$ and $L_z = -L\sin\varphi$. The value of the gyroscopic moment $\mathbf{M}_L = \boldsymbol{\omega}_z \times \mathbf{L}$ therefore results as

$$M_L = L\cos\varphi \cdot v/\rho \tag{7.30}$$

A transversely installed engine also contributes to the angular momentum by increasing or decreasing it according to the direction of rotation of the crankshaft. The resulting angular momentum of the engine can be compensated by masses rotating in the opposite sense (e.g. the electric generator or the flywheel and clutch).

In steady-state cornering, the angle of inclination $\varphi$ of the motorcycle follows from the condition of balance $M_a + M_L = M_G$ or, using equation (7.28) and the relationships $a_y = v^2/\rho$ and $\omega = v/R$ (R = tyre radius):

$$\tan\varphi = (-)v^2(mh + \Theta/R)/(\rho mgh) \tag{7.31}$$

The moment of inertia here is for both wheels, and for the rotating engine parts multiplied by the square of their transmission ratios. Wheels with high moments of inertia and engine parts rotating in the same direction require an increased angle of inclination $\varphi$ which may amount to 2° even in the higher gears.

To begin cornering by displacing the centre of gravity towards the inside of the corner, the rider of a motorcycle has no option but to apply a 'wrong' primary steering angle $\delta'$ directed to the outside of the corner – i.e. to swing out from the desired cornering direction, **Fig. 7.42**. With sufficient angular momentum of the front wheel, the steering angle input $\delta'$ generates a gyroscopic moment $M_{L,\delta'}$ which helps to lean the motorcycle. Every rider soon learns to lean or to 'unlean' the motorcycle with the aid of gyroscopic moments – e.g. for corners of varying radius or for avoiding the unexpected obstacle. Fig. 7.42 shows schematically the cornering process of a motor cycle: for a left-hand corner, the front wheel is temporarily steered by an angle $\delta'$ to the right, thus displacing the centre of gravity S to the inside of the corner, and the vehicle begins to lean to the inside, forced by the moment of the vehicle weight and by the gyroscopic moment. Then the steering angle $\delta'$ is removed and replaced by the steady-state steering angle $\delta$. This process is controlled intuitively by the rider.

Cornering stability of motorcycles cannot be judged by the same criteria as are valid for two-track vehicles. Lateral forces at the tyres are, of course, generated by tyre slip angles, but to a considerable extent also by wheel camber, unlike two-track vehicles. In relation to the tyre slip angle,

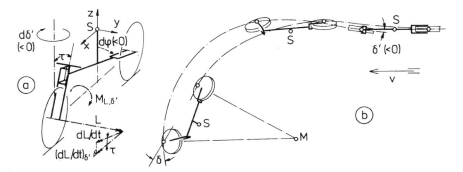

**Fig. 7.42** Steering input on a motorcycle

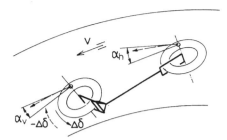

**Fig. 7.43**
Cornering stability of motorcycles

thoroughly 'understeering' and 'oversteering' motorcycles can be defined similarly to two-track vehicles. Two-track vehicles are self-stabilized if the tyre slip angle $\alpha_v$ at the front wheels increases more than the corresponding angle $\alpha_h$ at the rear wheels. For a motorcycle, though, this results in an increasing inclination angle that can be balanced only by increased lateral acceleration - i.e. by increasing the steering angle $\delta$ by $\Delta\delta$, or the vehicle speed, **Fig. 7.43**. Either of these measures will, of course, lead to a sliding front wheel when the adhesion limits are approached. Similar problems arise if the rear wheel begins to slide: removal of the steering angle by $(-)\Delta\delta$ will increase the inclination angle of the motorcycle, while increasing the steering angle by $\Delta\delta$ would, of course, bring it a bit towards the upright but, at the same time, reduce the cornering radius and increase the lateral acceleration. With a sliding rear wheel, though, there is the minor risk of the rider falling under the vehicle. In such emergency situations, the ability to use the gyroscopic moments correctly is very important; since riders are rarely confronted by gyroscopic moments in everyday life, there is a real need for training and experience.

# Chapter 8

# The Steering

## 8.1 Basic systems

The steering of road vehicles is achieved by changing the angles between the centre plane of the vehicle and the centre planes of several or all wheels. For this purpose, a rigid axle can be swivelled about its centre point ('fifth-wheel steering', the oldest method), **Fig. 8.1a**, or the front and rear halves of the vehicle can be swivelled against each other (articulated-vehicle steering - e.g. for work vehicles or agricultural tractors, Fig. 8.1b). The disadvantage of both methods is the reduction of the ground space embraced by the wheels (and, thus, of stability) with increasing steering angle, and, moreover, the lever arm of longitudinal disturbing forces which equals half the track width.

At the beginning of the 19$^{th}$ century, the cartwright Lankensperger invented axle-pivot steering (or kingpin steering) which makes for very small lever arms of disturbing forces, saves space in the vehicle and causes only negligible loss of embraced ground space. After a British patent was granted to his business partner Ackermann, who resided in England, this

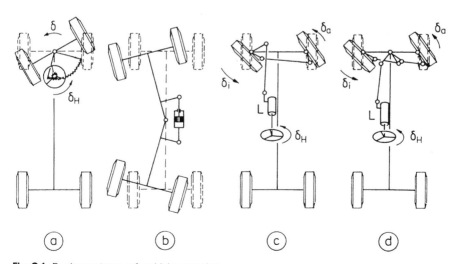

**Fig. 8.1** Basic systems of vehicle steering

steering system became known worldwide as 'Ackermann steering'. Fig. 8.1c shows kingpin steering as used on rigid axles, with a track rod connecting the two wheel carriers, and Fig. 8.1d a steering with 'divided' track rods for independent wheel suspension.

The steering angle is transmitted from the steering wheel to the wheels of the vehicle via the steering linkage (drop-arm, drag rods or links, track rods and steering arms) which is operated via the steering box L. The latter has an internal gear ratio $i_H$ to reduce the steering effort. The steering linkage, too, has a transmission ratio $i_L$ which normally varies with the steering angle. The total steering ratio between the steering-wheel angle $\delta_H$ and the outer and inner steering angles, respectively $\delta_a$ and $\delta_i$, is

$$i_S = (d\delta_H/d\delta_a + d\delta_H/d\delta_i)/2 \tag{8.1}$$

Kingpin steering is today the only method used on normal road vehicles. The swivel axis of the wheel carrier or 'knuckle' at the suspension is normally 'fixed' to the wheel carrier during a steering activity, but meanwhile there are also systems with variable axes ('virtual' kingpin axis - e.g. an instantaneous screw axis).

## 8.2 Steering boxes

The steering box transforms the steering angle $\delta_H$ applied to the steering wheel by the driver into a displacement of a joint of the steering linkage, e.g. into an angle $\delta_L$ of a drop-arm at the steering box, **Fig. 8.2**b-e. The steering-box ratio is then

$$i_H = d\delta_H/d\delta_L \tag{8.2}$$

The simplest mechanism to displace a track-rod joint is the rack-and-pinion steering box, Fig. 8.2a. Here the steering-box ratio can be defined only by the relationship between the steering-wheel angle $\delta_H$ and the travel h of the rack, which means that it is a 'dimensional quantity':

$$i_H = d\delta_H/dh \tag{8.3}$$

With a fixed rack-and-pinion ratio, the steering-box ratio follows from the number U of steering-wheel revolutions and the travel h as $i_H = 2\pi U/h$; special toothing geometries, however, nowadays allow variable steering ratios even for rack-and-pinion steering. The main advantages of the latter are simplicity, low cost and minimal space requirement in the vehicle, but not necessarily the direct and stiff connection of the steering wheel and the track rods. However, the rectilinear motion of the track-rod joints restricts

The Steering

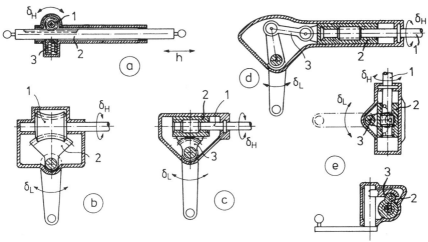

**Fig. 8.2** Steering boxes

the design possibilities of the steering geometry, which normally is three-dimensional.

Fig. 8.2b shows a steering box with an hour-glass worm 1 and a worm-wheel sector 2, where the steering-wheel angle $\delta_H$ is transformed into a drop-arm angle $\delta_L$ with a constant ratio $i_H$. To reduce friction, recent versions of this type have had a fork at the drop-arm shaft with a rotatably mounted profiled roller substituting for two teeth of the worm-wheel.

The worm-and-nut steering mechanism is a common design, today always with a recirculating-ball arrangement for lower friction. In Fig. 8.2c, the nut carries a rack segment that meshes with a gear segment on the drop-arm shaft. Alternatively the gears may be replaced by a crank mechanism, Fig. 8.2d, where the nut serves as piston. This steering-box type shows a variable transmission ratio with steering angle.

The steering-box systems of Fig. 8.2c and d are, as well as rack-and-pinion steering, well suited for hydraulic power assistance, due to the rectilinear movement of the driving element which can function as the servo piston.

Among the kinematically attractive variants of worm-and-nut steering, the relatively simple system of Fig. 8.2e is noteworthy: the ball at the nut 2 engages a cup-shape member 3 mounted on the drop-arm shaft. Due to the circular motion of the cup about the drop-arm shaft, the nut is rotated with increasing steering angle towards the shaft. The resultant leverage of the ball is thereby reduced, and the steering-box ratio $i_H$ decreases (this effect may be reversed, too, by modifying the mechanism). Hence, as the steering nut swivels, in one case in the same sense as the steering-box input shaft (the end of the steering column) and in the other in the opposite sense, the relative rotational angle between the parts - and consequently the axial shift of the nut - is lowered in the first case and increased in the other. This leads to a non-symmetrical diagram of transmission ratio versus steering-wheel angle, and to slightly different steering angles towards the left and right locks.

Rack-and-pinion steering is kept free of play over the full steering angle by a spring-loaded piston (3 in Fig. 8.2a) which presses the rack against the pinion. However, such compensation is not possible on several types of steering boxes, as for instance the worm-and-wheel or worm-and-nut designs. As wear occurs principally around the straight-ahead position, and any readjustment would cause restriction with greater steering angles, backlash at straight-ahead is deliberately made closer than towards the locks. The resulting greater play at large steering angles is acceptable as being unimportant in normal driving.

For reasons of space and/or of safety, the shaft (steering column) connecting the steering wheel to the steering box is frequently jointed - i.e. not in one piece. If the angles between adjacent portions of the shaft are relatively small, fabric-reinforced rubber joints will suffice, but preferably cardan joints are needed for greater deflection angles, since homokinetic joints based on balls and slideways have too much play if friction-free and too much friction if without play. A resultant 'gimbal error' will then cause an additional twice-per-revolution variation of the steering ratio. If the steering column has two cardan joints, though, the gimbal error can be avoided by a 'Z' or 'W' arrangement - see Chapter 3, Fig. 3.16. This is possible also with spatial distortion of the planes of deflection of the two joints by displacing the yokes on the intermediate column by the distortion angle of the planes. Elimination of the gimbal error will not, however, prevent a fluctuation of the torque in the intermediate section of the column, and with it a fluctuation of the efficiency level - sensible as a slight oscillation in the steering torque.

A hydraulic power-assisted steering box needs valves that control the oil pressure dependently on the steering torque. This is frequently achieved by a non-continuous steering column and transmission of the torque via a torsion bar, the torsional angle being a measure of the torque. The torsion bar is easily harmonized and, since its torque is generated by the driver, provides a 'feedback' of the forces and moments acting between the tyres and the road - necessary (and prescribed) on fast vehicles. On steering boxes with a nut having its torque supported by the housing (Fig. 8.2c and d), this reaction torque may also be used for pressure control.

# 8.3 Characteristics of steering geometry

## 8.3.1 Conventional definitions and physical interpretations

After the introduction of kingpin steering in motor vehicles, various special 'steering characteristics' have been established describing angular relationships at the vehicle wheels on the one hand and reactions of the steering to external forces or moments on the other.

The 'classic' kingpin steering of a rigid axle features an ordinary bolt for the kingpin which rotatably connects the axle beam with the wheel carrier or 'knuckle'. A similar method was applied on independent suspensions for a long time: the wheel carrier was not identical with the 'coupler' of the suspension mechanism but connected to it by a kingpin. When reliable ball-joints became available, the coupler of the suspension was itself used as the wheel carrier, **Fig. 8.3**. Now the line d between the ball-joints connecting the suspension links and the wheel carrier has taken over the role of the kingpin.

Normally the kingpin axis is inclined with respect to the vertical z-axis, in cross-section by the **kingpin inclination σ** and in side view by the **castor angle τ**. These definitions are valid also with steering angle - i.e. they are always measured in cross-sectional or in side view. The kingpin inclination and the castor angle are of essential importance for wheel-camber change relative to steering angle, something that will be discussed later.

The kingpin axis d intersects the road surface at a point D. The horizontal distance between the wheel centre-plane and D is the **scrub radius $r_S$** (or 'kingpin offset'); with steering, however, the tyre contact point A does not move on the road surface according to this radius, as will be shown later. In side view, the distance between D and A is the **castor offset n**. The corresponding distances measured at the height of the wheel centre M are the **wheel-centre offset $r_c$** and the **castor offset at wheel centre $n_\tau$**.

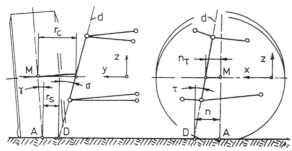

**Fig. 8.3** Diagram of steering characteristics with conventional steering geometry and a kingpin axis 'fixed' to the wheel carrier

Unlike the kingpin inclination and the castor angle, these four characteristics refer to the wheel coordinate system and not to that of the vehicle.

Every vehicle specialist knows the scrub radius $r_S$ to be the effective lever arm of a braking force, and the castor offset n to be the lever arm of a lateral force; with independent suspensions, the wheel-centre offset $r_c$ is usually regarded as being the lever arm of a traction force if the driving torque is transmitted to the wheel via a driveshaft with universal joints. The radius $r_c$ is also the lever arm of disturbing forces acting on the free-rolling wheel - as, for instance, impact forces, aquaplaning forces, etc. - because these forces can be transmitted to the wheel carrier only via the wheel bearing and the wheel centre M.

It is clear from Fig. 8.3 that the lever arms mentioned do not represent the true spatial distances from the kingpin axis d, because they are all defined in the road-surface plane or the x-y plane. However, it would not be worth the effort to 'enrich' the steering geometry by three-dimensionally defined characteristics, because those defined for the principal reference planes for the vehicle and the wheels are not only very well adapted for vector calculation but also provide accurate correlations between the forces acting and the steering when the definition of the steering ratio is considered, as will be explained in what follows.

Considering **Fig. 8.4**, the spatial moment $M_B$ of a braking force vector $F_B$ with its components $F_{Bx} = -F_B$, $F_{By} = F_{Bz} = 0$ about the intersecting point D of the kingpin axis d and the road follows from the radius vector $r_A$ with its components $r_{Ax} = -n$, $r_{Ay} = r_S$ and $r_{Az} = 0$ as $M_B = r_A \times F_B$ with the components $M_{Bx} = M_{By} = 0$ and $M_{Bz} = r_S F_B$. Obviously the scrub radius $r_S$

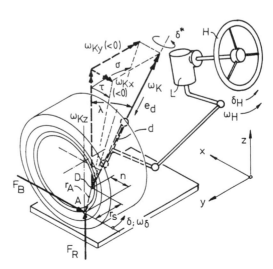

**Fig. 8.4**
Forces and velocities at the wheel carrier, where the kingpin axis is 'fixed'

# The Steering

and the braking force $F_B$ generate a moment about the z-axis and *not* about the kingpin axis d!

The same is valid for a lateral force acting on the wheel through the castor offset n. It should be pointed out here that n is the 'geometrical' castor offset, while the lateral force of a tyre resulting from a side-slip angle is displaced by the 'pneumatic trail' $n_R$ - see Chapter 4, Fig. 4.2 - and that the resultant lever arm will then be $n + n_R$.

While the steering-geometry characteristics mentioned in the foregoing have already been standardized, this is not yet true for another, less easily visualized characteristic which, however, is also of very great importance in respect of the relevant amount of the wheel load $F_R$ - i.e. the **wheel-load lever arm p (14)**. As long as the line of action of the wheel-load vector $F_R$ does not intersect the kingpin axis d, and the latter is not rectangular to the road, the wheel load exerts a moment about d. With the components $F_{Rx} = F_{Ry} = 0$ and $F_{Rz} = F_R$, and the radius vector $r_A$ as mentioned above, the moment vector is $M_R = r_A \times F_R$, and its components are $M_{Rx} = r_S F_R$, $M_{Ry} = n F_R$ and $M_{Rz} = 0$. The wheel-load lever arm p may be defined as positive if the wheel load acts in the restoring direction - i.e. by a moment $M_d$ about the kingpin axis that is opposed to the angular velocity $\omega_K$, meaning that the moment $M_d$ acts in the direction of the unit vector $e_d$. The resultant moment about the axis d is therefore

$$M_{dR} = \mathbf{M}_R \cdot \mathbf{e}_d \qquad (8.4)$$

The axis d is inclined with respect to the z-axis by a spatial angle $\lambda$ which follows from

$$\tan \lambda = \sqrt{1 + \tan^2 \sigma + \tan^2 \tau} \qquad (8.5a)$$

and the components of the unit vector $\mathbf{e}_d$ are

$$e_{dx} = \tan\tau / \tan\lambda \qquad (8.5b)$$
$$e_{dy} = \tan\sigma / \tan\lambda \qquad (8.5c)$$
$$e_{dz} = -1 / \tan\lambda \qquad (8.5d)$$

If the wheel-load lever arm is referred to the angular wheel-carrier velocity component $\omega_{Kz}$ (as with the characteristics already mentioned) according to their 'conventional' definitions, the theorem of virtual work leads to the condition

$$p F_R \omega_{Kz} = M_{dR} \omega_K \qquad (8.6)$$

and with $\boldsymbol{\omega}_K = -\omega_K \mathbf{e}_d$ and its z-component $\omega_{Kz} = \omega_K / \tan\lambda$, and the equations (8.4) to (8.6), the wheel-load lever arm (referred to the z-axis) is

$$p = r_S \tan\tau + n \tan\sigma \qquad (8.7)$$

The steering angles and attitudes of the two wheels of an axle assume non-symmetrical values with increasing steering-wheel angle, first because the inner steering angle leads the outer according to the different cornering radii, and second because the inclined kingpin axis d generates different camber angles at the two wheels. The total effect of the forces and moments at the two wheels must therefore be assessed by adding their respective moments at the steering box L or at the steering wheel H while considering the individual steering ratios of the two wheels with respect to the steering wheel. The steering ratio between a wheel with its steering velocity $\omega_\delta$ and the steering wheel with its velocity $\omega_H$ is

$$i_S = \omega_H/\omega_\delta \tag{8.8}$$

The steering velocity $\omega_\delta$ follows from the angular velocity $\boldsymbol{\omega_K}$ of the wheel carrier, the camber $\gamma$ and the steering angle $\delta$, see equation (7.3), Chapter 7. With $\omega_{Kx} = -\omega_{Kz}\tan\tau$ and $\omega_{Ky} = -\omega_{Kz}\tan\sigma$, the steering velocity is

$$\omega_\delta = \omega_{Kz}\{(\tan\tau\sin\delta - \tan\sigma\cos\delta)\tan\gamma + 1\} \tag{8.9}$$

and this means that normally the steering velocity $\omega_\delta$ is not equal to the z-component of the wheel-carrier velocity $\omega_K$. Only for a camber angle zero or for a steering angle $\delta = \mathrm{atn}(\tan\sigma/\tan\tau)$ - i.e. the wheel position where in plan view the wheel axis and the kingpin axis appear rectangular - does $\omega_\delta$ equal $\omega_{Kz}$.

Consequently, calculation of steering moments by aid of the 'conventional' characteristics mentioned, and division of these moments by the relevant steering ratio, and finally adding them at the steering wheel, leads to a physically incorrect situation, since the steering ratio refers to the steering velocity $\omega_\delta$ and not to $\omega_{Kz}$! In the following, therefore, the **steering velocity $\omega_\delta$** may be used to define the **characteristics of the steering geometry**, although the conventional definitions of the latter are factually based on $\omega_{Kz}$. The 'error' caused thereby is negligible, at least for typical steering angles of fast vehicles. **Fig. 8.5** shows the deviation of the steering velocity $\omega_\delta$ from the wheel-carrier velocity component $\omega_{Kz}$ versus steering angle $\delta$, assuming a kingpin inclination $\sigma = 8°$ and a castor angle $\tau = 5°$.

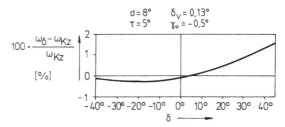

**Fig. 8.5**

Deviation between the steering velocity $\omega_\delta$ and the wheel-carrier velocity component $\omega_{Kz}$

# The Steering

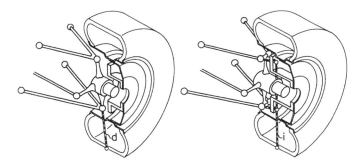

**Fig. 8.6** Space for a brake disc with fixed (d) and virtual kingpin axes (i)

Due to increasing engine performance and the consequent need for higher braking performance, dimensioning of the brake discs has become a real problem. The greatest possible brake-disc diameter is achievable by siting the disc alongside the drop centre of the rim; in practice, this results in a displacement of the disc towards the inner side of the wheel. Following the introduction of brake anti-lock systems, it proved necessary to provide a very small scrub radius $r_S$ (Fig. 8.4) to avoid the ill-effects of the pulsating braking forces on the steering. For this purpose, the kingpin axis d must be positioned near the wheel's centre plane. As the axis d is - at least on independent suspensions - preferably marked by two ball-joints connecting the wheel carrier to the suspension links, nowadays these ball-joints frequently occupy the space that was formerly available for the brake disc. The latter therefore has to be repositioned inside the drop centre - equivalent to a diameter loss of approximately 25 mm unless a larger rim diameter is adopted, **Fig. 8.6** (left side).

To avoid this, high-powered vehicles currently feature (apart from sophisticated solutions based on 'conventional' geometry with 'fixed' kingpin axes) suspension layouts with 'virtual' kingpin axes to enable the brake disc to occupy its optimum position and yet for the effective kingpin axis to be outboard of the disc. The right-hand drawing in Fig. 8.6 shows a wheel suspension based on the double-wishbone principle but with the upper and lower triangular links each divided into two rod links, each connected to the wheel carrier by ball-joints. The intersecting point of each pair of rods can be regarded as a 'virtual' ball-joint, and the line connecting these virtual joints is the virtual kingpin axis i.

The rods need not be arranged in common planes, as in Fig. 8.6, but may be freely positioned in space - **Fig. 8.7**. Such measures enable the steering geometry to achieve a 'spatial' character, and the rotation about the kingpin axis d may change into an instantaneous screw motion with an angular

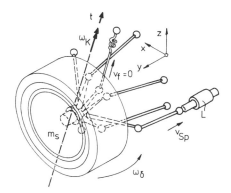

**Fig. 8.7**
'Spatial' steering geometry
with instantaneous screw

wheel-carrier velocity vector $\omega_K$ and a shift-velocity vector $t$ in the screw axis.

Since a screw axis, unlike a fixed kingpin axis, cannot be used as a reference axis for moments (and thus to define steering characteristics according to Fig. 8.3 and 8.4) the definitions of these characteristics have to be revised to obtain compatible physical meanings independently from the actual suspension systems. This would be possible, for example, by using 'instantaneous poles' as already shown in Chapter 3, Fig. 3.11 - but more elegantly by the application of the theorem of virtual work to a notional steering event, as will be explained in what follows.

### 8.3.2 Generalized definitions relating to spatial geometry

If the kingpin axis is variable with steering angle - maybe because it is a virtual axis as shown in Fig. 8.6 - the kingpin inclination $\sigma$ and the castor angle $\tau$ are not immediately usable for determining the working lever arms of forces, for instance the wheel-load lever arm p according to equation (8.7). The reason for this is that such a virtual axis may be an instantaneous screw axis and so cannot be considered as a reference axis for moments.

Purely angular relationships are not affected by this. The equations (7.1) and (7.3), Chapter 7, or (8.7) may therefore be used further on for the determination of the camber velocity $\omega_\gamma$ and the steering velocity $\omega_\delta$ in interdependence with the angular wheel-carrier velocity $\omega_K$. With $\omega_K$ as a vector in the direction of the instantaneous screw axis for a virtual steering event with a 'locked' spring, the **kingpin inclination** $\sigma$ and the **castor angle** $\tau$ follow from the projections of the vector $\omega_K$ in the cross-sectional or side view of the vehicle:

# The Steering

$$\sigma = -\operatorname{atn}(\omega_{Ky}/\omega_{Kz}) \tag{8.10}$$
$$\tau = -\operatorname{atn}(\omega_{Kx}/\omega_{Kz}) \tag{8.11}$$

An efficient and generalized definition of the steering characteristics scrub radius, castor offset, wheel-centre offset, castor offset at wheel centre and wheel-load lever arm must ensure that the physical interpretation of the respective characteristic is valid and compatible on a conventional steering system with 'fixed' kingpin axes, as well as on a generalized spatial system.

If, for instance, a 'scrub radius' is given for a conventional steering system, any chassis specialist will expect that a braking force $F_B$ acting on this scrub radius $r_S$ exerts a moment $r_S F_B$ that contributes, considering the steering ratio $i_S = \omega_H/\omega_\delta$, a moment $M_H = r_S F_B/i_S$ at the steering wheel. And if the same braking force exerts the same moment at the steering wheel of a spatial steering mechanism, and the steering ratio is the same too, this situation must have resulted from a scrub radius of the same amount.

For a notional steering event with an 'incompressible' spring ($v_f = 0$), the instantaneous state of motion of the wheel carrier K - which is characterized by the velocity $\mathbf{v}_M$ of the wheel centre M and the angular velocity $\boldsymbol{\omega}_K$ - can be determined according to the method shown in Section 3.4. From $\mathbf{v}_M$ and $\boldsymbol{\omega}_K$ we then get the virtual velocity $\mathbf{v}_A^*$ of the tyre contact point A, assuming the brake to be 'locked', according to equation (3.18).

As already recommended in Section 8.3.1, the steering characteristics in what follows refer to the steering velocity $\omega_\delta$ and not to the vertical component $\omega_{Kz}$ of the wheel-carrier velocity. Then a braking force $F_B$ will generate a 'power' with the steering velocity $\omega_\delta$ which follows from the product of $\omega_\delta$ and the moment $M_{\delta B} = r_S F_B$. On the other hand, the theorem of virtual work says that the power of the braking-force vector $\mathbf{F}_B$ of a hub-mounted brake is the scalar product of the force vector and the virtual velocity vector $\mathbf{v}_A^*$ of the tyre contact point, assuming the brake to be locked: $r_S F_B \omega_\delta = \mathbf{F}_B \cdot \mathbf{v}_A^*$ - **Fig. 8.8**. The vector $\mathbf{F}_B$ has the components $F_{Bx} = -F_B\cos\delta$, $F_{By} = -F_B\sin\delta$ and $F_{Bz} = 0$. This leads to $r_S F_B \omega_\delta = -v_{Ax}^* F_B \cos\delta - v_{Ay}^* F_B \sin\delta$ and to the **scrub radius**

$$r_S = -(v_{Ax}^*\cos\delta + v_{Ay}^*\sin\delta)/\omega_\delta \tag{8.12}$$

with $\omega_\delta$ according to equation (7.3), Chapter 7. Assuming a force $F_L$ in the medial plane of the wheel acting at the wheel centre M, the **wheel-centre offset** can be determined analogously, using the wheel-centre velocity vector $\mathbf{v}_M$:

$$r_c = -(v_{Mx}\cos\delta + v_{My}\sin\delta)/\omega_\delta \tag{8.13}$$

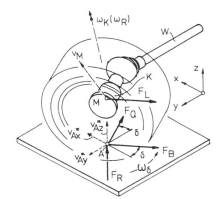

**Fig. 8.8**
Velocities and forces on a steered wheel with spatial steering geometry

Referring to the steering angle $\delta$, a lateral force $F_Q$ acting on the wheel by the lever arm of castor offset n generates a moment $nF_Q$, and from the condition of virtual work $nF_Q\omega_\delta = \mathbf{F}_Q \cdot \mathbf{v}_A^*$ we get (using the components $F_{Qx} = F_Q\sin\delta$, $F_{Qy} = -F_Q\cos\delta$ and $F_{Qz} = 0$) $nF_Q\omega_\delta = v_{Ax}^* F_Q\sin\delta - v_{Ay}^* F_Q\cos\delta$, or the **castor offset**

$$n = (v_{Ax}^*\sin\delta - v_{Ay}^*\cos\delta)/\omega_\delta \tag{8.14}$$

Though no lateral force normally acts at the wheel centre M, the **castor offset at wheel centre** can be determined by analogy to the castor offset n, using a notional lateral force:

$$n_\tau = (v_{Mx}\sin\delta - v_{My}\cos\delta)/\omega_\delta \tag{8.15}$$

As already mentioned, the wheel-load lever arm p may be defined as positive if the moment generated by the wheel load $F_R$ acts in the 'restoring' sense - i.e. reducing the given steering angle $\delta$. The 'power' of the moment $(-pF_R)$ at the steering velocity $\omega_\delta$ is therefore to be balanced by the power of the wheel-load vector $\mathbf{F}_R$, which has only a z-component, at the vertical component of the velocity of the tyre contact point A. As the latter is in plan view always on the extension of the wheel axis, its vertical velocity component is the same if the brake is 'locked' or if the wheel is freely rotatable; and as, moreover, the velocity vector $\mathbf{v}_A^*$ is already available due to the foregoing calculations, it may be used here, too, as a simplification. From $-pF_R\omega_\delta = \mathbf{F}_R \cdot \mathbf{v}_A^*$ follows the **wheel-load lever arm**

$$p = -v_{Az}^*/\omega_\delta \tag{8.16}$$

By multiplication with the time differential dt, equation (8.16) changes to

$$p = -dz/d\delta \tag{8.17}$$

# The Steering

or the wheel-load lever arm p is the derivative of the vehicle lift and the steering angle.

On spatial mechanisms, as already mentioned, the steering characteristics 'kingpin inclination' and 'castor angle' are responsible merely for angular interdependences. A very important interdependence for driving behaviour is the **camber change with steering angle** $d\gamma/d\delta$. With $\omega_{Kx} = -\omega_{Kz}\tan\tau$ and $\omega_{Ky} = -\omega_{Kz}\tan\sigma$, equation (7.1) of Chapter 7 leads to

$$\omega_\gamma = \omega_{Kz}(\tan\tau\cos\delta + \tan\sigma\sin\delta) \tag{8.18}$$

and with equation (8.9):

$$\frac{d\gamma}{d\delta} = \frac{\tan\tau\cos\delta + \tan\sigma\sin\delta}{\tan\gamma(\tan\tau\sin\delta - \tan\sigma\cos\delta) + 1} \tag{8.19}$$

In the straight-ahead position ($\delta = 0$), the gradient of camber against steering angle is clearly proportional to the camber angle $\tau$ (more precisely, for wheel camber $\gamma = 0$, because the denominator then assumes the value of unity). **Fig. 8.9** is an attempt to visualize this: in space, the wheel centre M moves along a circular path, while the intersecting point $D_M$ of the wheel axis and the kingpin axis remains fixed. Hence, the difference between the vertical coordinates of M and $D_M$ is proportional to the wheel camber $\gamma$. The 'pole' P of the wheel-centre motion in side view is the intersecting point of the kingpin axis and the vertical plane through M, and the circular path of M about P approximates to the elliptical projection of the spatial path, its tangent in the straight-ahead position obviously being inclined by the castor angle $\tau$. The radius of curvature $\rho$ is determined by the kingpin inclination; in cross-section, the distance M-P is equal to the quotient from the wheel centre offset $r_c$ and the tangent of the kingpin inclination, $(M-P'') = r_c/\tan\sigma$; from this, in side view, we get the radius $\rho = (M-P)''/\cos\tau = r_c/(\tan\sigma\cos\tau)$. The radius of curvature of the diagram $\gamma(\delta)$ is therefore

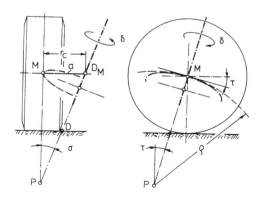

**Fig. 8.9**
Influence of kingpin inclination and castor angle on camber change with steering angle

proportional to the reciprocal of the kingpin inclination. Positive kingpin inclination σ bends the diagram towards positive camber with increasing steering angle δ, as is clear in Fig. 8.9.

The interdependence of camber change and castor angle is of special importance for the design of motorcycle steering geometry (which is, of course, limited to the characteristics castor offset and castor angle). Due to the considerable inclination ('roll angle') of the motorcycle during cornering, different effective camber angles with respect to the road occur normally at the front and the rear wheels. If the castor angle was 0°, a front wheel steered (theoretically) by 90° would show a camber angle about 0° independently from the inclination or camber angle of the rear wheel. It follows that the castor angle of a motorcycle must be adapted to the machine's purpose, as in the examples given below.

Fast touring motorcycles corner at high speed, with substantial lateral inclination but rather small steering angles. A moderate castor angle will here be sufficient to adapt the camber angle of the front wheel to that of the rear wheel.

Moto-cross machines are chased through tight curves with considerable lateral inclination and steering angle; hence they need a large castor angle to ensure a sufficient camber angle at the front wheel.

Trials motorcycles have to provide 'acrobatic' performance at relatively low speeds and with little lateral inclination of the machine; for this reason, their front forks have comparatively small castor angles.

For two-track vehicles, the exact functions of wheel camber $\gamma$, castor offset n and wheel-load lever arm p against steering angle δ are drawn in **Fig. 8.10** for four steering-geometry variations with conventional 'fixed' kingpin attitudes.

In comparison with the variants 1 and 2, the variants 3 and 4 show a threefold inclination of the tangent of the curve $\gamma(\delta)$ in the straight-ahead position; this corresponds to the relationships of the respective castor angles (9° and 3°). Due to the greater kingpin inclination σ (12° to 5°), curve 2 shows a considerably smaller radius of curvature than curve 1, and even achieves positive camber at the end of the outer steering angle (δ < 0). For all the variants, the castor offset n increases with the inner steering angle (δ > 0) and decreases with the outer steering angle (and even assumes negative values with the latter). The wheel-load lever arm p is defined so that the wheel load generates a restoring moment if p and the steering angle δ show the same sign - here always true for inner steering angles but only for greater outer steering angles. The reason is that all variants have a certain positive lever arm p in the straight-ahead position.

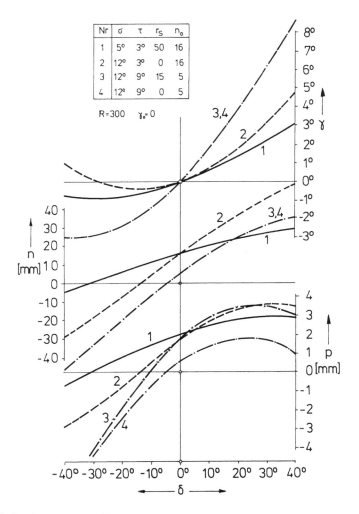

**Fig. 8.10** Camber, castor offset and wheel-load lever arm plotted against steering angle on several variants of steering geometry with a 'fixed' kingpin axis

The variants 1 and 2 have no castor offset at the wheel centre $n_\tau$, because the offset n is here the product of the tyre radius R and the tangent of the castor angle $\tau$, and consequently the kingpin axis runs through the wheel centre in side view. In this case, the extensions of the kingpin axis and the wheel axis in plan view coincide at an outer steering angle that follows from $\tan\delta = -\tan\tau/\tan\sigma$; both the castor offset n and the wheel-load lever arm p must therefore be zero in this position.

A castor offset n different from zero causes a lateral displacement of the tyre contact point with steering. In the straight-ahead position of the steering, this means a lateral displacement of the vehicle's front end. If the castor offsets of the inner and outer wheels become different with increasing steering angle, as on all four variants in Fig. 8.10, this causes respective constraint of the tyres and increased parking moments.

Due to their castor offset n ≠ 0 in the straight-ahead position, even the variants 2 and 4 with a scrub radius $r_S = 0$ will displace the vehicle with steering action. This is, of course, a pointer to the fact that the path of the tyre contact point A on the road is **not** described by the scrub radius! As already mentioned, the scrub radius is generally regarded as being the effective lever arm of a braking force generated by a hub-mounted brake, see equation (8.12). For this reason, it would be wrong and superfluous to analyse the path of the tyre contact points on the road surface with steering, and to define the radius of curvature of this path as the 'scrub radius': **Fig. 8.11** shows by a numerical example that this radius of curvature assumes a considerable value of about 40 mm on a suspension with considerable kingpin inclination and camber angle, despite a scrub radius that is always zero (the intersecting lines of the wheel centre planes with the road, here marked by arrows, all run through the intersection point D of the kingpin and the road, so braking forces will never generate a torque about the kingpin axis).

On a suspension with a 'virtual' and consequently variable kingpin axis, the scrub radius may vary too and for instance increase with the steering angle. It will then be advantageous to position its minimum in approximately the straight-ahead position by suitable arrangement of the suspension joints (**14**).

The sum of the castor offset n and the pneumatic trail $n_R$ is the lever arm of the lateral force $F_y$ acting on the steering in cornering. With increasing lateral acceleration, the outer wheels' lateral force becomes more and

**Fig. 8.11**

Path of the tyre contact point on the road with steering angle (scrub radius $r_S = 0$)

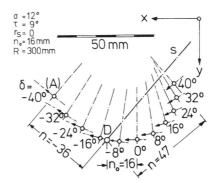

# The Steering

more predominant. Since the castor offset n of the outer wheel decreases with steering angle on most of the wheel suspensions (except for those with $\sigma = \tau = 0$), and since the pneumatic trail $n_R$ collapses at the adhesion limits, the torque at the steering wheel does not increase proportionally to the lateral acceleration. Naturally this non-linearity is desirable, because high steering-wheel moments lead to tired arm muscles, hindering sensible reaction by the driver – something that is especially important at the vehicle-dynamic limits. On the other hand, it must of course be ensured that the steering-wheel moment does not assume negative values.

Different wheel-load lever arms p at the two wheels generate a restoring torque too; the (higher) outer wheel load is, however, to be felt only if the outer lever arm p is negative – i.e. generates its own restoring torque at this wheel. However, at high speed that self-aligning torque from the vehicle's weight is considerably lower than the torque generated by the lateral forces.

Similarly to the castor offset n, the wheel-load lever arm p shows a gradient against the steering angle around the straight-ahead position, Fig. 8.10. The lever arm changes its sign at the particular steering angle where the tyre contact point lies in plan view on the extension of the kingpin axis and where, consequently, the wheel load cannot generate a moment about that axis (this is not valid for suspensions with a 'virtual' kingpin axis which is represented by an instantaneous screw). For this reason, the outer wheels of the suspension variants 4 and 1 do not generate their own restoring moments in the respective steering-angle ranges of 0° to -5° and 0° to -32°. The total restoring torque due to weight results from the difference between the inner and outer wheel-load lever arms and is obviously ensured in Fig. 8.10 over the full angular range of the steering.

In the straight-ahead position ($\delta = 0$), the derivative $dp/d\delta$ is the **restoring-force lever due to weight** (14)(16), which may be envisaged as the length of a pendulum that bears the weight acting on the front axle. If deflected by the steering angle $\delta$, the weight generates the restoring torque. A restoring torque due to weight is, consequently, always given if the diagram of the wheel-load lever arm p versus steering angle $\delta$ shows a positive gradient in the straight-ahead position. Using the restoring-force lever due to weight, the torque gradient on a wheel in the straight-ahead position is

$$dM/d\delta = F_z (dp/d\delta) \qquad (8.20)$$

On a suspension with conventional steering geometry (or with a 'fixed' kingpin axis), that restoring-force lever can be calculated (**16**) from

$$(dp/d\delta)_{\delta=0} = r_c \tan\sigma - n\tan\tau \qquad (8.21)$$

The restoring torque due to weight is the most important precondition for self-aligning of the steering in the straight-ahead position. However, as the steering angle and lateral acceleration increase, it becomes less influential than the moments generated by lateral forces.

The wheel-load lever arm p itself should be made as small as possible to avoid reaction torques at the steering caused by wheel-load fluctuation – e. g. on bad road surfaces.

The foregoing analysis has assumed that the spring is supported by one of the suspension links and that it acts on the wheel carrier by a joint connecting both. In this case, and on a suspension with conventional steering geometry, the kingpin remains 'fixed' in space during a pure steering event, with the spring assumed to be incompressible. The function $p(\delta)$ is then independent of the spring's position, and the restoring torque due to weight is unequivocally determined by the position of the kingpin axis and by equation (8.21). This is true in practice for most steered suspensions.

However, the restoring moment due to weight can also be made independent of the kingpin axis by connecting the spring immediately to the wheel carrier. **Fig. 8.12** shows schematically a torsion-bar-sprung steerable suspension where the lever of the torsion bar is connected to the wheel carrier by a short tension rod (a) which is out of line with the kingpin axis. As steering angle is applied, the tension rod will incline, thus lifting the vehicle and generating a torque about the kingpin axis.

On suspensions with a virtual (and then variable) kingpin axis, suitable choice of the suspension joint positions and the spring mounting allows the restoring moment due to weight to be influenced independently of the kingpin attitude (**14**). Hence, the equation (8.21) is no longer applicable on such suspensions.

Any kind of self-aligning due to weight is characterized by lifting of the vehicle with steering angle; the energy needed for this must, of course, be applied at the steering wheel.

If the wheel-load lever arms $p_a$ and $p_i$ of the outer and inner wheels are not of opposite signs and equal amount ($p_a = -p_i$) and if, therefore,

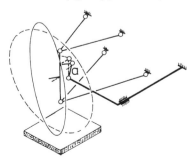

**Fig. 8.12**
Free choice of restoring torque
due to vehicle weight

the vehicle is lifted at the two wheels by different distances, the vehicle will be 'wound up' between its axles against the resistance of the springing system; the front and rear roll rates of the springs have to be treated as a 'series arrangement', see Chapter 5, equation (5.26). The spring deflection energy again has to be applied at the steering wheel. In terms of usual spring rates, however, this 'restoring torque due to spring deflection' is only a very low percentage of the restoring torque due to weight, and normally is negligible.

If the traction or braking torque is transmitted to the wheels via driveshafts, the effect of a traction or braking force on the steering is usually assessed by the wheel-centre offset $r_c$; on the other hand, it is well known that different deflection angles of the driveshafts on the two wheels of a driven front axle cause disturbing moments at the steering. Being constant with wheel travel on normal suspensions, though, the wheel-centre offset obviously cannot be regarded as an effective lever for traction or braking forces under all conditions.

In the early stage of vehicle development, when reliable homokinetic joints were not yet available, the problem of traction on steerable wheels was sometimes solved in a special manner: an intermediate shaft running coaxially with the kingpin, and carrying bevel pinions, transmitted the traction torque from a vehicle-side bevel gear to another at the wheel axle, **Fig. 8.13**. Remembering Fig. 6.16, Chapter 6, it will at once be clear that, with a steering motion of the wheel carrier K about the kingpin axis d, assuming a 'locked' torque support (the engine or the brake, consequently here the pinion shaft, too), the bevel gear at the wheel axle must swivel about the meshing point E of the gears, and that the effective axis of rotation $d^*$ of the wheel must run through point E and the centre point T of the bevel drive. The circumferential force at the tyre contact point A therefore acts on a lever defined by the intersecting point of $d^*$ with the road surface and the tyre contact point. This lever, $r_T$, may be introduced here under the designation 'traction-force radius' as a new 'steering characteristic' **(15)**.

Disadvantages of this particular solution are poor efficiency plus noise and wear of the bevel gears which have to move continuously; on the other hand, perfect homokinetic motion is provided.

It would obviously be easy to let the effective axis $d^*$ of rotation run through the tyre contact point A and, thus, to achieve a traction-force radius $r_T = 0$. The steering mechanism would then be free of any disturbance by traction forces in spite of a considerable wheel-centre offset $r_c$. In steering action with the vehicle at rest, the wheel would be able to roll freely on the road even with a 'locked' engine (this criterion would be generally valid, and applicable to all driven wheels independently of the system

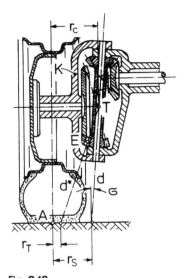

**Fig. 8.13**
Steerable wheel driven by bevel gears running about the kingpin axis

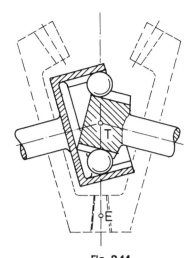

**Fig. 8.14**
Analogy between a homokinetic joint and a bevel-gear drive

and the torque-transmission method, though surely very difficult to prove by measurement in respect of the elastic properties of the suspension and the driveline).

In looking at the bevel-gear torque transmission in Fig. 8.13, not much imagination is needed to foresee what would happen with a normal torque transmission via driveshafts: for a given deflection angle of a homokinetic joint, the two joint halves can instantaneously be replaced by bevel gears (**17**), as in **Fig. 8.14**.

The method used in the foregoing to define steering characteristics – i.e. to determine effective lever arms of external forces by equalizing the 'power' generated by the force vectors and the virtual velocity vector at the tyre contact point and the 'power' of the steering moment at the steering velocity – allows us to calculate in a similar way the traction-force radius $r_T$, considering the complete driveline including possible hub-reduction gears: if the angular velocity $\omega_K$ of the wheel carrier for a notional steering motion with an 'incompressible' spring is known, the equations given in Section 3.6 enable the determination of the virtual velocity $\omega_R$ of the wheel, taking into account a 'locked' torque support (e.g. the engine), possible driveshafts and a possible hub-reduction gear. With $\omega_R$ and the wheel-centre velocity $v_M$, the virtual velocity $v_A^{**}$ of the tyre contact point A follows from equation (3.35). By analogy to equation (8.12), the

# The Steering

**traction-force radius** can be determined formally in the same manner, replacing, of course, the velocity $v_A^*$ by $v_A^{**}$:

$$r_T = -(v_{Ax}^{**}\cos\delta + v_{Ay}^{**}\sin\delta)/\omega_\delta \qquad (8.22)$$

Where traction torque is transmitted via driveshafts, this radius $r_T$ is the correct lever arm to replace the wheel-centre offset $r_c$, and in the (rare) case of braking-torque transmission it replaces the scrub radius $r_S$.

Application of this new steering characteristic $r_T$ leads to interesting and – with knowledge of the interdependences discussed in Section 3.6 – easily understandable results.

Assuming a 'locked' torque support – e.g. engine and, consequently, a locked inboard driveshaft end, too – an angular velocity $\omega_K$ of the wheel carrier K with steering causes a relative angular velocity $\omega_{R,K}$ of the wheel and the wheel carrier which is situated on the wheel axis, see Chapter 3. The virtual velocity vector $v_A^{**}$ of the tyre contact point A therefore differs from the velocity vector $v_A^*$ (for a hub-mounted torque support) merely by a circumferential component. As this component is rectangular to the vectors of the wheel load and of lateral forces at the wheel, the wheel-load lever arm p and the castor offset n are not affected by the transmission of braking or traction forces.

Analysis of a typical passenger-car suspension with conventional steering geometry ('fixed' kingpin axis at the wheel carrier) shows that the traction-force radius $r_T$ is almost constant with steering angle, similar to the radii $r_S$ and $r_c$. Much more interesting, however, is its behaviour with wheel travel.

While the scrub radius $r_S$ and the wheel-centre offset $r_c$ are practically invariable with wheel travel s, the traction-force radius $r_T$ in the 'normal position' is equal to wheel-centre offset, decreasing with wheel bump and increasing with rebound, **Fig. 8.15**.

Bearing in mind the statements made in Section 3.6, this behaviour is easy to explain – see **Fig. 8.16**. In the normal position, Fig. 8.16a, there is no deflection angle at the outer shaft joint $G_a$. For reasons of symmetry, the relative angular velocity of the wheel and the shaft W lies in the plane of symmetry which is here vertical. If $G_a$ is close enough to the kingpin axis d (desirable on suspensions with a 'fixed' kingpin axis to avoid superfluous lengthening or shortening of the shaft with steering), it does not move during the steering process, and the absolute angular velocity $\omega_W$ of the shaft is zero; the relative angular velocity of the shaft and wheel proves then to be also the absolute angular velocity $\omega_R$ of the wheel. $\omega_R$ differs from $\omega_K$ by a relative angular velocity $\omega_{R,K}$ which causes the wheel to rotate on its bearings. As the intersecting point H for the wheel axis and the kingpin

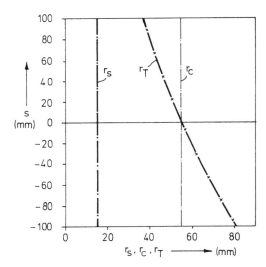

**Fig. 8.15**

The traction-force radius $r_T$ versus wheel travel s, in comparison with the wheel-centre offset $r_C$ and the scrub radius $r_S$ ('fixed' kingpin axis)

axis does not move with steering, the line $d^*$ parallel to $\omega_R$ and running through H is the effective axis of rotation of the wheel during an instantaneous steering event. In Fig. 8.16a, the axis $d^*$ is parallel to the centre plane of the wheel, so the traction-force radius $r_T$ is equal to the wheel-centre offset $r_C$. Obviously the latter may replace the traction-force radius in the particular position of the suspension where the outer driveshaft joint has zero articulation.

Fig. 8.16b shows the suspension with a certain wheel travel towards bump, causing a shaft deflection angle $\alpha$. The plane of symmetry at the outer shaft joint $G_a$ now inclines by $\alpha/2$ and with it the virtual angular velocity vector $\omega_R$ of the wheel and the effective axis $d^*$ of rotation. The latter meets the road at a point $D^*$ the distance of which, $r_T$, from the tyre contact point A is smaller than in Fig. 8.16a. With rebound travel, of course, $r_T$ will analogously increase.

Fig. 8.16 therefore offers a simple method of estimating the traction-force radius $r_T$ on a suspension with shaft drive: in cross-section, the line through the point H – which is parallel to the plane of symmetry of the shaft halves at the outer joint – intersects the road line at a distance $r_T$ from the tyre contact point A.

On front-drive vehicles, reduction of the traction-force radius (and, consequently, of the wheel-centre offset also) is desirable to minimize the influence of non-stationary traction forces at the steering, as may occur on an uneven road or with a variable friction coefficient. If the kingpin inclination $\sigma$ is positive, this leads to an even smaller scrub radius which

# The Steering

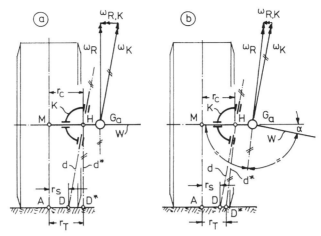

**Fig. 8.16** Angular velocities of wheel and wheel carrier with steering and torque transmission via a driveshaft

again is advantageous to avoid disturbances caused by brake-force fluctuation - e. g. with anti-lock brake systems. A considerable kingpin inclination, as typical for strut suspensions, may then lead to a 'negative' scrub radius which, however, must be kept within reasonable limits to avoid misleading information at the steering wheel about the true distribution of the braking forces at the wheels, and so to avoid a wrong reaction from the driver, especially at the beginning of the braking process.

Simple front-drive vehicles with transverse engines often have driveshafts of different lengths, due to the offset position of the final drive, **Fig. 8.17**.

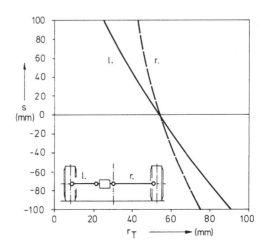

**Fig. 8.17**
Traction-force radii on a suspension with driveshafts of unequal length

With parallel travel of the two wheels, this causes different traction-force radii at the wheels, and consequent moments at the steering wheel even where the two traction forces are equal. This can be felt if, for example, the vehicle is accelerated or decelerated in the straight-ahead attitude while the suspensions are subjected to shaft deflection angles, or if such angles are caused by pitching.

In cornering, with opposing travel of the two wheels, the traction-force radius of the outer wheel becomes smaller than that of the inner wheel, even with driveshafts of equal length. For this reason, acceleration forces cause a moment that tries to return the wheels to the straight-ahead position (i.e. in the 'understeering' sense), while deceleration forces act in the opposite sense.

After the comments about the influence of deflected driveshaft joints as in Figs 8.15 and 8.16, it is easy enough to include the influence of hub-reduction gears as well. **Fig. 8.18** shows the traction-force radius $r_T$ versus wheel travel s for the same suspension as in Fig. 8.15 but with a hub-reduction ratio of 2, in comparison with the suspension without the hub reduction (i = 1, see Fig. 8.15). A negative reduction ratio means that the reduction-gear input shaft rotates counter to the wheel.

The curves $r_T(s)$ for i = 2 and i = -2 are approximately symmetrical to a vertical line corresponding to a radius $r_T = 15$ mm, and this is the value of the scrub radius $r_S$ of the suspension – see Fig. 8.15. The reason will become clearer if the motions of the driveshaft W, the wheel carrier K and the wheel itself are analysed, assuming (as previously) that the inner

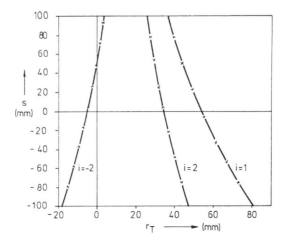

**Fig. 8.18** Traction-force radius versus wheel travel on a suspension with hub-reduction gear and different reduction ratios

The Steering

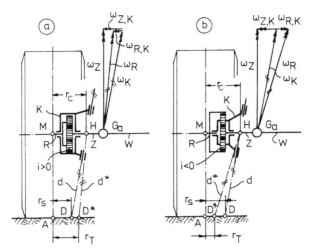

**Fig. 8.19** Angular velocities for a suspension with hub-reduction gear:
a) same direction of rotation of input and output shafts
b) opposite rotation of shafts

driveshaft end is locked and the outer shaft joint is near or on the kingpin axis. Consequently, the shaft W does not move during a virtual steering event, **Fig. 8.19**.

Then the angular velocity $\omega_Z$ of the input shaft Z is a vector in the plane of symmetry of the input shaft and the driveshaft W. The relative angular velocity $\omega_{Z,K}$ between the input shaft Z and the wheel carrier K generates a relative wheel velocity that is reduced by the gear ratio i: $\omega_{R,K} = \omega_{Z,K}/i$, and the absolute wheel velocity is the sum $\omega_R = \omega_K + \omega_{R,K}$.

In the normal position, without shaft deflection at the outer joint and with a hub-reduction ratio i = 2, the angular-velocity vector $\omega_{R,K}$ has the same direction as $\omega_{Z,K}$ but is of half its amount, Fig. 8.19a. The vector of the absolute wheel velocity $\omega_R$ is therefore inclined towards the kingpin axis d, and the virtual swivelling axis d* of the wheel meets the road surface at a distance $r_T$ from the tyre contact point A which is smaller than the wheel centre offset $r_c$ but still greater than the scrub radius $r_S$. With reversed rotation (i = -2), Fig. 8.19b, the relative wheel velocity $\omega_{R,K}$ is opposed to $\omega_{Z,K}$, and the resultant wheel velocity vector $\omega_R$ is inclined beyond the kingpin axis d. Here the traction-force radius $r_T$ is smaller than the scrub radius $r_S$.

With a wheel camber angle $\gamma$ near zero, and a shaft-to-wheel axle deflection angle $\alpha$ defined according to Fig. 8.16b, the geometrical relationships in Fig. 8.19 enable us to derive an approximation of the traction-force radius as

$$r_T \approx r_S + (R/i)[\sigma + \gamma - \alpha/2]$$

where R is the tyre radius and $\sigma$ the kingpin inclination (angles in radians).

If the reduction ratio were infinite, the vectors of the wheel velocity $\omega_R$ and the wheel-carrier velocity $\omega_K$ would coincide. With $i = \infty$, however, the driveshaft would no longer be able to force any rotation of the wheel which would then appear to be 'fixed' to its carrier. This represents the borderline case of a 'wheel-hub drive' where the scrub radius $r_S$ and the traction-force radius $r_T$ are identical.

With wheel travel, and varying driveshaft deflection angle, the attitude change of the virtual axis d* is reduced by the ratio i compared with the shaft joint deflection. Consequently the change of the traction-force radius $r_T$ with wheel travel is reduced, too, from what is shown in Fig. 8.16. Hence, a hub-reduction gear softens the effects of traction forces on the steering in cornering or with driveshafts of unequal length.

Hub-reduction gears are adopted mainly for two reasons, and therefore in different manners: first on off-road vehicles for better ground clearance (then preferably as cylindrical gearing with non-coaxial shafts, possibly without an actual reduction ratio); and second on heavy trucks to save weight in the driveline elements (then preferably as planetary gears with coaxial input and output shafts, and with a considerable reduction ratio).

On driven rigid axles, the deflection angle of the universal joint at the kingpin does not change with wheel travel. For such axles, the drawings in Fig. 8.19 consequently show the continuing state, and the traction-force radius is invariable with wheel travel. With a given kingpin inclination $\sigma$, different from zero, a suitable choice of the hub-reduction ratio then clearly enables the traction-force radius to be kept small and constant, with the advantage of lower stresses in the driveline parts and the steering linkage.

## 8.4 The steering linkage

### 8.4.1 Basic types

Steering angles applied to the steering wheel by the driver are transmitted to the wheels via the steering linkage. This linkage has to cope also with the bump and rebound wheel travel, except in special cases such as Dubonnet suspension. Independent wheel suspensions therefore usually

# The Steering

feature two transverse track-rods between the wheels and either the steering box or a medial linkage. The arrangement of the track-rods has to be suited to the suspension type and the desired kinematic bump-steer properties - see Fig. 7.30, Chapter 7.

Typical steering linkage layouts for independent wheel suspensions are shown in **Fig. 8.20**; due to the longitudinal compliance of today's passenger car suspensions, only transversely orientated track-rods come into question. In the example (a), the axes of rotation of the drop-arm at the steering box L and of the floating or 'slave' arm on the other side of the vehicle are parallel, and the two arms and the medial track-rod form a planar linkage. The transverse track-rods are connected to the arms independently from the medial rod. To achieve a symmetrical steering function and equal turns of the steering wheel in both directions, the medial linkage is a parallelogram. In the linkage of Fig. 8.20a, all six joints povide approximately the full steering angle (and thus contribute friction torques), and their elastic rates act in series.

By connecting the medial track-rod to the drop-arm and slave arm via turning joints, Fig. 8.20b, the medial rod is enabled to carry the inner joints of the track-rods (even out-of-line within certain limits). On one hand, more freedom is gained by this method for the spatial arrangement of the rod joints but, on the other hand, the possibility of geometrical variation is

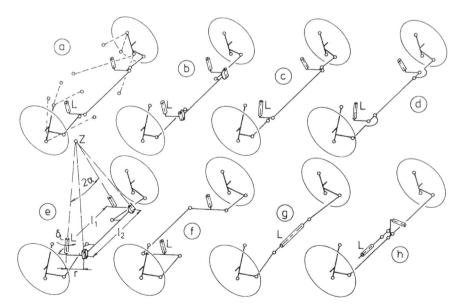

**Fig. 8.20** Steering linkage types for independent wheel suspensions

reduced, as all points at the medial rod (and hence the inner joints of the track-rods too) copy in space the circular movement of the drop-arm, which in the straight-ahead position moves exactly in the transverse direction. Only four joints are loaded with the full steering angle, which reduces friction in the linkage, and only the elastic rates of four joints are arranged in 'series', providing a stiffer connection between the two wheels.

If a wheel suspension features a very large kingpin inclination or castor angle (or both), a planar steering linkage is not very compatible with the spatially inclined paths of motion of the outer track-rod joints. For this reason, the axes of rotation of the drop-arm and the slave arm may also be inclined against each other - example (c). If the medial track-rod is guided by ball-joints (d), and has also to carry the inner joints of the transverse track-rods, these inner joints must lie on the axis of the medial rod to avoid free rotation about it. Another possibility is to guide the medial rod by turning joints whose axes intersect at the point Z of the drop-arm and slave arm axes (e); the medial linkage then forms a 'spherical' mechanism, see Section 3.3. To achieve equal maximum steering angles at the steering wheel and the drop-arm in both directions of turn, the linkage layouts of examples (c) to (e) must fulfil the condition that the distance $l_1$ between the centres of rotation of the joints of the medial rod is smaller than the rod length $l_2$: $l_2 = l_1 + 2r^2\sin^2\alpha(1 + \cos\delta_{L,max})/l_1$.

Drop-arms and slave arms that act like 'scale beams', example (f), are disadvantageous because of high reaction forces which cause increased friction and compliance.

Rack-and-pinion steering systems where the rack represents the medial track-rod (g) allow only linear movement of the inner joints of the transverse track-rods. In rare cases, rack-and-pinion steering boxes have been combined with lever arms (h).

On rigid axles, the wheel carriers are usually connected by a single track-rod 3 crossing the whole vehicle (so-called 'steering trapezium'), **Fig. 8.21**a, and a drag-rod 2 connected to the drop-arm 1 at the steering box L swivels one of the wheel carriers. The position of the drag-rod must be adapted to the suspension properties, see Section 7.5, and the linkage must be designed so that equal steering-wheel angles $\delta_H$ (and equal drop-arm angles $\delta_L$) generate first the outer and then the inner steering angle of that wheel carrier. A very rare layout comprises the use of 'divided' track-rods 3a and 3b, Fig. 8.21b, possibly combined with an intermediate lever at the axle.

Steerable tandem axles, Fig. 8.21c, need two steering trapeziums (track-rods 3a and 3b) and two drag-rods (2a and 2b), and an intermediate arm 5 that takes over the task of the 'second' drop-arm and is actuated by an

# The Steering

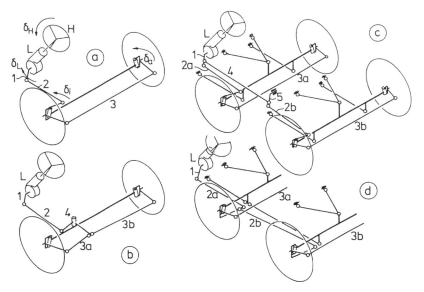

**Fig. 8.21** Steering linkages for rigid axles

intermediate rod 4. A very simple solution - which, however, requires axle suspensions featuring a linkage - is shown in Fig. 8.21d. The lower longitudinal links of the second axle's suspension are connected not to the chassis but to the first axle, and the second drag-rod 2b can therefore be connected immediately to the steering arm of the first axle. The upper triangular links which have to deal with the lateral forces should, of course, be connected immediately to the vehicle chassis.

Each steering linkage shows a steering ratio $i_L$ between the drop-arm of the steering box and the respective wheel carrier, this ratio normally varying with steering angle:

$$i_{La,i} = d\delta_L/d\delta_{a,i} \tag{8.23}$$

(a = outer and i = inner wheel), and with the steering-box ratio $i_H = d\delta_H/d\delta_L$ the overall steering ratio is

$$i_S = i_H(i_{La} + i_{Li})/2 \tag{8.24}$$

With rack-and-pinion steering, the steering ratio follows from the steering-wheel angle $\delta_H$ and the stroke h of the rack, $i_H = d\delta_H/dh$, the linkage ratio being

$$i_{La,i} = dh/d\delta_{a,i} \tag{8.25}$$

### 8.4.2 Steering geometry

A steering linkage has to ensure a certain relationship of the outer and inner steering angles $\delta_a$ and $\delta_i$, which must for example fulfil the condition that in plan view the axes of the two wheels intersect at the nominal centre K of the corner which lies on the rear-axle line, **Fig. 8.22**. This special steering requirement is called the 'Ackermann condition'.

On a suspension with conventional steering geometry, having a kingpin axis that appears fixed during a steering event, the relationship of the inner and outer steering angles can be determined using the intersection points of the kingpin axes and the wheel axes; with their distance $b^*$, Fig. 8.22a, and the distance of the axle's centre line from the cornering centre K (normally the wheelbase l), the 'Ackermann condition' can be obtained from the plan view (Fig. 8.22b):

$$\cot\delta_a = \cot\delta_i + b^*/l \tag{8.26}$$

There is an Ackermann condition for each steered axle on a vehicle. The relationship between the steering angles of the wheel carriers on the same side of the vehicle is defined by

$$l_1\cot\delta_1 = l_2\cot\delta_2 \tag{8.27}$$

On all-wheel-steered vehicles, the perpendicular line from the centre K of the corner to the vehicle's longitudinal axis represents the notional non-steered axis, and its distance from the steered axis has therefore to be taken for the 'wheelbase' in the equations.

If a suspension has a considerably inclined kingpin axis, and especially a virtual kingpin axis, it is better to depart from the simple Ackermann condition according to equation (8.26) and to use the actual wheel positions as shown in Fig. 8.22c. With the coordinates of points of the projections of the wheel axes in plan view - e. g. the wheel centres - the inner Ackermann angle $\delta_i$ (which corresponds to an outer angle $\delta_a$) follows from

$$\tan\delta_i = (x_i + l_h)/[(x_a + l_h)\cot\delta_a + y_a - y_i] \tag{8.28}$$

(where in Fig. 8.22c the coordinate $y_a$ is negative!). For a large kingpin inclination and/or castor angle, the error caused by application of the simple equations (8.26) and (8.27) may amount to 1-2° at full lock.

On heavy trucks, non-steerable tandem rear axles are frequently employed, Fig. 8.22d. Assuming equal wheel loads and equal tyre characteristics at all four wheels of the tandem axle system, and low speed of the vehicle without appreciable lateral acceleration, the distribution of lateral forces between the front axle and the two rear axles corresponds to the force

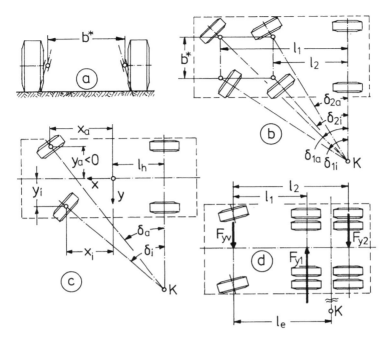

**Fig. 8.22** The Ackermann condition

distribution on a beam carrying three forces. The condition of balance of the lateral forces of the two rear axles is $F_{y1}l_1 = F_{y2}l_2$, where the forces act in opposite directions; the force $F_{y1}$ is the greater and therefore causes a greater sideslip angle than $F_{y2}$. From the relationship of the slip angles we get the effective wheelbase of a vehicle with non-steerable tandem axles:

$$l_e = (l_1^2 + l_2^2)/(l_1 + l_2) \tag{8.29}$$

and the centre K of the corner is slightly *behind* the medial line of the tandem assembly!

If the steering geometry is chosen exactly according to the Ackermann condition, the wheels will move at very low speed without tyre sideslip angles. With high lateral acceleration, however, these tyre sideslip angles occur on all wheels and the cornering centre K is displaced forward, see Chapter 7, Figs 7.37 and 7.39. The tyre slip angles $\alpha_{vi}$ and $\alpha_{hi}$ of the inner front and rear wheels will then be greater than the corresponding angles $\alpha_{va}$ and $\alpha_{ha}$ of the outer wheels, **Fig. 8.23**. On the other hand, the outer wheels will be capable of generating greater lateral forces than the

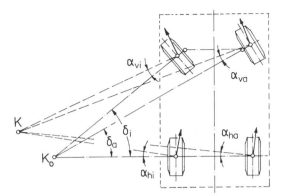

**Fig. 8.23** Tyre slip angles with high lateral acceleration and Ackermann steering

(less-loaded) inner wheels. For this reason, it is common practice to deviate from the true Ackermann condition towards a greater 'toe-in angle' – i.e. to more-nearly parallel steering angles of the two wheels – see also Chapter 7, Fig. 7.5. A further advantage of this measure is to save space in the vehicle due to the smaller inner steering angles, and in many cases the kinematic problems are reduced (see the later references to 'transmission angle'). Too much deviation from Ackermann, however, leads to problems of self-alignment of the steering on suspensions with large kingpin inclination and hence castor-offset change with steering angle, as will be discussed later.

Due to the three-dimensional character of steering geometry, the design of a steering linkage cannot be done by simplified planar methods and must take account of the kingpin attitude. Planar design can at best serve for a first approximation.

For rigid axles with a steering trapezium, literature offers methods for design (partly based on 'planar' geometry). However, bearing in mind the complicated procedures and often unsatisfactory results, the designer will do better in respect of the spatial paths of the track-rod joints by adopting trial-and-error methods with the compasses or at the computer!

For preliminary investigations into the basic form of a steering linkage – i.e. the four-joint chain of the drop-arm, the track-rod and the steering arm at the wheel carrier – planar kinematics offer a method of determining the joint positions for a selected pair of steering angles at the wheel and the drop-arm, **Fig. 8.24**. For three given positions $B_g$, $B_a$ and $B_i$ of a joint B (e.g. a track-rod joint at the straight-ahead position and at outer and inner steering angles) the straight-ahead position $A_g$ of a joint A is required, rotating about a centre $A_0$. The path of joint B does not have to be a circle

The Steering                                                                                                     227

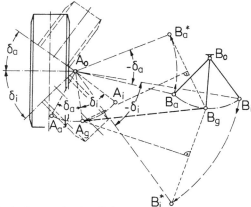

**Fig. 8.24** Planar method of designing a steering linkage

about $B_0$, as drawn, but may have any shape, as for instance the straight line that is typical for rack-and-pinion steering. With movement of $B_g$ into the positions $B_a$ and $B_i$, the lever $A_gA_0$ assumes steering angles $\delta_a$ and $\delta_i$. Assuming the (desired) linkage to be 'frozen' in the – still unknown – position $A_0A_aB_a$, and swivelled backwards about $A_0$ by the angle $\delta_a$, point $A_a$ must coincide with $A_g$, and $B_a$ comes into position $B_a^*$. Because of the constant (and still unknown) track-rod length $A_aB_a = A_gB_g = A_gB_a^*$, the wanted point $A_g$ must lie on the mid-perpendicular line from $B_gB_a^*$. Corresponding conditions are valid for the steering angle $\delta_i$, and the joint $A_g$ is at the intersection point of the two mid-perpendiculars.

Though the design of a steering linkage can be successful only if spatial geometry is utilized, some principal investigations of linkage properties may be made on a simplified planar model, as follows.

In a plane, the steering ratio $i_L = d\delta_L/d\delta$ can be determined by, for example, the method of rectangular velocity vectors (see Chapter 3, Fig. 3.1) or by using the effective levers $e_1$ and $e_2$ of the track-rod about the kingpin axis and the drop-arm shaft, **Fig. 8.25**: $i_L = e_1/e_2$.

The 'transmission angles' $\mu_1$ and $\mu_2$ are critical for the operating safety of the linkage. For $\mu_1$ or $\mu_2 = 0$, the linkage assumes an undetermined position. Especially the angle $\mu_1$ at the 'driven' member of the linkage (here the steering arm at the wheel actuated by the track-rod) must not become too small in order to avoid overstressing of the linkage in respect of its compliance properties, here symbolized by a track-rod spring rate c. The reserve of safety depends on the steering arm and track-rod lengths $r_1$ and l. For a particular example, the diagrams show – beginning with the upper one – the effective lever $e_1$ of the track-rod about the kingpin,

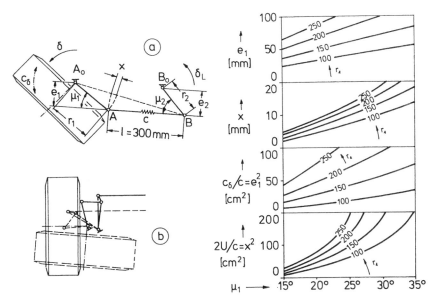

**Fig. 8.25** Transmission angles and effective levers of a steering linkage (planar model)

depending on $\mu_1$ and $r_1$, and the remaining overlap x of the steering-arm and the track-rod. The approximate rotational spring rate of the wheel about the kingpin is $c_\delta = c e_1^2$, and the energy needed for overstressing of the linkage is $U = c x^2/2$. Short steering-arms (rack-and-pinion steering) obviously require considerably greater transmission angles $\mu_1$ than the long ones of other steering linkages. Rigid axles on trucks must provide inner and outer steering angles of about 60° and 45° respectively; due to the greater stiffness of the steering trapezium of a rigid axle, transmission angles can be down to 10-12°. The multi-link (and rather flexible) steering linkages of passenger-car independent suspensions, however, require approximate minimum transmission angles of between 20° for long steering arms and 30° for short arms. In special cases - e.g. agricultural tractors - a series arrangement of two linkages helps to achieve large steering angles with acceptable transmission angles, Fig. 8.25b. If the steered wheels are not driven, auxiliary measures such as independently actuated rear-wheel brakes may be necessary.

Since $e_1 = r_1 \sin \mu_1$ and $e_2 = r_2 \sin \mu_2$, the steering-linkage ratio is $i_L = r_1 \sin \mu_1 / (r_2 \sin \mu_2)$, meaning that $i_L$ is infinite for $\mu_2 = 0$ and zero for $\mu_1 = 0$.

On a 'real' spatial steering linkage, the transmission angle does not occur only in the common plane of the steering arm and the track-rod. The linkage will always become unstable if the axis of the track-rod intersects

The Steering

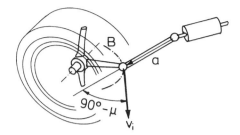

**Fig. 8.26**
Transmission angle on a spatial linkage

the kingpin axis, which occurs in Fig. 8.25 for $\mu_1 = 0$ but would in space also occur for a parallel attitude of the track-rod and the kingpin. This second feature becomes the more important the more obliquely the track-rod is inclined with respect to the kingpin axis on wheel bump or rebound.

Analogously to Fig. 8.25, the transmission angle $\mu$ on a track-rod joint i has to be defined in space by the angle between the track-rod a and the normal plane of the vector $v_i$ of its velocity which results from a steering process, **Fig. 8.26**. If the track-rod was rectangular to $v_i$, it would obviously not be able to fix the position of the joint i on its path B; the linkage would become unstable, and the transmission angle would be zero. Consequently, the angle between the velocity vector $v_i$ and the track-rod vector a is $90° - \mu$. According to equation (3.5) in Chapter 3, the scalar product of the vectors **a** and **v**$_i$ is also the product of its absolute values and the cosine of the included angle $(90° - \mu)$. From $\mathbf{a} \cdot \mathbf{v}_i = a \, v_i \cos(90° - \mu)$ and $\cos(90° - \mu) = \sin \mu$ we get the **transmission angle** through

$$\sin \mu = \frac{a_x v_{ix} + a_y v_{iy} + a_z v_{iz}}{\sqrt{a_x^2 + a_y^2 + a_z^2} \sqrt{v_{ix}^2 + v_{iy}^2 + v_{iz}^2}} \tag{8.30}$$

Typical properties of the common steering-linkage types, found especially on passenger cars, are shown in **Fig. 8.27** - again simplified by planar models. The linkages are all drawn to scale and are for inner steering angles of 45° corresponding to outer angles of 33° 20'. Moreover, the diagram contains the 'limiting curves' of the Ackermann condition A, determined by equation (8.26), and also the straight line P for parallel steering of both wheels.

Linkage 1 is frequently used (with the steering-arm at the wheel carrier pointing rearward and the drop-arm pointing forward), first because of its advantageous position near the front bulkhead and second because of the short 'safety steering column'. This linkage does not conform very well to the Ackermann condition A; it shows nearly parallel steering of both wheels over a wide range, while around the straight-ahead position the outer

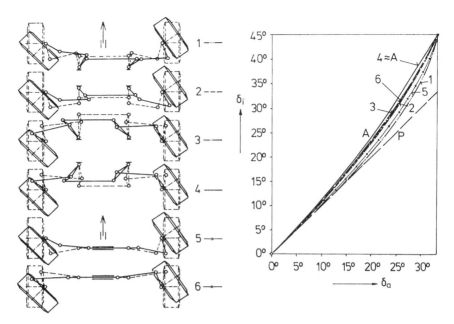

**Fig. 8.27** Typical steering characteristics of different linkage arrangements

steering angle may transitorily be even greater than the inner one. Not until approaching full lock does the characteristic become increasingly curved and finally intersects the Ackermann line. At the inner wheel, the track-rod approaches close to an unstable position and the inner transmission angle decreases rapidly, the converse occurring at the outer wheel. This behaviour may explain, too, the conspicuous curvature of the steering characteristic. It is easy to see that on the inner wheel the steering-arm and the track-rod swivel in opposite directions, thus leading to an increasing loss of transmission angle.

Linkage 2, with all the levers lying behind the axle and rotating in the same sense, shows a considerably better approximation to Ackermann, and slightly better transmission angles.

Essential improvement in the approximation to Ackermann as well as in the transmission angle is achieved by systems with steering arms pointing forward, as linkage 3 and especially linkage 4 which shows properties opposed to those of linkage 1. Unobjectionable transmission angles and a nearly constant transmission ratio, and a steering characteristic that practically coincides with the Ackermann curve (or that is, at least, proportional to it) are, however, purchased at the expense of a steering box in the extreme forward position.

# The Steering

The Ackermann characteristic is not steadily curved in the same sense but tends towards a point of inflection, because the inner and outer wheels would change their roles if the steering angle approached 90°. Only with linkage 4 is the Ackermann characteristic theoretically feasible up to the fifth derivative in the straight-ahead position (**9**).

A rack-and-pinion steering corresponds to a system with a drop-arm of infinite length, and its properties are therefore between those of steering linkages with steering arms pointing respectively forward and rearward – see the curves 5 and 6.

For all linkage types with steering-arms pointing forward, and with conventional 'fixed' kingpin axes, the steering-arms must be inclined towards the wheel centre planes (and so may foul the brake discs), in order to achieve a good approximation of Ackermann. With a 'virtual' kingpin axis, which is then normally variable at the wheel carrier, the kingpin displacement leads to 'shortening' of the effective steering-arm for one direction of steering and to 'lengthening' for the other direction and so may be used to influence the harmonization of the steering angles. Skilful application of this effect may reduce the previously mentioned brake-disc problem.

From linkage types 1 to 4, the steering linkage ratios $i_{La}$ and $i_{Li}$ of the outer and inner wheels become more and more well-balanced. The total ratio $i_L = (i_{La} + i_{Li})/2$ is symmetrical with steering to the left and to the right, and it increases with growing steering angle if the sum $\delta_a + \delta_i$ of the outer and inner wheel steering angles is smaller than the total angle $2\delta_L$ of the drop-arm (as typical for passenger cars), but decreases if it is greater (on trucks and buses). With rack-and-pinion steering, the total steering ratio normally decreases as the steering angle increases.

The foregoing comments are not intended to judge the linkage types mentioned but merely to call attention to their special properties and possible problems.

## 8.4.3 Steering returnability

The steering wheel of a fast vehicle is expected to return automatically to the straight-ahead position after any steering event; moreover, the steering-wheel moment should correspond to the external forces acting on the wheels. This self-centring property of the steering can be achieved by suitable tuning of the steering characteristics and the steering linkage. In fast straight-ahead driving, only the restoring torque due to weight is of essential influence, see equations (8.20) and (8.21).

In cornering, too, a steering-wheel moment is required that depends on the external forces and that informs the driver about the actual driving situation. The skilful harmonization of the steering-wheel moment and lateral acceleration is of great importance to the overall 'driving impression'. However, the steering moment should not increase proportionally to lateral acceleration, because this would lead to high forces at the steering wheel and hence to cramped muscles and loss of sensitivity for steering corrections that might be needed at the vehicle's dynamic limits. As power-assisted steering is today nearly standard, the characteristics of the control valve can also have an influence on the steering moment.

In fast cornering only small steering angles are applied. Due to the tyre sideslip angles, the centre K of the corner moves ahead of the rear-axle line — see also Fig. 8.23. **Fig. 8.28** shows schematically, in plan view, a vehicle undergoing high lateral acceleration, and the forces acting on its wheels.

Considering the tiny (and here drawn overscaled) steering angles of the wheels, it does not matter if the latter are steered according to the Ackermann condition or in the parallel sense. The position of the cornering centre K therefore has a primary influence on the distribution of the tyre sideslip angles.

The moment exerted on the steering system by the front wheels is caused mainly by the lateral forces, and only to a small degree by the wheel load $F_z$ acting on the wheel-load lever arms p. With positive wheel camber with respect to the road at the outer front wheel, its camber force $F_{\gamma va}$ tries to increase the steering angle.

The tyre slip forces $F_s$ are determined by the tyre characteristics and the parameters of wheel load, friction coefficient etc., and increase degressively with wheel load and slip angle, see Chapter 4: $F_s = F_s(F_z, \alpha)$; in contrast with the camber forces $F_\gamma$, they act displaced behind the tyre contact point by the pneumatic trail $n_R$.

Possible traction forces $F_{Tva,i}$ at the outer and inner front wheels act at the traction-force radii $r_{Ta,i}$ which normally have different values with increasing roll angle — see Fig. 8.15.

With these forces and their working lever arms as shown in Fig. 8.28, the restoring torque at the steering box L is

$$M_L \approx [F_{sva}(n_{va} + n_{Rva}) - F_{\gamma va}n_{va} - F_{zva}p_{va} - F_{Tva}r_{Ta}]/i_{La}$$
$$+ [F_{svi}(n_{vi} + n_{Rvi}) + F_{\gamma vi}n_{vi} + F_{zvi}p_{vi} + F_{Tvi}r_{Ti}]/i_{Li} \quad (8.31)$$

where va and vi apply to outer and inner front wheels respectively. The torque at the steering wheel then follows from $M_H = M_L/i_H$.

Caused by the wheel-load transfer to the outer side of the corner, the wheel load $F_{zva}$ of the outer wheel is greater than the corresponding $F_{zvi}$ of the inner one. As already explained in Chapter 7, the load transfer can

The Steering

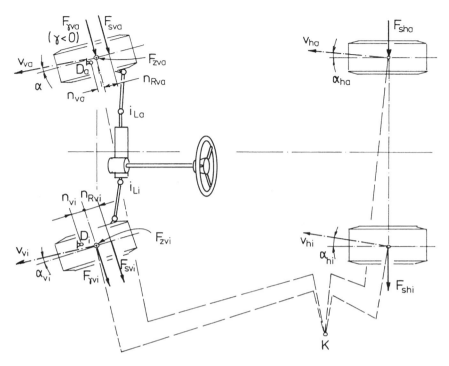

**Fig. 8.28** Steering returnability with high lateral acceleration

be influenced by the front and rear roll-centre heights and by the distribution of the roll spring rates to the axles. A further factor that should not be underestimated is the anti-dive property of the front suspension - i.e. the amount of the support angle $\varepsilon^*$ and its interdependence with wheel travel, as already shown in Fig. 7.4, Chapter 7. The longitudinal component $F_{sx}$ of the lateral force $F_s$ in Fig. 7.4 loads the suspension like a braking force. Therefore the 'anti-dive' characteristic of a suspension may generate a 'stabilizing' roll moment at the vehicle, thereby changing the driving behaviour and the torque at the steering wheel. Consequently, the support angle must be chosen in the light of the cornering behaviour of the vehicle, too.

The tyre sideslip forces act on a lever that is the sum of the castor offset n and the pneumatic trail $n_R$. On almost all wheel suspensions the castor offset n increases with growing inner steering angle and decreases with increasing outer angle (and may ultimately assume a negative value), see Fig. 8.10. Moreover, while the tyre slip forces assume their maximum values with growing lateral acceleration, the pneumatic trail $n_R$ tends back towards zero. Hence, the contribution of the lateral force of the outer wheel to the restoring torque decreases more and more and may eventually become negative. This effect is, however, partially cancelled by the fact that the steering-linkage ratio $i_L$ usually increases with the outer steering angle and decreases with the inner angle, thus reducing the influence of the moments generated at the outer wheel.

The outer front-wheel load which increases with lateral acceleration contributes a positive restoring moment only if the outer wheel-load lever arm $p_a$ is negative.

In contrast with the restoring moment at high lateral accelerations, the steering returnability at very low speed – e.g. on turning into a side street or during a parking manoeuvre – is influenced by a considerably greater number of parameters. In the extreme case of zero driving speed and lateral acceleration, of course, the wheel-load transfer and dynamic lateral forces vanish. The essential forces, lever arms and moments acting during extremely slow cornering are schematically represented in **Fig. 8.29**.

The two front wheels are steered according to any steering-angle layout that deviates from Ackermann towards more 'parallel' steering angles (or, so to speak, to increased 'toe-in'); consequently the extensions of the front-wheel axes intersect in plan view *behind* the rear-axle line.

When cornering without any lateral acceleration, the vehicle will move about a cornering centre K which may, unlike the case of fast cornering, lie throughout behind the rear-axle line, as drawn; the reason will be given at the end of this section.

In the following, the outer and inner front and rear wheels and the steering characteristics and forces attached to them are indicated respectively by the subscripts va, vi, ha, and hi.

With the coordinates of the tyre contact points and the cornering centre K, the tyre sideslip angles at the wheels are

$$\alpha_{va} = - \text{atn}[(x_{va} - x_K)/(y_K - y_{va})] + \delta_a \qquad (8.32a)$$
$$\alpha_{vi} = \text{atn}[(x_{vi} - x_K)/(y_K - y_{vi})] - \delta_i \qquad (8.32b)$$
$$\alpha_{ha} = \text{atn}[(x_{ha} - x_K)/(y_K - y_{ha})] \qquad (8.32c)$$
$$\alpha_{hi} = \text{atn}[(x_{hi} - x_K)/(y_K - y_{hi})] \qquad (8.32d)$$

and from these slip angles we get the tyre sideslip forces, according to Fig. 4.2, Chapter 4.

The camber forces $F_\gamma$ are drawn in Fig. 8.29 when they act with positive camber angles – i.e. towards the outside of the vehicle.

With the wheel loads $F_z$ and the rolling-resistance coefficient $f_R$, the rolling-resistance forces are

$$F_{Wv,h} = f_R F_{zv,h}$$

# The Steering

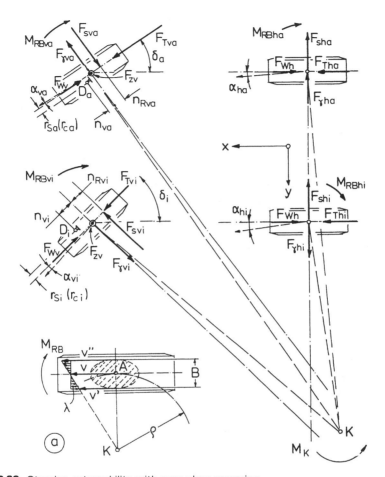

**Fig. 8.29** Steering returnability with very slow cornering

Even extremely slow cornering requires some engine power. For generalized investigation, all-wheel drive is assumed here, and with the share $\chi$ of the front axle on the total traction force, and the differential locking rates $\eta_{Sp(v,h)}$, the total traction force $F_T$ is distributed to the four wheels as follows:

$$F_{Tva} = \chi F_T(1 - \eta_{Spv})/2 \qquad (8.33a)$$
$$F_{Tvi} = \chi F_T(1 + \eta_{Spv})/2 \qquad (8.33b)$$
$$F_{Tha} = (1 - \chi)F_T(1 - \eta_{Sph})/2 \qquad (8.33c)$$
$$F_{Thi} = (1 - \chi)F_T(1 + \eta_{Sph})/2 \qquad (8.33d)$$

The tyre sideslip forces $F_S$ act on the steering by the resultant lever arm $n + n_R$, while the camber forces $F_\gamma$ act through the castor offset n only. The rolling-resistance force $F_W$ acts at the wheel-centre offset $r_c$, and the traction force $F_T$ at the traction-force radius $r_T$.

The forced movement of the tyres on a circular path about the cornering centre K causes a 'scrub torque' $M_{RB}$ on the tyres in their contact areas on the road; the different velocities v' and v" at the inner and outer edges of the tyre contact area - Fig. 8.29a - lead to a linear distribution of circumferential slip $\lambda$ over the width B and thus to a restoring torque $M_{RB}$ that increases with increasing wheel load, width B and increasing gradient of circumferential force and slip (see Fig. 4.3, Chapter 4), but decreases with increasing cornering radius $\rho$. With the radius of the wheel's path

$$\rho_n = \sqrt{(x_n - x_K)^2 + (y_n - y_K)^2} \tag{8.34}$$

(n = va, vi, ha, hi), the scrub torque of the tyre is $M_{RB} = M_{RB}(F_{zn}, \rho_n)$.

The forces and moments already defined allow us to establish the balance condition about the (still unknown) cornering centre K; in this, however, we can easily neglect the castor offset n, pneumatic trail $n_R$, traction-force radius $r_T$ and wheel-centre offset $r_c$, because these lever arms are tiny compared with the dimensions of the vehicle and will therefore not exert any sensible influence on the balance of moments. The sum of all moments about the cornering centre must be zero:

$$\begin{aligned} M_K = &\ [(F_{Tva} - F_{Wv})\sin\delta_a + (F_{sva} - F_{\gamma va})\cos\delta_a](x_{va} - x_K) \\ &+ [(F_{Tva} - F_{Wv})\cos\delta_a - (F_{sva} - F_{\gamma va})\sin\delta_a](y_K - y_{va}) \\ &+ [(F_{Tvi} - F_{Wv})\sin\delta_i - (F_{svi} - F_{\gamma vi})\cos\delta_i](x_{vi} - x_K) \\ &+ [(F_{Tvi} - F_{Wv})\cos\delta_i + (F_{svi} - F_{\gamma vi})\sin\delta_i](y_K - y_{vi}) \\ &- (F_{sha} + F_{\gamma ha} + F_{shi} - F_{\gamma hi})(x_h - x_K) \\ &+ (F_{Tha} - F_{Wh})(y_K - y_{ha}) \\ &+ (F_{Thi} - F_{Wh})(y_K - y_{hi}) \\ &- M_{RBva} - M_{RBvi} - M_{RBha} - M_{RBhi} = 0 \end{aligned} \tag{8.35}$$

The cornering centre K will assume a position where the required traction power reaches its minimum:

$$F_{Tva}\rho_{va} + F_{Tvi}\rho_{vi} + F_{Tha}\rho_{ha} + F_{Thi}\rho_{hi} = \min. \tag{8.36}$$

which can be investigated by a simple iteration program.

With the coordinates of the cornering centre K, the tyre sideslip angles of all wheels and all the forces and moments can now be determined. Then, the moments generated at the steering linkage are

The Steering                                                                                   237

$$M_{va} = F_{sva}(n_{va} + n_{Rva}) - F_{\gamma va}n_{va}$$
$$\phantom{M_{va} =} - F_{Tva}r_{Tva} + F_{Wv}r_{cva} - F_{zva}p_{va} + M_{RBva} \quad (8.37a)$$
$$M_{vi} = - F_{svi}(n_{vi} + n_{Rvi}) + F_{\gamma vi}n_{vi}$$
$$\phantom{M_{vi} =} + F_{Tvi}r_{Tvi} - F_{Wv}r_{cvi} + F_{zvi}p_{vi} + M_{RBvi} \quad (8.37b)$$

(these moments act in the opposite sense to the steering if their signs are positive) and, with the relevant steering-linkage ratios of the front wheels, the total returning torque at the steering box is

$$M_L = M_{va}/i_{La} + M_{vi}/i_{Li} \quad (8.38)$$

With the steering-box ratio $i_H$, the returning moment at the steering wheel is

$$M_H = M_L/i_H \quad (8.39)$$

On a wheel suspension with conventional kingpin axis 'fixed' to the wheel carrier, the traction-force radius $r_T$ and the wheel-centre offset $r_c$ remain more or less constant with steering angle $\delta$, so their respective inner and outer values are equal. For this reason, the influence of the traction forces (at least if the differential locking rate is low) and of the rolling-resistance forces on the returning torque is negligible.

The wheel-load lever arm of the outer wheel is normally smaller than that of the inner, and the wheel load therefore acts in the restoring sense. The contribution of the restoring torque due to wheel load is, however, of minor importance also.

On the other hand, the moment caused by the tyre sideslip forces is of predominant importance because on normal suspensions the castor offset n varies considerably with the steering angle $\delta$. While the outer wheel acts on a significantly smaller castor offset than the inner (or even a negative one), the pneumatic trail $n_R$ may be assumed for very low speed to be equal on both wheels; for this reason, the moments generated by the tyre slip forces tend to increase the steering angle, provided that the sideslip forces of both front wheels are (as shown in Fig. 8.29) directed towards the medial plane of the vehicle, according to the pre-assumed deviation from Ackermann of the steering geometry.

However, the 'scrub torque' of the tyre acts in the restoring sense and, since it assumes similar values to the moment of the sideslip forces mentioned above, it will in many cases 'save' the returnability of the steering.

Obviously the returnability of the steering depends on the difference between two very large moments which may, moreover, both be affected by considerable tolerances of different kinds: tyre wear and pressure, temperature, manufacturer, etc. To escape this problem, it may be advantageous to provide a steering geometry not too far different from the

Ackermann condition, accepting, of course, a rather great steering angle of the inner wheel (and consequent loss of interior space) and, probably, a reduced transmission angle (see Figs 8.25 and 8.26) at that same wheel.

As already mentioned at the beginning of this section, the cornering centre K is drawn in Fig. 8.29 behind the rear-axle line. The reason is that, with large steering angles, front-wheel lateral forces that act towards the middle of the vehicle (as they do if the steering layout deviates considerably from Ackermann) show lines of action that are considerably displaced and thus generate a moment at the vehicle acting against the direction of the corner. This moment must be offset by the lateral forces at the rear wheels, which are then consequently directed towards the outside of the corner (as in Fig. 8.29).

Obviously, the method frequently used - at least in former times - to save space in the vehicle by approximately parallel steering of the front wheels, and at the same time to avoid kinematic problems (transmission angle - see Fig. 8.25) and achieve a small turning circle, suffers limitations on vehicles with wide tracks and short wheelbases. This is especially true on wheel suspensions that give considerable castor-offset change with steering angle - e.g. those with a large kingpin inclination, as is typical for today's strut systems, due to the requirement of a minimal scrub radius.

### 8.4.4 Steering-system vibrations

Two principal types of vibrations occur in a steering system. One of them is in the region of the natural frequency of the 'unsprung masses' - i.e. about 10-15 Hz (the so-called 'wheelfight') and the other around the natural rolling frequency of the vehicle - i.e. about 2 Hz (the 'rolling oscillation').

Wheelfight is caused by unbalanced masses of the tyres and the wheel, and by elastic non-uniformity of the tyres, especially radial force fluctuations. The latter excite the steering system via the wheel-load lever arm but the unbalanced masses work via the wheel-load lever arm as well as the wheel-centre offset.

In a simplified view, the rolling oscillation is an oscillation of the steering wheel against the whole vehicle mass, influenced by dynamic tyre properties and the vehicle's springing and damping system. On a passenger car, the polar moment of inertia of a steering wheel amounts to only one-tenth that of a vehicle wheel but, when 'reduced' to the kingpin (i.e. multiplied by the square of the steering ratio) it assumes a considerable value in the range of the yaw moment of inertia of the vehicle (**6**). A steering wheel with a low moment of inertia increases the natural frequency of the rolling oscillation, but on the other hand it transmits the wheelfight moments more

# The Steering

noticeably to the driver's hands. In general, so-called 'steering dampers' avoid rolling oscillation but not wheelfight because they respond too late for this purpose.

For reasons of important non-linear parameters, neither type of vibration is suitable for simplifying methods of calculation.

## 8.5 Self-aligning steering systems

In the preceding sections, we have considered only steering systems that are actuated by the driver via the steering wheel. On articulated vehicles such as semi-trailer trucks or articulated buses, however, the trailer may be provided with steerable axles the attitude of which results from the relevant driving situation and cannot be influenced immediately by the driver.

**Fig. 9.30** shows schematically the self-aligning steering system of the trailer portion of an articulated bus. The front vehicle V and the trailer N are connected by a joint K. Depending on the deflection angle between the halves of the vehicle, the wheels of the trailer's axle are steered via a steering linkage connected to the front half V by a joint S.

The same principle is applied to semi-trailer trucks with only one trailer axle, but preferably using fifth-wheel steering - i.e. by swivelling a complete rigid axle (see Fig. 8.1a), instead of the kingpin steering system shown in Fig. 8.30. The 'steering geometry' of the trailer wheels has to be established on a variable 'wheelbase' that decreases with increasing deflection angle between the vehicle halves, because the perpendicular line a from the cornering centre M to the centre-line of the trailer represents, so to speak, its (non-steerable) 'front axle'.

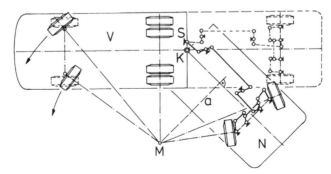

**Fig. 8.30** Steerable trailer axle on an articulated bus (according to documents of the M.A.N. company)

In the case of tandem axles, the maximum legal loading of two single axles is allowed only if a certain wheelbase is provided between them. In narrow corners this leads to considerable scrubbing of the tyres. To spare these and the road surface, self-aligning tandem-axle aggregates are then employed.

The tandem aggregate of **Fig. 8.31**a consists of two 'trailing' drawbar axles of the same type, the lateral attitude of which is controlled by leaf springs able to swivel about joints with vertical axes of rotation. If the path of the wheels on the road is not parallel to the central line of the trailer, both axles are displaced sideways by the lateral forces which work in opposed senses (see also Fig. 8.22d). The leaf springs act like scale-beams and force opposing lateral displacements and steering angles of equal amounts. Consequently the tyres move without side-slip; the cornering centre remains on the line of symmetry m of the tandem aggregate, and the resultant steering angle is zero.

In contrast, Fig. 8.31b shows a tandem axle with significant steering effect. The first axle is guided by three leading rod links, and the second by three trailing links. The two lower links of the first axle intersect in plan view at a point $P_v$ behind the axle, and those of the second axle at a point $P_h$ ahead of the axle, while the distance of $P_v$ from the first axle is greater than that of $P_h$ from the second axle. In cornering, the leaf springs again act like scale-beams and force opposing and equal lateral displacements of the axles. The latter swivel about their 'poles' $P_v$ and $P_h$ and the

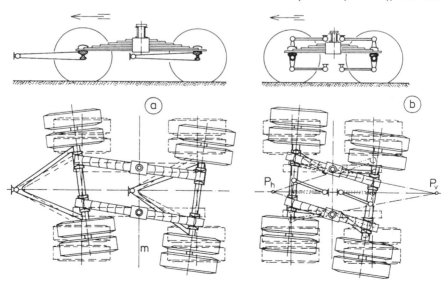

**Fig. 8.31** Self-aligning tandem axles

The Steering                                                                 241

second one, owing to its shorter distance from its pole $P_h$, assumes a greater steering angle than the first. As both axles are steered here in the same direction of cornering, unlike example (a), their intersecting point (i.e. the cornering centre of the trailer) is displaced forward, thus simulating a shorter wheelbase of the trailer. Hence, this aggregate saves space in cornering in the same way as the semi-trailer bus of Fig. 8.30. In contrast to the bus, however, the trailer axles' steering angle results not from the deflection angle of the vehicle halves but from the lateral forces acting between the axles and the road.

It should be pointed out here, too, that both the tandem-axle systems shown in Fig. 8.31 meet the conditions for well-balanced braking-force distribution - see Fig. 6.32 in Chapter 6.

For official approval, road vehicles have to conform to maximum dimensions prescribed to secure adequate mobility in mixed traffic. For example, the vehicle must be able to negotiate an annular section of road with an outer radius $r_1$ and an inner radius $r_2$ without any parts of it protruding - see **Fig. 8.32**a. With a given width B, the maximum distance of the rear axle from the front end may then be

$$L = \sqrt{r_1^2 - (B + r_2)^2}$$

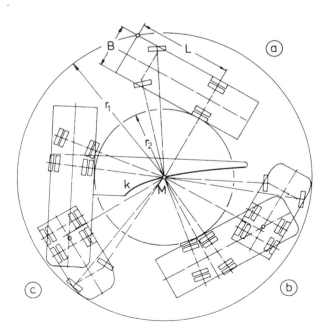

**Fig. 8.32** Road area requirement in cornering

To gain more length, the front ends of long vehicles are therefore frequently rounded, as indicated on the tractors of examples b and c.

Fig. 8.32b shows a semi-trailer truck with a tandem axle according to Fig. 8.31a; the latter does not serve to increase the vehicle length but only to achieve higher permissible axle loading. Neglecting the tyre slip angles that would occur with high lateral acceleration, the trailer will assume a position where the line of symmetry of its rear axles passes through the cornering centre M, thus behaving more advantageously than a trailer with non-steerable tandem axles - see Fig. 8.22d and equation (8.29).

For the tandem-axle system of Fig. 8.31b, the 'steering geometry' is defined by the geometrical locus k of the intersecting points of the two axles, Fig. 8.32c. If the curve k is assumed to be fixed to the trailer, the latter will assume a position where k touches the cornering centre M which is defined by the axles and steering angles of the tractor. This axle system obviously saves a lot of space in the corner and so allows a longer trailer to be used.

For the required tangential entry to the annulus, there are prescribed limits for the lateral protrusion of the vehicle's tail. Every vehicle with rear overhang is bound to protrude laterally when entering the circle. With semi-trailer vehicles, the deflection angle between the front and the rear portion, or a deflection angle of the trailer's centre-line from the actual path of its rear axle, does already occur when the trailer or its rear axle is still moving along the straight section of the road. If the rear-axle aggregate is of the self-steering type, as on the bus of Fig. 8.30 or the trailer of Fig. 8.31b, the lateral protrusion will be increased and the vehicle may have to show a warning notice at its rear end.

# Chapter 9

# Elasto-Kinematics

## 9.1 Principal considerations

The wheel suspension is excited by vibrations, from road irregularities or tyre non-uniformity, especially around the natural frequency of the 'unsprung masses'. Single obstacles, such as pot-holes or transverse joints in the road surface, show a wide frequency range which can come into the audible region. The same is true for natural vibrations of the tyre. Since modern tyres have a high circumferential stiffness and therefore respond minimally to circumferential force fluctuations, the disturbances mentioned above generate in the tyres circumferential forces that may be of the same order as the wheel loadings.

The rubber joints of the suspension links allow the system to comply with these forces and thus they reduce the effects of impact forces. They also serve for noise insulation and have the additional advantages of being maintenance-free, friction-free, capable of recovering after temporary overload and – last but not least – cost saving.

On vehicles intended to be comfortable, those impact forces require a longitudinal compliance in the wheel suspension of as much as ±15 mm. Such compliance would cause unfavourable attitude changes of the wheels – i. e. deviations from the desired 'kinematic' function (above all, elastic steering angles) – unless suitable countermeasures are taken.

The term 'elasto-kinematics' means a conscious harmonization of the spring rates of the suspension joints (and possible elastic rates of any chassis elements) and of the spatial arrangement of the suspension links, with the aim of compensating the elastic displacements that occur under external loads, or even of converting them into 'wanted' displacements.

Because even in quasi-static cases the different force groups – like wheel load, braking force, traction force and cornering force – must all be embraced by the elasto-kinematic attunement of the suspension, it is clear that, in the pre-computer era, such attunement was possible only by empirical processes, and that suspensions consciously designed for elasto-kinematic functioning could not be developed before the early 1970s. Multi-link suspensions offer more scope for variety and hence of elasto-kinematic

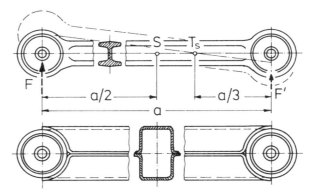

**Fig. 9.1** Simple rod link and its percussion point

harmonization; this, apart from saving weight and space, is the main reason for their increasing application even on low-cost vehicles.

Suspension vibrations cause inertia forces in the links, and consequently reaction forces at their joints. The latter may, at least theoretically, be avoided by choosing the suspension-link joints for mutual 'percussion points' (**1**) (**5**) meeting equation (5.23) in Chapter 5 (so-called 'decoupling of masses' on the suspension link). **Fig. 9.1** shows a rod link with constant-profile cross-section; here the percussion point $T_s$ of each joint is one-third of the length 'a' from the other joint. An impact force F on one of the joints will therefore cause a reaction force $F' = F/2$ at the other.

Suitable positioning of the percussion point is the more important the greater the 'unsprung masses', as for example on drive assemblies where the engine is mounted to the wheel carrier (so-called 'drive-unit wheel carriers') – see Fig. 6.6a, Chapter 6.

A similar 'percussion' problem is the 'steering kickback', visualized in schematic form in **Fig. 9.2** on a 'planar' model. The centre of gravity S of the wheel with the wheel carrier and brake is usually displaced inboard from the centre-plane of the wheel by a distance 'e'; therefore a longitudinal force $F_x$ will try not only to push the wheel backwards by a travel $v_x$ but at the same time to swivel it by a steering angle $\delta$. If there was no wheel suspension, and the wheel was freely movable in space, it would swivel about the percussion point $T_s$ whose distance from S follows from the lever e of the force $F_x$ and the radius of inertia i, according to equation (5.23), as $p = i^2/e$. If the suspension resists movement of the wheel about $T_s$, this will cause reaction forces at the suspension links (and hence at the track-rod Sp, too). However, the suspension in Fig. 9.2 shows a transverse link Q and a track-rod Sp with their axes intersecting at the

Elasto-Kinematics 245

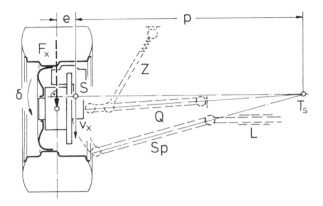

**Fig. 9.2** Steering kickback shown schematically

percussion point $T_s$; and since the tension rod Z may be assumed to have quite a flexible mounting at the vehicle body (to achieve good 'longitudinal compliance') it will not generate a noticeable reaction force at the start of the wheel displacement. Thus, the suspension offers no resistance to the combined shift-and-steer movement of the wheel.

On this simple planar model, the steering kickback could obviously be avoided if the centre of gravity S of the 'unsprung' masses lay on the centre-plane of the wheel or, if this is not possible, by tolerating steering angles under impact forces (and then, consequently, under braking or accelerating forces too).

A 'historical' solution of a longitudinally compliant suspension is shown in **Fig. 9.3**. The double-wishbone system itself is made as stiff as possible,

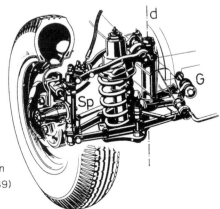

**Fig. 9.3**
Longitudinally compliant front suspension of the Mercedes-Benz type '170 S' (1949) (courtesy Daimler-Benz AG)

and is attached to a bracket that is mounted to the chassis rotatably about an axis d and supported by a rubber joint G. The track-rod Sp runs roughly parallel to the plane that is defined by the axis d and the kingpin axis of the suspension, thus allowing longitudinal displacement of the wheel without any steering angle. This is, of course, not yet an 'elasto-kinematic' assembly in the modern sense, but one can already see the intention of avoiding elastic 'wind-up' of the wheel carrier under braking forces in spite of the longitudinal compliance of the wheel.

If loaded with high-frequency forces, the suspension links are excited to sympathetic vibrations. The resonant bending frequencies of a rod link with a constant cross-section like that of Fig. 9.1 are

$$\omega_n = n^2 \pi^2 \sqrt{E I_B / m l^3} \tag{9.1}$$

where E is the elasticity modulus, $I_B$ the geometrical moment of inertia, m the mass, l the length and n the order of the resonant vibration. With the cross-sectional area A and the radius of the geometrical moment $i_B = \sqrt{I_B/A}$, and the volume $V = Al$ and the density $\rho = m/V$, equation (9.1) changes to

$$\omega_n = (n^2 \pi^2 i_B / l^2) \sqrt{E/\rho} \tag{9.2}$$

The steels and aluminium alloys that are generally used for suspension parts do not show significant differences of the quotient $E/\rho$; therefore a high resonant bending frequency of a suspension link is best achieved by a big radius $i_B$, as for example on a thin-wall hollow component – see Fig. 9.1 (lower drawing).

Large surfaces with thin walls may be excited to transverse vibrations of these walls and are therefore often provided with two-directional bowing to achieve higher inherent stability.

As already mentioned, elasto-kinematics is the appropriate harmonization of the effective spring rates and the geometrical positions of the links. A wheel suspension that meets kinematic demands (as for instance having a desired roll-centre height or braking-force support angle) does not necessarily also lend itself to good elasto-kinematic attunement. Using a simple 'planar' model, **Fig. 9.4** serves to remind us of several truisms about the limitations of elasto-kinematics.

If a suspension is elastically displaced, it stores energy delivered by the external force causing the displacement. The latter must therefore have a component that is directed in the same sense as the force vector. Consequently, the displacement vectors v caused by a lateral force $F_q$ or a longitudinal force $F_l$ (examples a and b) may, according to the positions of the

Elasto-Kinematics 247

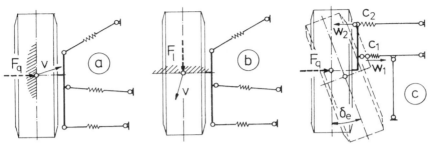

**Fig. 9.4** Directions of forces and of accompanying displacements

principal spring axes of the suspension mechanism (see Chapter 5, Figs 5.17 and 5.18), lie within a sector of ±90° on each side of the respective force vectors, but not in the hatched regions.

Referring to the two transverse links, the lateral force $F_q$ in Fig. 9.4c acts in a 'floating' manner at the wheel carrier, thus compressing the nearer link by a deflection $w_1$ and extending the farther link by $w_2$; this leads to a resultant transverse displacement $w_1 - w_2$ of the wheel carrier and to an elastic steering angle $\delta_e$. Obviously this steering angle could be avoided if the displacements $w_1$ and $w_2$ were of the same direction and amount, this requiring suitable arrangement of the transverse links on *both* sides of the force vector and, in addition, an appropriate choice of the resultant spring rates $c_1$ and $c_2$ of the links. If enough links are available, this will always be possible.

However, the elasto-kinematic properties may change considerably with wheel travel, depending on the kinematic characteristics of the suspension. The five-link design in **Fig. 9.5** comprises three transverse and two longitudinal links. It could be designed for 'neutral' cornering behaviour in the position shown - i.e. to generate no elastic steering angle under a lateral force F which acts displaced rearward from the tyre contact point A by the pneumatic trail $n_R$ (see Chapter 4). This has been achieved here in an especially simple manner by positioning the two lower transverse links in plan view at equal distances ahead of and behind the line of action of the force F - providing them with equal effective spring rates - and by positioning the upper link in the cross-sectional plane through that line of action.

The suspension shows, though, a 'longitudinal pole' L in front of the wheel and quite close to it, to counter braking dive (see Chapter 6). With wheel travel, the wheel carrier (and with it all joints) will consequently swivel about that pole in side view. The point of 'neutral' lateral force transfer is therefore swivelled about the pole L on a path $g_N$, while the

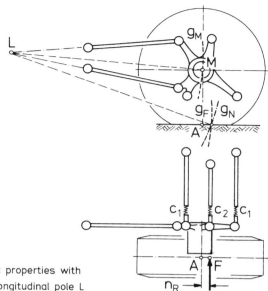

**Fig. 9.5**

Change of elasto-kinematic properties with wheel travel caused by a longitudinal pole L

wheel centre M travels along its path $g_M$ about L. The tyre contact point A travels on a parallel path to the wheel centre, of course, and as the point of application of the force F is always offset from A by $n_R$, its path $g_F$ is parallel to that of the wheel centre too. Consequently the point of application of force F is displaced forward, with respect to the 'neutral' point, with wheel bump and displaced rearward with rebound. This means that, in cornering, the lateral force at the outer wheel generates an elastic toe-in angle with increasing wheel travel as does that at the inner wheel (where, of course, the lateral force acts in the opposite direction to the drawn force F). So the wheels assume increasing, and perhaps considerable, elastic toe-in as the vehicle rolls. With a rear suspension this is equivalent to elastic understeer due to the predominant influence of the outer wheel. A front suspension would, of course, require a pole L *behind* the wheel to achieve elastic understeer.

More difficult is the elasto-kinematic treatment of longitudinal forces, because all construction parts of the suspension are unilaterally positioned with respect to such a force which therefore generates a resulting moment at the suspension. When loaded with a braking force $F_x$, the wheel carrier in **Fig. 9.6**a (which is guided by a trapezoidal link) will be swivelled by an elastic angle $\delta_e$, as the rubber joints at both the wheel-end and the vehicle-end axes of revolution are pulled outward at the front and pushed inward at the rear. Theoretically, the steering angle could be avoided by a skew

# Elasto-Kinematics

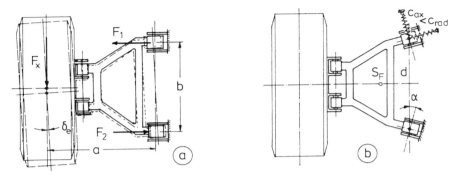

**Fig. 9.6** Trapezoidal-link suspension loaded with a longitudinal force

arrangement of the axes of the rubber bushes with respect to the axes of revolution under an angle $\alpha$, Fig. 9.6b, and thus to generate an 'elastic centre' of the suspension (see Chapter 5, Fig. 5.17) which lies on the line of action of the force (or the wheel centre-plane). The angle $\alpha$ indicated in the drawing would, however, hardly be sufficient to position the elastic centre in that plane, applying realistic radial and axial spring rates to the rubber joints (quite apart from the 'coning' angles with wheel travel caused by this measure, which would impose narrow limits on the attunement – see Fig. 5.41).

Wheel suspension systems are often mounted on a subframe which may also carry the final drive and which is connected to the vehicle body through rubber mountings. This arrangement is useful if, for example, the suspension itself does not enable the best elasto-kinematic attunement; it also provides better insulation of the body structure against tyre or gearbox noise as well as simplifying the installation of the axle on the production line.

The elastic centre $S_F$ of such a subframe normally lies in the medial plane of the vehicle, **Fig. 9.7**a. While steering angles under lateral forces

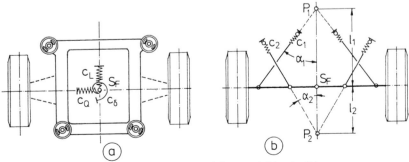

**Fig. 9.7** Compliant mounting of a subframe (a) or a rigid axle (b)

can always be avoided, in theory, by suitable elasto-kinematic attunement of the subframe's mountings, one-sided longitudinal forces will always cause a slight steering angle of the whole assembly.

The same is true on all rigid-axle suspensions, Fig. 9.7b. With the angles of attack $\alpha_1$ and $\alpha_2$ of the upper and lower links, and the spring rates $c_1$ and $c_2$ of the links, the resultant lateral spring rate of the lower links at their intersecting point $P_1$ is $c_{Q1} = 2c_1 \sin^2 \alpha_1$ and that of the upper links at their intersecting point $P_2$ is $c_{Q2} = 2c_2 \sin^2 \alpha_2$. The resultant lateral spring rate is $c_Q = c_{Q1} + c_{Q2}$ or

$$c_Q = 2(c_1 \sin^2 \alpha_1 + c_2 \sin^2 \alpha_2)$$

If the elastic centre $S_F$ is desired to be in the wheel axle line, as drawn, it must meet the condition

$$c_{Q1} l_1 = c_{Q2} l_2$$

and the rotational rate of the axle suspension about the z-axis is

$$c_\delta = c_{Q1} l_1^2 + c_{Q2} l_2^2$$

Through its lever arm about the vehicle's centre of gravity – i.e. the half track width – a one-sided longitudinal force exerts a yawing acceleration on the vehicle. For a front axle the steering angle of a compliantly mounted subframe or rigid axle will be in the same direction as the yaw angle, and for a rear axle it will be in the opposite direction. Clearly, a compliantly mounted rear-axle subframe or rigid rear axle can help to stabilize the vehicle under one-sided longitudinal forces. However, while a certain elastic *steering angle* is related to a certain force (independently of the vehicle speed) and will therefore generate different yaw velocities with different speeds, the moment of the one-sided force generates a certain yaw *acceleration* independently of the speed. Hence, only approximate compensation for yawing is possible.

Best suited for optimum elasto-kinematic attunement is of course the multi-link suspension. Due to three-dimensional and interacting events with different patterns of external forces, a proper analysis of the elastic behaviour of a wheel suspension can be done only by computer calculation. The metal parts of the suspension – such as the suspension links, possible subframes and the relevant structure of the vehicle body – generally show compliance of a similar order to that of the rubber joints; a good program for elasto-kinematic analysis must therefore be able to work with the stiffness matrices (which are generated, for example, by the application of finite-element methods) of those metal parts. Complicated assemblies

Elasto-Kinematics 251

respond differently from a single rubber joint or a suspension link, depending on the combination of forces – something that will be discussed later.

In the following section, typical effects on compliant suspensions and suitable design measures will be shown, using a simple multi-link model.

## 9.2 Elasto-kinematics of independent wheel suspensions

### 9.2.1 Elastic behaviour of the suspension mechanism

To visualize the principal problems of the elasto-kinematic harmonization of a wheel suspension, it is helpful first to consider a simplified multi-link suspension with respect to different and typical quasi-static cases of external forces, **Fig. 9.8**. The suspension consists of two triangular links one above the other, compliantly mounted to a vehicle body which is assumed to be rigid, and of a rod link (e.g. a track-rod on a steerable suspension). The suspension may form a 'planar' system, and the axes of the turning joints of the triangular links are directed parallel to the x-axis. The two ball-joints of the triangular links guiding the wheel carrier are in the cross-sectional plane through the tyre contact points and the wheel centres. This suspension could be applied in principle to front or rear wheels. It is here

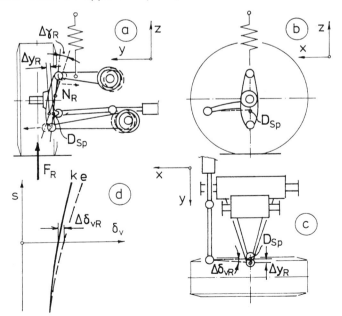

**Fig. 9.8** Compliant double-wishbone suspension with wheel loading

deliberately chosen in rather simple form in order to discuss its elastic behaviour only by visualization and without the need for equations and complicated calculations.

Because the spring is supported by the upper link, Fig. 9.8 a, and the ball-joint of this link is displaced from the wheel centre-plane, the wheel load $F_R$ generates a moment at the wheel carrier that tries to extend the lower triangular link and to compress the upper. The wheel carrier is therefore inclined by an elastic 'negative' camber angle $\Delta\gamma_R$, and swivels in cross-section about a 'neutral' point $N_R$, the position of which depends on the spring rates of the rubber mountings of the links.

The rod link or track-rod is situated in front of the suspension and relatively low. On the 'rigid' suspension (without rubber joints), its position could have been determined for the drawn 'kinematic' toe-in curve k against wheel travel s. Because of its position below the neutral point $N_R$, the centre of rotation $D_{Sp}$ of the outer track-rod joint on the 'kingpin' axis is displaced towards the outside of the vehicle by $\Delta y_R$, while the track-rod itself remains unloaded due to the wheel-load force acting in the cross-sectional plane through the kingpin axis. Consequently the wheel assumes a 'toe-in' angle $\Delta\delta_{vR}$, Fig. 9.8c. Hence, the 'elastic' toe-in curve e starts displaced with respect to the 'kinematic' curve k by the angle $\Delta\delta_{vR}$, and the displacement increases with bump travel according to the rising wheel load and, of course, joint deflections.

In comparison with the kinematic curve, the elasto-kinematic toe-in curve in general shows a displaced point of origin and a different gradient. The displacement can be compensated by (for instance) readjustment of the track-rod on the production line, and the gradient deviation and perhaps the curvature by a deliberate 'kinematically wrong' location of the track-rod in the vehicle – see Chapter 7, Fig. 7.30.

Unlike the elastic steering angle $\Delta\delta_{vR}$, the unavoidable elastic camber angle $\delta\gamma_R$ is of minor importance if limited to a fraction of an angular degree.

This first example has already made it clear that elasto-kinematic attunement of a wheel suspension is not a task for the later road-test stages but begins in the planning phase of the suspension mechanism by alternating and iterative kinematic and elasto-kinematic analyses. These analyses have, of course, to cover not only the (just discussed) case of wheel load but the cases considered in the following too; naturally this requires the elastic properties of the relevant constructional parts to have been estimated as early as possible.

The suspension has to withstand not only vertical wheel loading but other external forces resulting from driving events, such as lateral forces

Elasto-Kinematics

in cornering or during a steering input, and braking or traction forces. The elastic response of the suspension to these forces has an essential influence on the driver's overall impressions about handling and driving safety.

In **Fig. 9.9**, the suspension shown in Fig. 9.8 is loaded with a lateral force $F_Q$ which acts towards the vehicle's centre-plane – i.e. as on an 'outer' wheel in cornering. The suspension may already be preloaded by the wheel-load force, and the elastic displacements caused by this may already be covered by the toe-in curve k which therefore serves as the basic curve of the following considerations.

In cross-section, Fig. 9.9a, $F_Q$ causes a compression of the lower triangular link and an extension of the upper one, swivelling (contrary to Fig. 9.8) the wheel carrier by a 'positive' camber-change angle $\Delta\gamma_Q$ about a 'neutral' point $N_Q$. The centre of rotation $D_{Sp}$ of the outer track-rod joint is therefore displaced towards the inner side of the vehicle by $\Delta y_Q$. And although the tyre contact point is in the cross-sectional plane through the kingpin axis, a moment arises in the suspension caused by the pneumatic trail $n_R$ (see Chapter 4) which displaces the line of action of $F_Q$ behind the tyre contact point. This moment loads the track-rod with a tensile force and extends it (and its compliant joints) by an elastic deflection $\Delta l_{SpQ}$. The displacements $\Delta y_Q$ and $\Delta l_{SpQ}$ both lead to a toe-out angle $\Delta\delta_{vQ}$, Fig. 9.9d.

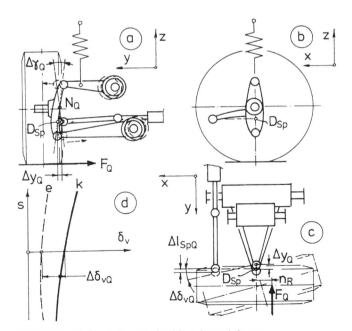

**Fig. 9.9** The suspension of Fig. 9.8 loaded with a lateral force

In the case of a front suspension, this toe-out angle – if within reasonable limits – would be advantageous, because it acts in the 'understeering' sense and softens the vehicle's reaction to a steering input. On a rear suspension, however, the toe-out angle would mean 'oversteering' and so is unfavourable.

It is easy to see in Fig. 9.9a that the share of the toe-out angle resulting from the lateral displacement $\Delta y_Q$ depends essentially on the height co-ordinate of the track-rod with respect to the neutral point $N_Q$. If the track-rod was above $N_Q$, the steering tendency would be reversed, and the wheel would assume a toe-in angle with the track-rod ahead of the axle (or toe-out with it behind the axle). Hence, if a front suspension is required to react in an understeering manner, a track-rod in the lower region of the suspension should be positioned *in front* of the axle and one in the upper region should be *behind* it. The latter situation often applies on front suspensions of front-drive cars having the steering box above the gearbox.

The pneumatic trail $n_R$ has a maximum value at a tyre sideslip force of zero, and it drops down to zero at the adhesion limits. For this reason the line of action of the lateral force $F_Q$ is displaced forward at the wheel with increasing lateral acceleration, and that by an elasto-kinematically remarkable travel of 25–40 mm on passenger cars. Consequently the moment about the kingpin axis in Fig. 9.9 decreases with increasing lateral acceleration, and hence the track-rod extension $\Delta l_{SpQ}$ also, and with this its share of the elastic steering angle $\Delta \delta_{vQ}$. This behaviour affects all compliant wheel suspensions, whether their elasto-kinematic attunement is good or bad: the elastic understeering tendency decreases on front suspensions with growing lateral acceleration and increases on rear suspensions. Accordingly the steering torque caused by the self-aligning moment of the tyres increases in a degressive manner on steerable front wheels. To ensure a 'restoring' torque at the steering wheel, rear wheels – if steered according to Fig. 7.38b, Chapter 7 – need a suspension with a 'negative' castor offset that is at least of the amount of the maximum pneumatic trail (for zero lateral acceleration) of the tyre; the steering torque will then increase in a progressive manner with lateral acceleration.

The triangular links of the suspension in Figs 9.8 and 9.9 are arranged above and below the wheel centre by roughly equal distances. They may be assumed to have equal longitudinal spring rates so as to provide good longitudinal compliance of the wheel. If the wheel is driven via a transverse shaft with universal joints, **Fig. 9.10**, the traction force $F_A$ acts on the wheel centre and thus loads the triangular links with equal longitudinal forces. In side view, the wheel carrier will then move forward by a shift $\Delta x_A$, and with it the kingpin axis and the centre of rotation $D_{Sp}$ of the outer

Elasto-Kinematics

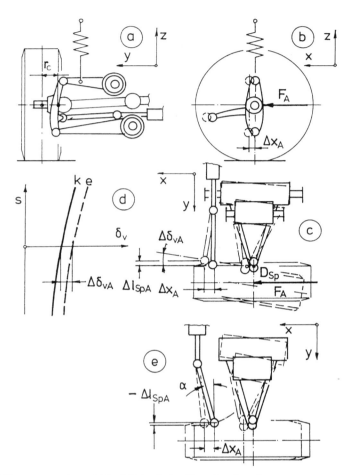

**Fig. 9.10** The suspension of Fig. 9.8 loaded with a traction force

track-rod joint. With the transverse position of the track-rod, as drawn, this situation will not cause a steering angle. However, the traction force $F_A$ acts on a lever arm $r_c$ (the 'wheel-centre offset' on a steerable suspension) and generates a moment about the 'kingpin' axis that leads to compression of the track-rod mountings and extension of those of the triangular links, with a resultant displacement $\Delta l_{SpA}$. The wheel thereby assumes a toe-in angle $\Delta \delta_{vA}$, Fig. 9.10c and d. The 'elastic' toe-in curve e will run roughly parallel to the 'kinematic' curve k.

A wheel that assumes toe-in under a traction force will swivel towards toe-out if the accelerator pedal is released. On an outer rear wheel this would be equivalent to an 'oversteering' angle and would for instance

256                                                                    Road Vehicle Suspensions

aggravate the vehicle's reaction to a sudden power change as might occur in cornering, especially in an emergency situation. For this reason it is preferable to set-up a rear suspension for elastic toe-out under a traction force and so to get an 'understeering' toe-in reaction with power change, at least at the outer wheel which is more important for cornering stability than the inner. On the suspension of Fig. 9.10, this effect can be achieved by positioning the track-rod with an angle of incidence $\alpha$ (see Fig. 9.10e) and compensating the lateral deflection $\Delta l_{SpA}$ by skew displacement $(-)\Delta l_{SpA} = \Delta x_A \tan\alpha$ of the outer track-rod joint. The angle $\alpha$ follows then from $\alpha = \mathrm{atn}(\Delta l_{SpA}/\Delta x_A)$.

This arrangement represents a typical solution of 'elasto-kinematic' attunement as it is applied today.

What has just been demonstrated for the track-rod is, of course, possible at any suspension link. A skew arrangement of two transverse links against each other will always generate steering angles if the suspension is longitudinally compliant.

There is one more typical case of an external force - braking by means of a brake mounted on the wheel carrier and acting on this by a force $F_B$ at the tyre contact point, as shown in **Fig. 9.11**.

Since the entire suspension is positioned above the road surface, and thus above the braking force, the lower triangular link is pulled rearward by

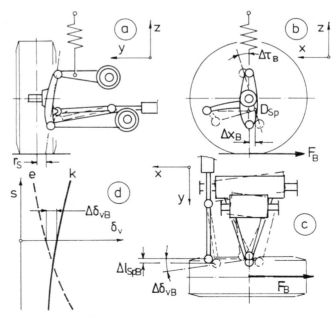

**Fig. 9.11** The suspension of Fig. 9.8 loaded with a braking force

a very large force while the upper one is pushed forward by a considerably lower force. The centre of rotation $D_{Sp}$ is displaced rearward by $\Delta x_B$ and, in addition, the wheel carrier is 'wound up' by an elastic castor-angle change $\Delta \tau_B$, Fig. 9.11b. The braking force $F_B$ acts on the lever arm $r_S$ (the 'scrub radius' on a steerable suspension) and generates a moment about the 'kingpin' which tries to lengthen the track-rod and its mountings by $\Delta l_{SpB}$, thus causing a toe-out angle $\Delta \delta_{vB}$, Fig. 9.11c. Moreover, and in contrast to Fig. 9.10, the elastic castor change $\Delta \tau_B$ leads to downward displacement of the outer track-rod joint, Fig. 9.11b, which thus assumes a considerably 'wrong' vertical coordinate in relation to the inboard joint and so to the previous kinematic attunement – see also Chapter 7, Fig. 7.30. Consequently the 'elastic' toe-in curve e is not only displaced with respect to the 'kinematic' curve k by $\Delta \delta_{vB}$ in the 'normal' position of the suspension, but is also inclined towards increasing toe-out with wheel bump (if the track-rod were, contrary to Fig. 9.11, to be sited behind the axle, its outer joint would be displaced upward, with the same result).

The gradient change of the toe-in curve could be welcomed because it is equivalent to an 'understeering' angle in cornering, at least on an outer front wheel (which has travelled against bump) and possibly on an outer rear wheel (which has travelled against rebound). However, the inclined toe-in curve e is valid also for straight-ahead driving where it is less desirable.

The toe-out angle $\Delta \delta_{vB}$ that occurs even in the normal position during a braking event is, in any case, disadvantageous on a rear wheel since it reinforces the yawing of the vehicle under a one-sided braking force and could, of course, be compensated by a measure already described for Fig. 9.10 – i.e. by a suitable angle of incidence of the track-rod, or of any transverse suspension link. However, because the angles of incidence of the suspension links can be chosen only once, and unequivocally, this measure may perhaps not be available.

Fortunately, we still have another kinematic measure that was not yet used in the foregoing: the elastic steering angles $\Delta \delta_{vA}$ (Fig. 9.10) and $\Delta \delta_{vB}$ (Fig. 9.11) depend on the moments caused by the traction and braking forces and the lever arms $r_c$ and $r_S$ (the 'wheel-centre offset' and 'scrub radius' on a steerable suspension). Hence, adequate choice of the lengths of these two lever arms allows us to fix the ratio of the moments so that both steering angles can be compensated by a single setting of the angles of incidence of the suspension links (**7**). Changing the radii $r_c$ and $r_S$ for elasto-kinematic reasons means, of course, changing the attitude of the 'kingpin' axis, too (which may cause problems with steerable suspensions in view of the requirements of steering geometry).

In reality, the elasto-kinematic effects discussed for Figs 9.8 to 9.11 in a simplifying manner occur in superimposition. The suspension chosen for our example was simplified for easier visualization insofar as the important joints of the triangular links lie in the cross-sectional plane through the tyre contact point. In general, this is not true for a steered front suspension because of its castor angle and castor offset, and even more so for a multi-link rear suspension. The principal effects and measures discussed before will then appear superimposed. To take this into consideration would needlessly have hindered the intended main investigations and their visualization.

It is quite clear that the deliberate elasto-kinematic design of a wheel suspension is easier the more links and joints are available. The primary reason for the increasing adoption of multi-link suspensions is the desire for well-balanced elasto-kinematic properties.

The example of a braked wheel in Fig. 9.11 showed that a longitudinally compliant suspension may lead to elastic wind-up of the wheel carrier. In this case, free choice of longitudinal compliance is not possible.

The wind-up can be reduced if only one link of the suspension is loaded with longitudinal forces, and therefore may alone be provided with longitudinally 'soft' joints.

With a double-wishbone suspension this is achievable by positioning the lower triangular link at approximately the same height as the wheel centre and, to ensure a sufficient length of base for the support of the braking torque, by positioning the upper link above the tyre tread, **Fig. 9.12**a. A further advantage of this layout is that the kingpin inclination can be chosen within desired limits and, with this, a limited wheel-centre offset and very small scrub radius. This is in fact the preferred form of today's double-wishbone suspensions. On a strut suspension, on the other hand (Fig. 9.12b), higher positioning of the triangular link would cause increasing bending moments at the piston rod of the damper; the considerable distance between the triangular link and the strut's upper mounting, however, allows sufficient longitudinal compliance of that link even if it is positioned nearer to the road.

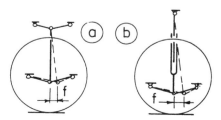

**Fig. 9.12**
Reduction of elastic 'wind-up' under braking force

# Elasto-Kinematics

Another method of escaping from the problem of elastic wind-up is to mount the whole suspension on a subframe that is itself compliantly connected to the vehicle body; the suspension can then be equipped with stiffer joints. Such a measure is quite common for driven rear suspensions, because a subframe is needed for insulation of the final-drive gear noise, and because the installation of the complete suspension and subframe assembly in the vehicle body, via three or four rubber mountings, makes production easier. However, such an assembly is subject to certain elastic steering angles under one-sided longitudinal forces, see Section 9.1.

Considering the relationships explained for Figs 9.8 to 9.12, it would certainly be advantageous if each suspension on one side of the vehicle could react sufficiently to longitudinal forces without influencing its fellow on the other side.

More freedom of elasto-kinematic harmonization is achievable, clearly, if the elastic wind-up angle ($\Delta\tau_B$ in Fig. 9.11b) can be avoided. This wind-up is caused by the longitudinally compliant reaction of the braking torque via two suspension joints one above the other (i.e. the joints of the triangular links in Figs 9.8 to 9.11). According to this, a suspension would be improved by having a single suspension link to react the braking torque, provided this component could be mounted with longitudinal compliance. This is one of the reasons for the adoption of trapezoidal-link suspensions in several variations – **Fig. 9.13** (see also Fig. 2.10 in Chapter 2).

In its simplest form, the suspension needs an additional rod link only, example (a). However, the turning joints of the trapezoidal link must have a relatively high resistance to angular deflection to avoid toe-in change under longitudinal forces (see also Fig. 9.6), and the desired longitudinal compliance can be achieved only by low axial spring rates of the joints, something that is difficult to achieve with normal rubber components. Some ('wrong') toe-in change – namely toe-out with braking or toe-in with traction – must be accepted here. Moreover, the simple system offers only a limited possibility of deliberate kinematic attunement because of its close similarity to 'spherical' mechanisms, see Chapter 3, Fig. 3.7.

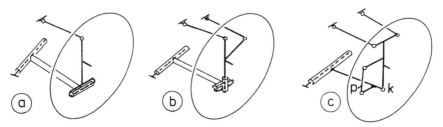

**Fig. 9.13** Trapezoidal-link suspensions

By replacing the wheel-side turning joint by a 'cardanic' connection, example (b), the wheel carrier gets an additional degree of freedom which allows rotation about a vertical axis, similar to the kingpin mounting of a 'knuckle' on a rigid axle. This degree of freedom has to be controlled by an additional (and second) rod link. Since the suspension now contains two rod links, it is possible to give them an angle of incidence (see Fig. 9.10) and thus to exert an influence on the elastic steering angles. Due to its more elaborate componentry, this suspension may show a true 'spatial' geometry and would, moreover, be suitable for steerable front wheels.

If the steering angle is of a minor amount, as on a rear suspension, the cardanic connection between the trapezoidal link and the wheel carrier can be simplified, replacing it by a ball joint k and a pendulum p - Fig. 9.13c. Choosing the spatial attitude of the latter offers an additional possibility of kinematic and elasto-kinematic attunement.

Similar effects to those with a trapezoidal link can be achieved by a 'wind-up-resisting' connection of the wheel carrier to the suspension links in side view, **Fig. 9.14**, as for instance by the longitudinal arm of a spherical 'double-wishbone' system (a). If the two longitudinal links of a multi-link suspension (b) are joined in a single chassis-side mounting (thus forming a triangular link), they may be replaced by a leaf or 'sword' of sheet metal that is flexible about the vertical axis. Another possibility is to support the two longitudinal links by a shackle with a rotatable mounting that allows longitudinal compliance of the wheel (c). All the solutions shown in Fig. 9.14 require some space in the vehicle and are therefore not always applicable, unlike those of Fig. 9.13.

The metallic components of the suspension and the subframes show compliance properties in the same range as the rubber joints'. Due to requirements of space, clearance and crash-testing, these parts cannot be designed for optimum stiffness in any case. Consequently their elastic deflections must be involved, too, in the elasto-kinematic harmonization of the suspension links and joints.

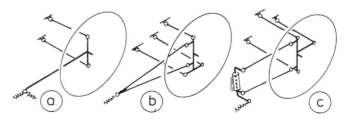

**Fig. 9.14** Wheel suspensions with good resistance to braking wind-up

## 9.2.2 Compliantly mounted subframes

If wheel suspensions are attached to subframes that are compliantly mounted to the vehicle body, one possible reason is that the suspensions themselves may not be suitable for optimum elasto-kinematic attunement; on the other hand, a subframe provides better noise insulation and facilitates handling of the axle assembly on the production line.

Elasto-kinematic harmonization of a subframe is more difficult on a front axle than a rear one because on the former steering-linkage requirements must be considered in addition. There are two principal cases.

In the first, the steering box is mounted on the subframe. Then the elasto-kinematic attunement of the subframe is carried out analogously to that for a rear axle. A one-sided longitudinal force will cause a steering angle of the axle which, unlike that of a rear axle, reinforces the vehicle's yaw reaction. Therefore the angular spring rate of the subframe about the vertical axis must be as high as possible, and this of course requires a wide-based arrangement of its mountings, taking up a lot of space in the vehicle.

In the second instance the steering box is attached to the vehicle body. Elastic displacement of the subframe under lateral forces will then cause elastic steering angles, track-rods in front of the axle leading to 'understeering' angles. Steering angles under longitudinal forces can be avoided by skilful arrangement of the track-rods - e.g. providing suitable angles of incidence (see also Fig. 9.3 and 9.9).

Because of the considerable space requirement in the front portion of the vehicle, compliantly mounted subframes for complete front suspensions are rare. Elasto-kinematic analysis of a front suspension *without* a subframe must, however, include the correct simulation of the elastic properties of the often extensively subdivided (and therefore compliant) structure of the vehicle's front end. Some aspects of subframe design are considered in what follows.

The five-link suspension of **Fig. 9.15** comprises three transverse links and a diagonal link connected to the subframe, while the longitudinal link is mounted immediately to the vehicle body. The subframe is designed like the arch of a bridge, perhaps because it carries a final-drive gearbox and so must provide clearance for the driveshaft - and maybe for the exhaust pipes too. In Fig. 9.15a, the two transverse links that control the toe-in angle (or the steering angle) of the wheel are positioned low-down in the suspension; they therefore load the subframe in its 'open' region where it is least stiff against lateral forces. A longitudinal force $F_L$ extends the front transverse link and compresses the rear one, thus bending the lower

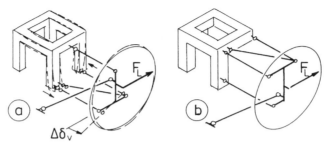

**Fig. 9.15** Compliance of the link-mounting points on a subframe

front portion of the subframe outward, and its rear portion inward. This causes an elastic 'toe-out' angle $\Delta\delta_V$. If the longitudinal link is mounted compliantly, the toe-out angle can be compensated by arranging the two lower transverse links to have an angle of incidence - see Fig. 9.10e. The lower the subframe's stiffness, the greater this angle of incidence must be, but the greater, too, the influence of the spring-rate tolerances of the suspension's rubber joints, and that of the elastic properties of the subframe. Good rubber joints show Shore hardness tolerances about $HS = \pm 3$ and consequently of about $\pm 10\%$ of their nominal spring rate. The metallic parts behave no better: the influence of sheet thickness tolerances of about $\pm 10\%$ on the flexural strength of profiles is linear on the webs of these sections and of the square order on the flanges, while the usual forging-die tolerance of around 2 mm causes (in view of the relatively small dimensions of forged profile sections) even greater variations of flexural strength.

It is, of course, very helpful if the design of a suspension allows elastic steering angles to be compensated by counteracting measures such as incidence angles of suspension links. However, in general it is certainly preferable to avoid such conflicting situations altogether.

Fig. 9.15b shows that the (basically unchanged) suspension can be preserved from elastic deflections of the subframe by interchanging the two lower links and the upper link, thus moving the two links that control the steering angle into the stiffest region of the subframe; this considerably reduces the angles of incidence for compensating the subframe's elastic deflection and makes the suspension much less sensitive to tolerances of the elastic properties.

As already mentioned, the elastic displacement of the point of action of an external force on such a complex spatial component as a normal subframe is not simply proportional to that force but depends on the combination of all the forces acting on the part.

Elasto-Kinematics

If there were a fixed proportionality, the spatial part could be replaced for calculation purposes by a system of virtual springs whose rates could be treated like those of the rubber joints of the suspension, and could consequently be regarded as acting in a 'serial arrangement' with the latter. This is true for the piston rod of the strut suspension shown in **Fig. 9.16**a. A transverse force F deflects the rubber mounting of the strut by a travel $f_G$ which follows from the force F and the spring rate $c_G$ of the rubber mounting as $f_G = F/c_G$. The same force acts, of course, on the end of the piston rod and exerts a bending moment that results in an effective spring rate $c_S$ and a deflection $f_S = F/c_S$ of the rod end. In practice, $c_S$ may throughout be lower than $c_G$! The resultant displacement of the point of action of F is therefore $f = f_G + f_S$, and the effective spring rate of this serial arrangement is $c = F/(f_G + f_S) = c_G c_S/(c_G + c_S)$ - see also Chapter 5, equation (5.26).

In contrast to the piston rod of the strut, however, the subframe in Fig. 9.16b and c shows different reactions depending on the combination of the external forces. The subframe may carry the trapezoidal links of two wheel suspensions. In case (b) both wheels are symmetrically loaded with longitudinal forces F which exert a constant bending moment at the medial transverse section of the subframe and linearly increasing moments at its longitudinal arms. The medial section will, therefore, assume a circular bending curvature, thus swivelling the connected ends of those arms by an angle $\alpha_1$, while the bending lines of the arms are parabolic. From the effective deflection $f_1^*$ of the free end of each longitudinal arm under the force

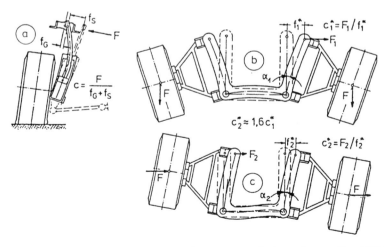

**Fig. 9.16** Effective spring rates for a complex component, depending on the combination of the external forces

$F_1$ of the front joint of the triangular link, a spring rate $c_1^* = F_1/f_1^*$ for the subframe can be derived. In Fig. 9.16c, however, both wheels are loaded with cornering forces F which act in the same direction, thus loading the free ends of the longitudinal arms with forces $F_2$. The curvature of the medial section of the subframe now becomes s-shaped and thus shows smaller deflection angles $\alpha_2$ at the corners to the longitudinal arms (see also the twice-supported leaf spring in Fig. 5.29, Chapter 5). Consequently the effective spring rate $c_2^* = F_2/f_2^*$ at the free ends of those arms is increased. If their length is the same as that of the medial section, and the cross-section of the subframe profile is constant, the relationship of the effective spring rates is $c_2/c_1 \approx 1.6$. Figs 9.16b and c showed a 'planar' frame; for a spatial one, the elastic behaviour will be more complicated and can be analysed only by calculation, for instance using finite-element methods.

While the foregoing considerations have dealt mainly with *avoiding* the consequences of elastic displacements of a subframe, it is worth finally describing a situation where these displacements can be used to advantage.

**Fig. 9.17** shows a rear suspension with driven wheels. The suspension consists of two longitudinal links one above the other, immediately connected to the vehicle body, and three transverse links, the upper two connected to a subframe in its stiffest region (and thus minimizing the influence of the subframe's compliance on the toe-in angle). The final-drive gearbox is mounted to the subframe, too.

When loaded with a traction force $F_A$ which exerts a moment $M_A$ at the final-drive gearbox, and consequently at the subframe too, the latter will

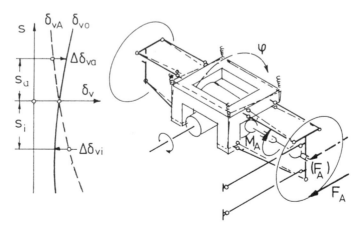

**Fig. 9.17** Stabilizing effect with power change in cornering

# Elasto-Kinematics

be inclined in its rubber mountings to the vehicle by an angle $\varphi$, and thereby will displace the joints of the front transverse links upward and those of the rear links downward. As the longitudinal links are immediately connected to the vehicle body, the wheel carrier does not take up the inclination of the subframe. This results, though, in a change of the relative attitude of the upper transverse links which is, on the other hand, of critical importance in determining the relationship between the toe-in curve and the wheel travel $\delta_{v0}(s)$. For this reason, the toe-in curve $\delta_{vA}(s)$ under driving torque is inclined with respect to the original curve $\delta_{v0}(s)$ – see also Chapter 7, Fig. 7.30 – and has a gradient that means 'oversteer' on a rear axle. The deviation increases with wheel travel s. With parallel wheel travel in straight-ahead driving, this causes a symmetrical toe-in change of the whole axle and normally is harmless though not useful.

In cornering, however, the outer wheel is displaced by a bump travel $s_a$ and the inner by a rebound travel $s_i$. If the driver suddenly releases the accelerator pedal (power change, see Chapter 7, Section 7.6), the subframe swivels back into its unladen position and both rear wheels revert to their original positions on the curve $\delta_{v0}(s)$; the outer wheel is steered by an angle $\Delta\delta_{va}$ in the toe-in direction, and the inner by an angle $\Delta\delta_{vi}$ in the toe-out direction. This means that both wheels on the axle counteract the natural oversteer reaction of the vehicle by assuming understeering angles. The amount of these angles depends on the wheel displacements $s_a$ and $s_i$ (and, consequently, on the roll angle and the lateral acceleration) and on the angular displacement $\varphi$ (and hence on the applied engine torque); in this way, the rear axle's elasto-kinematic reaction is best adapted to the power-change event.

Though having a detectable influence on a vehicle's driving behaviour, such steering angles take place merely over a few angular minutes. As with all measures on a 'passive' system, as represented by an elasto-kinematically optimized wheel suspension, achievable advantages must be weighed against possible functional disadvantages. In the case of influencing the power-change behaviour by the measures visualized by Fig. 9.17, the question may arise as to what will happen if the driver accelerates in cornering. The answer is that a driver always accelerates consciously and so knows about the yaw reaction of the vehicle when utilizing its high performance (which reaction is, of course, slightly increased by the measures described), while abrupt power withdrawal occurs especially in emergency situations where any aid to cornering stability is welcome.

In the foregoing, only quasi-static elasto-kinematic events have been considered. However, a compliant wheel suspension represents a vibrating system too, and that (in view of the limited displacements and consequently very progressive spring characteristics of the joints) a non-linear system with a broad spectrum of natural frequencies. An intrinsic and important problem is to avoid resonances with other natural frequencies, as for instance the vertical wheel vibration on the tyre whereby - among other phenomena - stick-slip braking effects may be excited.

This section should give the reader an idea of the manifold elasto-kinematic effects, problems and possible solutions that are available during the development of a wheel suspension. Conscious development of a multi-link suspension requires knowledge about kinematic and elasto-kinematic relationships and, even in this age of 'computer-aided design', a good spatial imagination - not only in the designer but also in the testing engineer and the calculation specialist. Successful suspension design is impossible without close cooperation between these groups, and this mutual interdependence has made work on modern suspension systems notably arduous but also rewarding.

## 9.3 Statically over-constrained systems

Occasionally, rigid-axle suspensions represent statically over-constrained mechanisms - e.g. showing five rod links instead of the sufficient four. This feature may be chosen for a better installation, perhaps to avoid a single longitudinal link in the medial plane of the vehicle beneath the luggage compartment. Another reason might be to avoid torques on the axle tube which could load the final-drive housing with elastic deflection and thus cause oil leakage.

**Fig. 9.18**a shows a rigid-axle suspension with four longitudinal links and one transverse link or 'Panhard rod'. In side view, the longitudinal links are inclined towards each other to provide a 'longitudinal pole' L for braking anti-lift or traction anti-squat properties. With antimetric wheel travel - e.g. in cornering - each side of the axle would try to swivel about its respective pole L. This, however, would require a distortion of the axle tube. The tube carries torsional stress, and the rubber joints of the suspension links are mutually constrained, so the work of deformation must be effected by the rolling moment of the vehicle, and the axle suspension therefore acts like a stabilizer spring or anti-roll bar.

These constraining forces are largely avoidable if the longitudinal links are of equal length and parallel in the normal position of the vehicle, and arranged in pairs, one above the other.

If the links are parallel in the normal position but of unequal length, as in Fig. 9.18b, their suitable arrangement may also avoid too much constraint:

Elasto-Kinematics 267

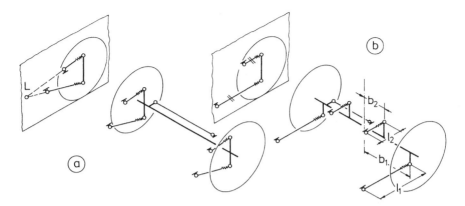

**Fig. 9.18** Statically over-constrained rigid-axle suspensions

with the vehicle rolling, the swivel angles of the links are in side view proportional to the roll angle $\varphi$ and to the quotients $b_1/l_1$ and $b_2/l_2$ of their distances from the medial plane and their lengths, and the longitudinal displacements of their axle-side joints are proportional to their lengths and to the square of the swivel angle. Equal longitudinal displacement of all axle-side joints – and thus, pure longitudinal translation (and minimum distortion) of the axle body with the vehicle rolling – is achieved for equal quotients of the squares of the distances b and the lengths l for all links: $b_1^2/l_1 = b_2^2/l_2$.

Over-constrained suspensions require particularly good maintenance of manufacturing tolerances to prevent pre-constraint even in the normal position.

On an over-constrained suspension, kinematic 'characteristics' such as a support angle $\varepsilon^*$ or an instantaneous roll axis $m_W$ follow from interaction of its kinematic and elasto-kinematic properties, and from application of strictly vertical (virtual) parallel or antimetric forces at the tyre contact points that serve only to move the mechanism; serious errors would be caused by superimposing braking, traction or cornering forces (see also the considerations in Chapter 6, Section 6.3.8!).

# Chapter 10

# Synthesis and Design

## 10.1 General remarks

For the choice of the suspension type and its kinematic harmonization, widely different criteria may be important – as for instance the purpose of the vehicle, its class, size and price, the manufacturer's tradition and corporate identity, experience with preceding models, further development of major assemblies of those models, available production facilities, modular concepts (i.e. component sharing on different models), manufacturing costs, possibility of quality control, reliability and cost of ownership. Last but not least come *new* requirements that can no longer be met by suspension systems used hitherto.

The customer today expects good driving behaviour of his new vehicle; however, he will not in any case be aware of money spent on the chassis, since the evidence of it is normally hidden, unlike the exterior and the internal equipment of the vehicle. Hence, it is not easy for a chassis designer to carry through new solutions that require fresh investment, and probably increased cost. Moreover, the chassis designer's intentions almost always conflict with those of other development areas such as body design (passenger compartment, luggage compartment, fuel system, ease of entry/exit and spare-wheel stowage) and drive-train development (engine, transmission and exhaust system). Willing cooperation of all design and research functions and the ability for convincing argument and accepting compromise is therefore a prerequisite for successful design work!

Every road vehicle has at least one front and one rear axle. The latter is as important as the former, or – due to its independence of the steering system and thus of any influencing by the driver – perhaps even more so: the occasionally heard opinion that the type of rear axle is immaterial since it simply follows the front axle is nothing but a joke. Both axles (together with the springing and damping system, and with such essential properties as dimensions, masses and moments of inertia), influence the driving behaviour and must be skilfully adapted to each other. This, however, is less a matter of the suitable choice of the suspension type than of fixing the kinematic and elasto-kinematic properties – as for example the roll centre, bump-steer behaviour, braking and traction anti-pitch measures and steering geometry.

The kinematics of a rigid-axle suspension are normally easy to consider and to determine, even by drawing. A great variety of compound suspensions is on offer, so there is little point in giving advice for their design (except the 'H-frame' layout - see Chapter 7, Figs 7.25 and 7.36). The following considerations are therefore restricted mainly to independent wheel suspensions.

## 10.2 Planar wheel suspensions

Planar mechanisms are today rarely found on multi-link independent wheel suspensions; the double-wishbone suspensions of former times, however, were almost always planar mechanism chains (where the axes of all joints run parallel): metallic turning joints resistant to any 'coning' deflection of their axes were used even at the wheel side, and the wheel carrier was not also the 'coupler' of the kinematic chain but was connected to the coupler by a real kingpin. Not until after reliable ball-joints became available was the wheel carrier directly connected to the suspension links.

With the axes of all joints arranged in the straight-ahead direction, only the wheel motion in the transverse direction can be influenced, **Fig. 10.1**. From the desired roll-centre height $h_{RZ}$ and the desired camber change against wheel travel $d\gamma/ds$ (which is equal to the reciprocal of the pole distance q) we get the position of the 'lateral' pole Q. With given joints 1 and 2 at the wheel carrier, the pole Q defines the lines of action of the transverse links. The distance q' of the centre of curvature of the path of the tyre contact point A follows from the desired roll-centre height change against wheel travel $dh_{RZ}/ds$ by the condition

$$dh_{RZ}/ds = - (b/2)/q' \qquad (10.1)$$

With q', the centre of curvature A' of the path of the tyre contact point A can be determined on the polar ray A-Q. If one of the inner joints of the transverse links is given, here for example the joint 1', Bobillier's method (see Chapter 3, Fig. 3.2) locates the inner joint 2' of the other link. On a steerable suspension, the attitude of the track-rod also has to be determined by this method. Since the circular path of the outer track-rod joint deviates from the theoretical path of the relevant point at the wheel carrier with increasing wheel travel even on a planar suspension, toe-in change will occur with that travel (see Chapter 7, Fig. 7.30). For this reason, any steerable suspension is, strictly speaking, a 'spatial' mechanism (except for special types like the Dubonnet system mentioned earlier).

# Synthesis and Design

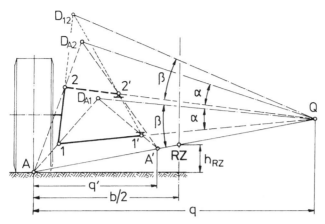

**Fig. 10.1** Planar double-wishbone suspension

The introduction of double-wishbone suspensions at the beginning of the 1930s was, besides other reasons (like weight and space-saving), caused by investigations into the 'shimmy' vibration of steerable wheels on rigid axles; these studies revealed that camber and track changes with wheel travel were disadvantageous because of gyroscopic coupling of the moments of inertia, and a roughly vertical and rectilinear path of the tyre contact point appeared to be desirable. This means, of course, that the roll-centre height remains constant with wheel travel. In the parallel position of the transverse links, **Fig. 10.2**, Bobillier's method leads to the condition that the distance e of the polar ray A-Q from one of the links must be equal to the distance of the links' relative pole $D_{12}$ from the other link. With parallel transverse links, the camber change with wheel travel $d\gamma/ds$ is instantaneously zero. This is not necessarily desired in the normal position of the suspension. However, Fig. 10.2 provides the rule - as a bit of advice for designing spatial suspensions too - that a rectilinear path of the tyre contact

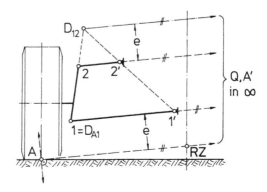

**Fig. 10.2**
Rectilinear path of
the tyre contact point

point is almost always achievable if the ratio of the lengths of the links is reciprocal to the ratio of their distances from the polar ray A-Q (or from the road surface).

The roll-centre height change with wheel travel is obviously determined by the ratio of the lengths of the transverse links. Normally the upper link is made shorter than the lower, something that also saves space in the cross-section of the vehicle.

If the upper link is considerably shorter than the lower, the camber change with wheel travel $d\gamma/ds$ is non-linear and highly progressive, **Fig. 10.3**. Progressive camber increase with bump is frequently demanded by test engineers in order to achieve some increase in the lateral camber force at the outer wheel in cornering while retaining an acceptable camber angle for the stationary fully loaded vehicle, and thus to delay the progressive increase of the side-slip angle. Owing to this demand it is possible to overlook the fact that certain suspensions with these properties (e.g. the double-wishbone layout) show a small roll-centre height change with wheel travel (see the considerations about Fig. 10.2) and thus promote 'jacking-up' in cornering (see Chapter 7, Figs 7.11, 7.12 and 7.20); the desired high negative camber at the outer wheel may therefore be achieved rather late in the travel or not at all, so the measure may prove useless or even disadvantageous. Moreover, several suspension types with progressive camber change - again such as the double-wishbone suspension - generate a progressive 'negative' camber with respect to the vehicle coordinate system at the (rebound) inner wheel too; this is superimposed on the roll angle of the vehicle and acts as 'positive' camber on the road, forcing the inner tyre to ride on its shoulder instead of the tread.

The preceding considerations about planar suspensions were made in order to emphasize several features of the suspension parameters of transverse dynamics that are analogously valid for spatial suspensions, too.

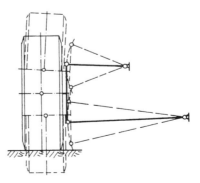

**Fig. 10.3**
Typical camber change with wheel travel on an unequal-length double-wishbone suspension

Synthesis and Design                                                                 273

For non-driven wheels the traction-force support angle is of no interest, so spherical or planar suspension mechanisms will be sufficient. Spatial mechanisms may at best be justified by elasto-kinematic requirements.

Since planar mechanisms are special cases of spherical mechanisms, they have the same level of kinematic potential. By a spatially skewed arrangement of the (though still parallel) joint axes, too, any planar wheel suspension can be provided with non-linear movement in all three coordinate planes of the vehicle; in contrast to spatial suspensions, of course, there is no free and independent choice of kinematic properties in each of the planes.

If spherical or planar suspensions are applied to driven wheels, a compromise is necessary between the five suspension 'characteristics' - roll centre, bump steer, camber change, braking-force support angle and traction-force support angle (the amount of which is normally equal to the wheel-travel angle, too) - and mostly it is the traction-force support angle that must be sacrificed.

## 10.3 Spatial wheel suspensions

In the foregoing chapter it was already made clear that the design of a modern wheel suspension is carried out consciously in respect of its elasto-kinematic function and, of course, the self-evident optimizing of its kinematic properties. This statement does not in any way reduce the importance of kinematic synthesis; however, it may be recalled that an infinite number of variants is imaginable to meet desired kinematic properties, but that only a small number of them is suitable also for elasto-kinematic attunement. For example, a suspension similar to that of Fig. 9.4c, Chapter 9, can readily be dimensioned for characteristics of longitudinal or lateral geometry, if regarded as a rigid system, but any attempt to avoid elastic steering angles in cornering will be useless since all the transverse links are arranged on one side of the lateral-force vector.

Consequently, consideration must be given right at the start of a suspension development to its ability to meet the elasto-kinematic requirements with adequately dimensioned rubber joints and link positions. The planning process for a new wheel suspension therefore occurs in practice by alternating iteration of kinematic synthesis or analysis plus elasto-kinematic investigations, while at least the standard requirements for wheel load, cornering force, braking and perhaps traction force must be tested.

The design of a wheel suspension has, of course, to take account of additional conditions such as the space available in the vehicle. Suspension joints – especially the rubber variety – require space too. The design of the latter depends on decisions about assembly, as will be discussed later, and the well-timed estimation of their approximate dimensions (allowing freedom for the almost inevitable modifications) is an essential precondition for effective design work.

It is, of course, very helpful if a proved preceding suspension of the same type can be used as the basis of a new design.

In the rather improbable situation of there being no background at all for making a first attempt at the arrangement of the suspension links for a desired kinematic function, the following method is based on the theorem of 'instantaneous screw' as discussed in Chapter 3. It is applicable at least to suspension types having links that immediately connect the wheel carrier with the vehicle body or possibly a subframe, and do not incorporate intermediate couplers as shown in the layout in Fig. 2.13b, Chapter 2.

The term 'instantaneous screw' describes the state of motion of a spatial body (here the wheel carrier of a suspension) and is defined by the vectors of the angular velocity $\boldsymbol{\omega}_K$ of the wheel carrier and of the velocity of a reference point on it. From this state of motion can be derived the kinematic 'characteristics' of the suspension, as for example the roll centre, as shown in Chapters 5 to 8. By the same token, of course, the instantaneous screw of a planned suspension can be derived from the desired values of its characteristics in a certain position (e. g. the 'normal' straight-ahead position). As most of the characteristics refer to the tyre contact point A, this point – or, better, the point on the wheel carrier that at the moment coincides with A – is the most suitable one to be chosen as the reference point. According to the practice used in this book, its (virtual) velocity vector has consequently to be named $\mathbf{v}_A^*$, thus demonstrating its close affinity with the definitions of the suspension characteristics.

From the six components of the vectors $\boldsymbol{\omega}_K$ and $\mathbf{v}_A^*$, one can be given – for instance the vertical wheel velocity $v_{Az}^*$ (e. g. $v_{Az}^* = 1$). The five remaining components will then be referred to this. With the simplifying presumption that the wheel camber angle is really negligible in the normal position, the wheel centre M will lie in the same side-view plane as the tyre contact point A. Then the vertical velocity components of both points are equal: $v_{Az}^* \approx v_{Mz}$. The x- and y-components of $\mathbf{v}_A^*$ follow from the equations (6.7) and (7.15), and from the desired support angle $\varepsilon^*$ and the desired roll-centre height $h_{RZ}$ as

$$v_{Ax}^* = \pm v_{Az}^* \tan \varepsilon^* \qquad (10.2a)$$

$$v_{Ay}^* = v_{Az}^* h_{RZ}/y_A \qquad (10.2b)$$

Synthesis and Design

(the upper sign is valid for front wheels), and the x-component of the velocity $v_M$ – according to equation (5.63), with the abovementioned assumption $v_{Mz} \approx v_{Az}^*$, and with the desired wheel-travel angle $\varepsilon$ – is

$$v_{Mx} = -v_{Az}^* \tan\varepsilon \tag{10.3}$$

As shown in Chapter 6, the wheel-travel angle $\varepsilon$ is equal to the support angle $\varepsilon^{**}$ on a suspension having its wheel driven by a transverse driveshaft with universal joints, provided there is no hub-reduction gear; then $v_{Mx}$ may also be defined by the support angle $\varepsilon^{**}$.

From $v_{Mx}$, $v_{Ax}^*$ and the tyre radius R we get the y-component of the angular wheel-carrier velocity $\omega_K$, **Fig. 10.4**, as $\omega_{Ky} = -(v_{Ax}^* - v_{Mx})/R$ or

$$\omega_{Ky} = v_{Az}^*(\mp \tan\varepsilon^* - \tan\varepsilon)/R \tag{10.4a}$$

and $\quad \omega_{Ky} = v_{Az}^*(\mp \tan\varepsilon^* \pm \tan\varepsilon^{**})/R \tag{10.4b}$

The components $\omega_{Kx}$ and $\omega_{Kz}$ are derivable from the desired camber change against wheel travel $d\gamma/ds$ and the bump-steer gradient $d\delta/ds$. If in equation (7.1), Chapter 7, the term $\omega_\gamma$ is modified to $\omega_\gamma = v_{Az}^*(d\gamma/ds)$, and $\omega_{Ky}$ is replaced according to equation (10.4), the desired toe-in angle $\delta_v = -\delta$ leads to

$$\omega_{Kx} = -(v_{Az}^*/\cos\delta)[(\sin\delta/R)(\mp\tan\varepsilon^* - \tan\varepsilon) + d\gamma/ds] \tag{10.5}$$

Accordingly, in equation (7.3) the term $\omega_\delta$ is modified to $\omega_\delta = v_{Az}^*(d\delta/ds)$, and the terms $\omega_{Kx}$ and $\omega_{Ky}$ are replaced according to the equations (10.5) and (10.4); suitable treatment of the angular functions leads at the end to

$$\omega_{Kz} = v_{Az}^*[d\delta/ds - (d\gamma/ds)\tan\gamma\tan\delta - (1/R)(\mp\tan\varepsilon^* - \tan\varepsilon)(\tan\gamma/\cos\delta)] \tag{10.6}$$

Now the state of motion of the wheel carrier is known, the velocity vector $v_i = v_A^* + \omega_K \times r_i$ of any point i on the wheel carrier can be determined, with $r_i$ as the connecting vector from the tyre contact point A to the point i, whose components are $r_{ix} = (x_i - x_A)$ etc. The components of $v_i$ are as follows:

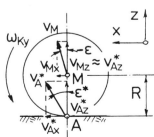

**Fig. 10.4**
Angular wheel-carrier velocity about the y-axis

$$v_{ix} = v_{Ax}^* + \omega_{Ky}(z_i - z_A) - \omega_{Kz}(y_i - y_A) \qquad (10.7a)$$
$$v_{iy} = v_{Ay}^* + \omega_{Kz}(x_i - x_A) - \omega_{Kx}(z_i - z_A) \qquad (10.7b)$$
$$v_{iz} = v_{Az}^* + \omega_{Kx}(y_i - y_A) - \omega_{Ky}(x_i - x_A) \qquad (10.7c)$$

A rod link or triangular link connected to the point i, the position of which conforms to the state of motion of the wheel carrier, must lie in the normal plane of the vector $\mathbf{v}_i$ – **Fig. 10.5**. The intersection lines $g_{xz}$ and $g_{yz}$ of this plane in the side-view plane and in the cross-sectional plane through i are rectangular to the respective projections of the vector $\mathbf{v}_i$.

If the x- and y-coordinates of a wanted second point n on the link are given, its z-coordinate can be determined by combining the elements of the lines $g_{xz}$ and $g_{yz}$, as indicated in Fig. 10.5, and the line i–n is a possible rod link (or arm of a triangular link) that meets the kinematic demand of the suspension. The gradients of the lines $g_{xz}$ and $g_{yz}$ are defined by the components of the vector $\mathbf{v}_i$. According to the graphical solution in Fig. 10.5, the z-coordinate of the wanted point n can be calculated by

$$z_n = z_i - (x_n - x_i)(v_{ix}/v_{iz}) - (y_n - y_i)(v_{iy}/v_{iz}) \qquad (10.8)$$

This method leads to possible **attitudes** of links that meet the given kinematic conditions in the given position of the suspension - e.g. to track-rod attitudes for zero gradient of toe-in change against wheel travel. Optimizing the **lengths** of the links (which is important for what happens to the suspension characteristics with increasing wheel travel) has, however, to be done by iterative calculation on a computer, and may for instance require changes of joint positions for elasto-kinematic or - more down-to-earth - clearance reasons.

Proposals have been made to set out the characteristics of a suspension in the normal position *and* their courses with wheel travel, so as to get a complete solution in one go. However, because of the always tight space

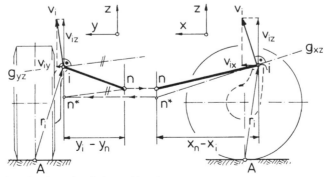

**Fig. 10.5** Determination of a link position for a given instantaneous screw

# Synthesis and Design

limitations, suspension design is in reality simply a hunt for millimetres, and it is doubtful if a computer program could propose favourable link positions. On the other hand, the calculation time required for programs that apply, for instance, to the methods given in Chapters 3 to 9 has fallen from a lot of minutes to hundredths of a second, since the era of punched-card-fed computers, and therefore no longer impedes iterative calculations.

In the course of kinematic and elasto-kinematic optimizing (which occur simultaneously, as mentioned earlier), a lot of geometrical changes will be required which become progressively smaller and eventually reveal the definitive form of the suspension. Often it will turn out during design work on the suspension elements that a minor change of a joint or link position will allow simpler or cheaper construction (e.g. by the standardization of brackets, replacement of deep-drawn parts by folded ones, etc.). In such a case, a new pass through kinematic and elasto-kinematic analyses may be worthwhile, having regard to the long time a new suspension system will be in series production.

Since the designer is responsible for the correct dimensioning of the suspension parts, it is very useful - even at the beginning of the planning phase - to carry out at least estimating calculations of strength or stiffness in order to become familiar with the dimensions of the parts and to provide suitable clearances. Changes that are found necessary at an advanced stage of the development process may force undesirable and expensive remedies for reasons of available space.

Many suspension components are standardized for physical reasons or to facilitate mass production: the weight and volume of a spring are determined by the required resilience energy, and they will increase if the need arises for 'softer' springing. The length of a damper follows from the required stroke and from certain dimensions fixed by the manufacturer. The steering box and steering linkage always compete for space with other assemblies such as the engine and gearbox. On the other hand, the spatial attitude of a track-rod is critical for the toe-in curve and the steering geometry (see Chapter 8). A power-assisted rack-and-pinion steering system with the gearing and power sections in line requires a considerable length of rack and leads to rather short track-rods (and, consequently, to short suspension links too). Far-sighted estimation of the required space and observance of suppliers' standards, and the notification of expected problems to all departments involved in the project, will promote smooth and cost-saving development.

## 10.4 Design considerations

With multi-link suspensions it is usual to fix certain parameters by adjustment at the end of the production line, allowing economically justifiable tolerances to be met.

Normally the toe-in angle $\delta_V$ and often the wheel camber angle $\gamma$ are adjusted in a defined vehicle position (e.g. in the 'normal' position but, bearing in mind the unladen state of the vehicle at the end of the production line, perhaps in the 'empty' position, too). It is important to consider the possibilities of adjustment as early as the *planning* phase of suspension development. Depending on the suspension type, only certain mounting positions come into question for the measures involved.

The suspension in **Fig. 10.6**, with three transverse links, is very well adapted for adjustment of the toe-in and camber angles, since the two lower links are at much the same height and the upper link is positioned directly above one of the lower links. With adjustment of the camber angle $\gamma$ by means of the upper link, the wheel swivels about the axis $a_1$ through the joints of the lower links, almost without causing any toe-in change, while adjustment of the toe-in angle $\delta_V$ by means of the front lower link swivels the wheel about the axis $a_2$ and has no influence on the camber angle. Adjustment by means of the rear lower link would swivel the wheel about $a_3$ thus leading to simultaneous toe-in and camber changes without the practicability of influencing either of these values separately.

As the tolerance of the wheel camber angle $\gamma$ is usually about 10-30 angular minutes, and that of the toe-in angle $\delta_V$ about 2-10 minutes, generally the camber angle has to be adjusted first and the toe-in angle second.

Obviously, the need for adjustment may pose additional conditions for the choice of link attitudes in the suspension layout.

On suspensions with triangular links, the attitudes of the axes of the revolute joints can be used to achieve far-reaching kinematic effects and, at the same time, to save space, as visualized for a front suspension in

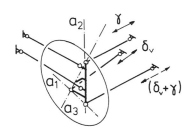

**Fig. 10.6**
Suitable positions for the adjustment of the camber angle $\gamma$ and the toe-in angle $\delta_V$

Synthesis and Design 279

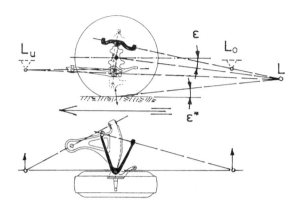

**Fig. 10.7** Arrangement of the triangular links of the front suspension of the Mercedes-Benz type '600' (1963)  (courtesy Daimler-Benz AG)

**Fig. 10.7**: the vehicle-side axis of rotation of the lower link intersects with the side-view plane through its ball-joint at the wheel carrier at a point $L_u$ which can be regarded as the 'pole' of the path of the ball-joint in side view. The same applies for the upper link and the point $L_o$. As the pole $L_o$ coincides in side view with the extension of the axis of rotation of the upper link, it is exactly the centre of curvature of the elliptical path of the upper ball-joint. The wheel suspension acts in side view like one with 'virtual' longitudinal links that are connected to the vehicle body at the points $L_u$ and $L_o$, and the intersecting point L of these virtual links is the suspension's 'longitudinal pole' which defines the wheel-travel angle $\varepsilon$ and the support angle $\varepsilon^*$ (neglecting, of course, that L is in the vertical plane through the ball-joints and not in the medial plane of the wheel!).

In former times, such considerations were essential means for suspension design by drawing, and may even today serve for improved appreciation of the spatial effects of triangular links.

It has already been mentioned that estimation of the required properties and dimensions of joints, and especially rubber joints, is very useful at the beginning of the suspension development, in order to avoid undesirable surprises during the design process. Rubber joints on suspension links are mostly of the cylindrical type because of its good ability to react angular deflections; there are numerous variants in respect of complexity, cost, function and means of assembly.

The simplest manner of making a cylindrical rubber joint is to vulcanize the rubber to the inner bush and to press it into a bore, using a slip additive

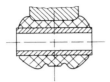

**Fig. 10.8**
Simple cylindrical rubber joint
pressed into a bore
(lower section, before mounting)

**Fig. 10.9**
Cylindrical rubber joint
with stop-collar

that evaporates within a short time - **Fig. 10.8**. The shape of the rubber as supplied (lower section) provides optimum preload compression of the rubber after installation (upper section). It is obvious that such a joint shows quite large tolerances regarding the spring rate and permissible angular deflection. When excessively loaded in torsion, the joint counters by sliding in the bore, with consequent distortion and possibly changed spring rates. This low-cost type is therefore rarely found today in significant suspension designs.

Cylindrical rubber joints for high duty always feature an additional outer bush, and the rubber is vulcanized to both it and the inner bush.

If a rubber joint has to react considerable axial forces in addition to radial forces, an axial stop-collar may be provided as in **Fig. 10.9**. The axial spring characteristic depends on the shape of the collar and the washer AF, but with substantial tolerances. In combined torsional and axial loading, the rubber of the collar may slide on the washer with risk of wear. It must be appreciated, too, that the axial creep of cylindrical rubber joints is more extensive than radial creeping. For this reason, in the case of steady axial load and if axial creeping would damage the long-term quality of the suspension (e. g. because of toe-in change resulting from axial deflection), it is better to reject this type of rubber joint in favour of a true ball-and-socket joint, **Fig. 10.10** (left side), or at least a rubber ball-joint (right side).

The dimensions of a rubber joint are essentially influenced by the decision of whether its inner bush shall be mounted in double-shear or single-shear

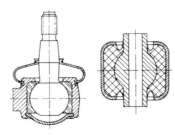

**Fig. 10.10**
Ball-and-socket joint (left)
and rubber ball-joint

Synthesis and Design                                                                 281

('floating'). This depends on the necessary design properties of the relevant parts - something that will be discussed later.

Double-shear mounting allows a thinner bolt, **Fig. 10.11**a. The required preload of the bolt is equal to at least half the radial force acting on the joint, divided by the friction coefficient. If considerable torsional angles occur, additional preload is necessary for the frictional support of the rubber torque. To ensure correct pressure at the faces of the inner bush, the bracket must have at least one flexible flange. The diameter of the bolt determines the inner diameter of the inner bush, while the permissible tension of the bush material and the preload force determine its outer diameter. The dimensions of the outer bush follow from the outer diameter of the inner bush and the desired radial and torsional spring rates.

Fig. 10.11b shows a simple variant of the cylindrical rubber joint, where the ends of the inner tube are compressed and form flanges for mounting by radially arranged bolts; this variant needs no flexible flanges on a bracket.

A single-shear or floating mounting, Fig. 10.11c, requires a thicker bolt not only because the preload force acts on one contact surface only, but also because the radial force at the joint acts on a lever of half the length of the inner bush and generates a moment that tries to tilt the bush off the contact surface and may cause gaping (with danger of corrosion or even bolt failure). Consequently, the inner bush needs a greater diameter, as does the outer bush also, but the length of the joint can of course be smaller. If failure of the rubber can lead to a total separation of the inner

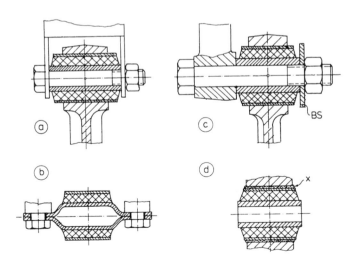

**Fig. 10.11** Assembly possibilities of cylindrical rubber joints

and outer bushes, a washer BS is necessary with a diameter greater than that of the outer bush.

On cylindrical rubber joints for mass production, normally the outer bush is provided with a thin rubber coating, x in Fig. 10.11d. This coating simplifies the equalization of tolerances and helps to avoid contact-corrosion of the metal components.

For easy assembly of a double-shear joint connection, the bracket needs a clear span that exceeds the greatest possible length of the inner bush, depending on the latter's tolerances. In practice this may require deflections of 0.5 mm or even more. Sheet flanges do not normally cause problems, but forged or cast parts will resist the establishment of the preload force owing to their greater stiffness. Because a double-shear connection requires parallel surfaces and some compliance of the flanges, forged or cast parts usually have to be machined, **Fig. 10.12**. At least one of the flanges must be milled down to a sheet-like thickness to allow correct bedding of the connection, Fig. 10.12a. To avoid fracture, the milled surface needs 'transition' radii towards the unmachined surface. In heavy-duty vehicles, the problem is occasionally bypassed by creating one 'fixed' and one 'floating' connection – see Fig. 10.12b. The fitted bush PB allows – at least in the new state – correct application of the bolt's preload force.

Both these solutions are better avoided on wheel suspensions for large production volumes, and a single-shear connection similar to that in Fig. 10.11c may be preferable.

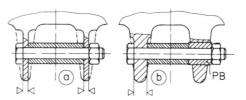

**Fig. 10.12**
Double-shear connections on forged or cast parts

Double-shear connection may be appropriate, though, for a suspension link made by an 'open' U-section beam which perhaps supports a coil spring, **Fig. 10.13**. Then the mating component, e.g. the wheel carrier, must contain the bore for the outer bush of the joint, the axis of which is (in this instance) disadvantageously in the same direction as the parting line FT of the die. Since this line runs across the bore, milling of the faces may prove necessary as well as of the bore which, of course, is produced by machining anyway. If the wheel carrier is cast, some of the machining can be avoided by using a mould core. Should the inner end of the link be made in the same manner as the outer (as shown in the drawing), this will require at the

Synthesis and Design

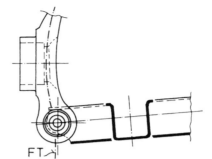

**Fig. 10.13**
Simple suspension link and a more expensive mating component

vehicle body or subframe a bracket containing a bush in a bore which perhaps also has to be machined – even less desirable than for the bore in the wheel carrier. Such a link should then have a tube welded to its inner end so as to permit a chassis-side bracket with two flanges.

These considerations about Fig. 10.13 serve to remind us that it may be unhelpful to design one part as simply as possible while swinging extra cost on to the mating component.

Suspension links like that of Fig. 10.13 often have to sustain additional loading: due to the spatial movement of the suspension, angular deflections may occur in any direction, and the axes of the joints at the ends of a link may be distorted relative to each other, leading to 'coning' deflection angles at the joints (see Chapter 5, Fig. 5.41). The moments caused by the coning angles load the link in torsion, **Fig. 10.14**. A link with an 'open' section has little resistance to torsion and reacts by distorting its cross-section, and the contact areas of the inner bush of a joint are deflected against each other by a (minute) angle $\alpha$, Fig. 10.14a. The deflection stresses the frictional connection of the link and the bush, and may lead to micro-slip with the

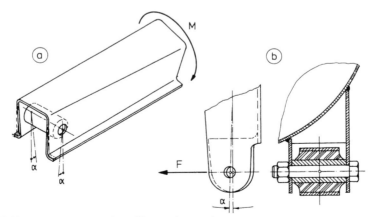

**Fig. 10.14** Warping of 'open' profiles under torsion

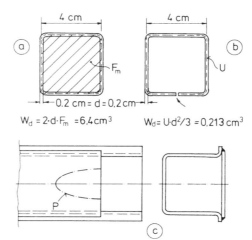

**Fig. 10.15**
Cross-section parameters of hollow and open profiles

danger of corrosion and/or gradual loosening of the connection. Therefore, such an open-section link is better closed to a hollow one with high torsional stiffness – at least in the region where no joints have to be mounted. Another possibility is to relieve the link of torsion by incorporating a proper ball-and-socket joint at one end.

A similar problem may arise on a bracket if its flanges have different resistances against a force F at the joint, Fig. 10.14b. In the drawing, the right flange is longer than the left and, consequently, more compliant to the force. This leads to a mutual distortion of the contact areas at the bush by an angle $\alpha$. Suitable design or reinforcement measures are therefore necessary to ensure equal deflection of both flanges.

On a suspension link, hollow sections are possible only in those regions where no joints are to be attached in double-shear. Due to the considerable difference between the section moduli of open and hollow profiles, the transient region between the two types of section must withstand a rapid change of tension loading. Tension peaks that especially endanger welded seams must be avoided by suitable design. The section moduli W of hollow and open profiles can be determined by the well-known Bredt formulae which are set out in **Fig. 10.15**. For the numerical example of hollow and open profiles of equal shape and equal volume or weight, the ratio of the section moduli is about 30:1. Sudden transition from the hollow to the open section, as shown in Fig. 10.15c, causes considerable danger of failure at the welded seams, and the notching-out of a 'relieving parabola' (P) is of only limited benefit.

Much more efficient, and not necessarily more expensive, is to effect a gradual transition from a hollow to an open profile, as shown in **Fig. 10.16**.

Synthesis and Design

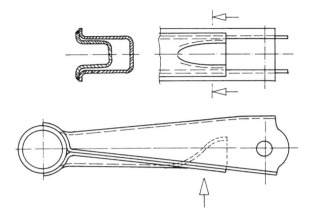

**Fig. 10.16** Advantageous transition from a hollow to an open link section

At its right end, a suspension link has two flanges for the double-shear connection of a joint, while a tube is welded to the left end to accomodate another joint. To avoid distortion of the link, and endangering the frictional double-shear connection on the right-hand side, the link is designed as a hollow beam welded from two U-shape sheet sections. In the transition region, one of the profiles is gradually reversed so as to approximate the shape of the other, namely an 'open' though double-wall profile.

Forged or cast components may not in simplification be regarded as rigid in any case. On the forged rigid-axle beam in **Fig. 10.17**, a second part is attached at two points a certain distance apart. If this part is sensitive to elastic deflection (e. g. because it is a gear housing), one of the mountings should be made compliant. The same is valid for any multiple connections of large constructional items − e. g. the housing of rack-and-pinion steering and the chassis structure. Constraint-free connection is of especial importance for structural parts with different coefficients of thermal expansion, considering the wide temperature range that occurs in vehicles.

If a curved beam made of sheet sections receives a bending moment, the direct stress should not be a problem if there is space enough to give

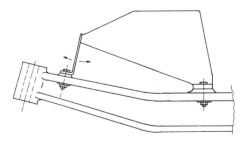

**Fig. 10.17**
Constraint-free connection
of two constructional parts

**Fig. 10.18**
Secondary bending moments on a curved hollow beam

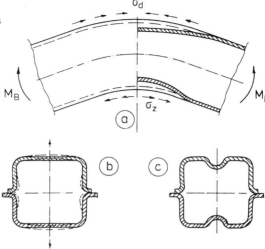

the beam the necessarily wide cross-sectional area. Secondary stress and a possible impairment of dimensional stability arise from the curvature, however. **Fig. 10.18** a shows the curved region of a hollow beam made from two sheet-metal half-sections. The bending moment $M_B$ tries to bend the beam towards a greater radius of curvature. The vectors of the compressive stress elements $\sigma_d$ and the tensile stress elements $\sigma_z$ follow the curvature of the beam and generate resultant components rectangular to the sheet surface – Fig. 10.18b. In the medial areas of the bottoms of the profiles, the sidewalls do not provide sufficient support, and the sheet is

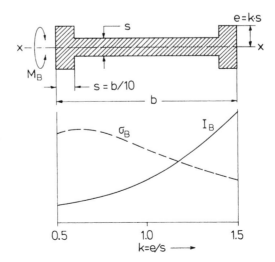

**Fig. 10.19**
Improvement of dimensional stability by flanges

bowed outward, causing secondary bending of the bottoms and of the sidewalls, and the latter try to swivel against each other about the weld seams with the danger of fracture. If the bending moment were inverted, the bottoms of the sections would be bowed inward and the welding flanges outward. Dimensional stability can, for example, be improved by beads in the bottoms of the profiles, Fig. 10.18c.

Beads or ribs for the reinforcement of thin walls must, however, have a certain minimum height to achieve greater stiffness without at the same time increasing the stress. **Fig. 10.19** shows the cross-section of a flat beam with its resistance against bending improved by longitudinal flanges. With increasing height of the flanges, the geometrical moment of inertia $I_B$ increases in a progressive manner, of course, but since initially the edge distance e increases more quickly than $I_B$, the stress $\sigma_B$ increases too until it decreases as desired (for $k = 0.5$, the profile has no flanges!).

If the bending moment on a beam is caused by a transverse force F, **Fig. 10.20**, the beam's cross-section is loaded with not only the linear bending stress $\sigma$ but also with the shear stress $\tau$ which assumes its

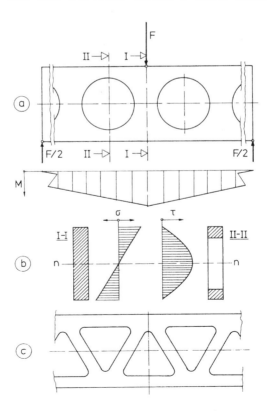

**Fig. 10.20**
Weight-saving by notched apertures

maximum value at the 'neutral line' n, Fig. 10.20b. Hence it is not very advantageous to save weight by notching-out circular holes along the neutral line, Fig. 10.20a; a cross-sectional area II across a hole cannot transmit the shear stress whereby stress peaks are generated around the hole. A better solution is shown in Fig. 10.20c, with triangular apertures that allow the part to act like a lattice beam.

If suspension parts cannot be constructed from sheet-metal profiles, recourse to forged or cast parts is necessary. The reason may be lack of space or, as often occurs, that considerable cranking is necessary to clear other constructional parts.

Due to the higher tensile strength of the material, forged suspension links can be of quite small section. Hence, a possible axial stress caused by a longitudinal force is no longer negligible in comparison with the bending stress. To avoid non-symmetrical stress distribution, and consequently bad material utilization or increased weight, it is sensible to use profiles that allow a free arrangement of the 'neutral line'.

In **Fig. 10.21**a, a cranked suspension link made by forging is loaded with a longitudinal force $F_L$. In the cranked region the link shows a T-profile having its geometrical centre S displaced towards the force's line of action. The edge distances $e_1$ and $e_2$ are therefore different, and with them the section moduli. From $F_L$, the cranking distance p and the geometrical moment of inertia $I_x$ follow the tensile (z) and the compressive (d) bending stresses:

$$\sigma_{zB} = F_L\, p\, e_1 / I_x$$
$$\sigma_{dB} = F_L\, p\, e_2 / I_x$$

and the axial stress follows from the cross-sectional area A:

$$\sigma_{dL} = F_L / A$$

The best material utilization and the minimum weight require equal tensile and compressive stresses:

$$\sigma_{zB} - \sigma_{dL} = \sigma_{dB} + \sigma_{dL}$$

and this leads to the condition for the equal-stress cranking distance p:

$$p = \frac{2\, I_x}{A\,(e_1 - e_2)} \tag{10.9}$$

This equation helps us to design the cross-section profile for optimum material utilization, if the cranking distance p is given.

Examples (b) to (d) show other possible profile types that allow a desired position of the neutral line to be fixed. Types (c) and (d) are 'solid sections' and therefore not quite as weight-saving as the profiles (a) and (b), but

Synthesis and Design

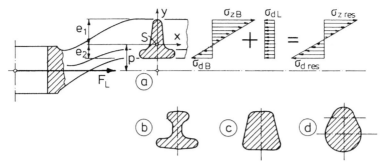

**Fig. 10.21** Optimized stress distribution on a cranked beam under longitudinal force

they have the advantage of not showing 'undercut' in any direction and so can be swivelled deliberately into the directions of maximum bending moment without impeding the forging process.

Beams with 'open' cross-sectional profiles, even forged or cast parts with their usually thicker walls, are not very resistant to torsion, but occasionally it is possible to avoid torsional stresses by suitable design of the profile.

In **Fig. 10.22**, the end of a forged suspension link is cranked and loaded with a transverse force $F_q$. According to the cranking, the connecting line of the geometrical centres S of the cross-sections is s-shaped, and the correct determination of the stressing in the cross-section requires that the cross-sectional planes be chosen rectangularly to the line S.

In the region of full cranking the connecting line S is parallel to the longitudinal axis of the link, and in a cross-section I the force $F_q$ acts on

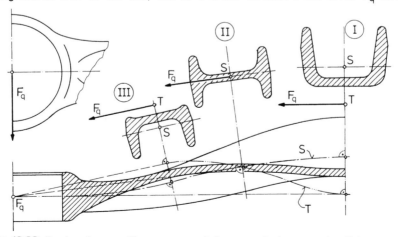

**Fig. 10.22** Torsion-free profile arrangement for a cranked suspension link

the intersection point of the link's longitudinal axis, displaced from the geometrical centre. However, the cross-section is not torsionally loaded because it is U-shaped, and its shear centre T is on the extension of the line of action of $F_q$.

In section II, the tangent of the line S runs through the point of action of $F_q$, and the extension of $F_q$ acts through the geometrical centre S of the profile. Here a symmetrical I-section is the best solution.

In the region between section II and the end of the link, the line S is curved in the opposite direction, and in any cross-section the line of $F_q$ appears on the opposite side of the longitudinal axis of the link. To avoid torsion, the web of the section is displaced to the same side, too, and generates a shear centre T on the extension of $F_q$.

With this method, the link can be designed with an 'open' profile in spite of being cranked and carrying a transverse force. The variable position of the web does not cause any forging problems.

The foregoing considerations underline the fact that, at the beginning of a suspension development programme, a far-sighted analysis of the problems and possible solutions is very important. By the wise choice of design principles and methods, the designer determines essential features of the project which are difficult to change at a later stage.

The constructional parts of a modern wheel suspension are often too complicated to be accessible to 'classical' calculation methods. In spite of this, the designer, due to his close cooperation with the production departments, is the best-suited person to make decisions about the principal features of a design and should not expect the calculation or testing specialist to sort them out for him. Though virtually all constructional parts today are tested and investigated by competent specialists, it is always very helpful for the parts to be based upon the most suitable design.

The measures discussed in the foregoing serve mainly to solve clearance problems. On a wheel suspension, clearance must be provided for all possible wheel positions - e.g. bump and rebound travel and full-lock steering angles for steerable wheels. Apart from interference with other assemblies of the vehicle such as the engine, gearbox, exhaust system etc., the most critical constraints occur between the wheel itself and the suspension parts or the wheel opening of the body; every stylist tries to make the wheel opening as elegant as possible.

The wheel with the tyre is a rotating body the clearance of which is hard to establish by drawing methods but less so by computer-aided design techniques. Due to the rotary property of the wheel, it is very useful to know the rotational contours of relevant assemblies in the neighbourhood around the wheel axis.

Synthesis and Design

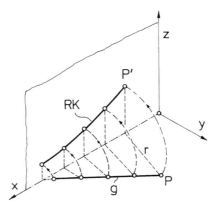

**Fig. 10.23**
Rotational contour RK
of a straight line g
about an x-axis

A rotational contour is the intersection curve of the envelope surface of a revolving part with a plane through the axis of rotation. For example, the rotational contour RK of a straight line g that runs skewly to an x-axis is a hyperbola - **Fig. 10.23**.

For any spatial curve or body, a rotational contour about a given axis can be determined - as for example for the contour curve of a wheel opening in **Fig. 10.24** about the wheel axis $\xi$. For each point of the

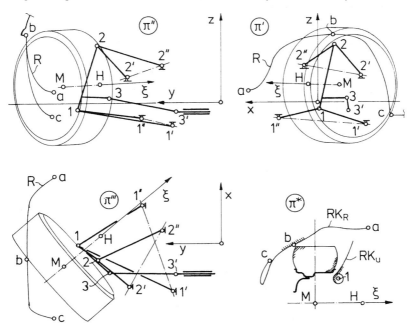

**Fig. 10.24** Rotational contours about the wheel axis (wheel opening and suspension link)

wheel-opening contour curve R, its distance from the wheel axis $\xi$ is determined and drawn in a plane $\Pi^*$ through the axis $\xi$. The rotational contour $RK_R$ may show little similarity to the original curve R. By the same method, rotational contours of other parts can be drawn, such as the curve $RK_u$ of the lower suspension link 1-1'. If corresponding points of the original curve and of the rotational contour are marked (here for example the points a, b and c), it is easy to determine in the plane $\Pi^*$ the points on the curve R that are most critical with respect to tyre clearance.

In view of the variety of possible relevant wheel positions in a suspension system, it is quite clear that a comprehensive envelope of the rotational contours will be necessary.

Establishment of rotational contours about the wheel axis is very useful for the design phase as well as the production phase: questions about changes of tyre or rim equipment, for example - as often arise during a vehicle's period of production - can then quickly be answered (e.g. by the superimposition of transparent drawings).

# Chapter 11

# Motorcycle Suspensions

The wheel suspensions for single-track vehicles - i.e. motorcycles and motor-scooters - always represent 'planar' mechanisms, whose geometrical properties can be described completely in the side view of the vehicle. Lateral displacements and camber change of the wheels affect the directional stability adversely; for this reason, suspensions with transverse links have been unsuccessful, even on motorcycles with sidecars. With planar suspension mechanisms, of course, only the parameters of wheel-travel angle, support angle, castor angle and castor offset can be influenced.

On a motorcycle, the corresponding phenomenon to the 'roll oscillation' of a two-track vehicle (see Chapter 8, Section 8.4.4) is 'wobble oscillation', an oscillation of the front fork with the front wheel against the rear part of the vehicle, which essentially depends on the moment of inertia of the fork about the kingpin axis or 'steering pivot'. It is therefore important to keep the fork's moment of inertia as low as possible.

**Fig. 11.1** shows a selection of front-wheel suspensions schematically, and in some instances without representation of the springs.

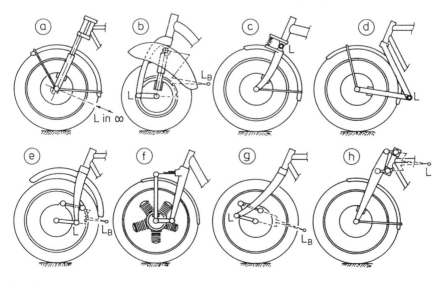

**Fig. 11.1** Front-wheel suspensions for motorcycles and motor-scooters (schematically)

The telescopic fork (a) is the most popular today, not only because of its elegant shape but also because of its relatively low moment of inertia about the steering axis and its acceptable stiffness. The 'longitudinal pole' L of the sliding guidance is at infinity and, due to the roughly parallel attitude of the sliding guidance and the steering pivot, the wheel-travel angle is positive (and allows easy response of the springing) but the braking-force support angle is negative (and forces substantial diving of the steering head under braking). A rotatably mounted brake support, **Fig. 11.2**, combined with a reaction rod to the lower fork bridge, can be used to provide a braking pole $L_B$ and a positive braking-force support angle $\varepsilon_B$ – see also Chapter 6, Fig. 6.8.

A short trailing-link to support the brake would cause even heavier diving under braking, since its longitudinal pole L would be closely in front of the wheel carrier. If a rotatably mounted brake support is carried on a strut from the steering pivot (Fig. 11.1b), the suspension changes into a 'slider-crank mechanism', generating a 'braking pole' $L_B$ behind the wheel which exerts a progressive 'anti-dive' effect (Vespa, for instance).

An articulated fork (Fig. 11.1c) which was occasionally used, especially on motorized bicycles (and the Opel motorcycles of the 1920s), provides longitudinal compliance rather than vertical springing and swings rearwards under braking.

Nearly '100% anti-dive' and, in spite of this, very small castor-offset change with wheel travel, can be achieved by a 'leading-fork' or 'Earles fork' (Fig. 11.1d). However, these advantages are offset by a higher moment of inertia about the steering pivot, which can be reduced by short leading links, Fig. 11.1e. Such an arrangement, of course, requires a rotatably mounted brake support and a reaction rod to produce a 'braking pole' $L_B$ far enough away from the wheel to avoid jacking-up of the vehicle under braking force.

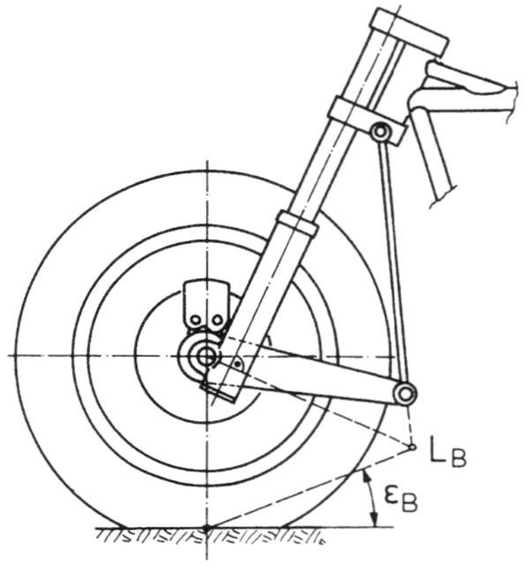

**Fig. 11.2**
Braking anti-dive measure on a telescopic fork

Motorcycle Suspensions 295

**Fig. 11.3** Earles-type suspension of the BMW model 'R 69' (1955) (courtesy BMW AG)

A front-wheel-hub engine was a characteristic feature of the 'Megola' motorcycle of the 1920s (Fig. 11.1f); the rotary five-cylinder radial engine was guided by lower leading links and an upper quarter-elliptic leaf spring.

Suspensions with short trailing links, again with a rotatably mounted brake and a reaction rod, have been used too, as for instance on an NSU racing motorcycle (1952) – Fig. 11.1g.

The four-link trapezoidal fork or 'girder fork', Fig. 11.1h, was virtually the standard front-wheel suspension in former times. It has quite a low moment of inertia about the steering axis but some lack of lateral stiffness. By setting the upper and lower pairs of links at different angles, anti-dive properties can be achieved.

Swinging-fork suspensions were widely used in the 1950s and 1960s because of easy response of the springing and because of their anti-dive effect which allows relatively soft springing. **Fig. 11.3** shows the leading swinging-fork of the BMW motorcycles of that time.

A leading-link suspension with rotatably mounted brake support and reaction link is shown in **Fig. 11.4**, and a girder fork in **Fig. 11.5**. On both suspensions, a relative inclination of the links for braking 'anti-dive' purposes is recognizable – see also Figs 11.1e and h. For constant braking-force support angle with wheel travel, the ratio of the lengths of the links should be the reciprocal of the ratio of their distances from the road (see also Chapter 10, Fig. 10.2).

**Fig. 11.4**
Short-leading-link front suspension of a Honda motorcycle of the 1950s (Photo by the author)

**Fig. 11.5**
Girder fork of an NSU motorcycle of the 1940s (Photo by the author)

Motorcycle Suspensions

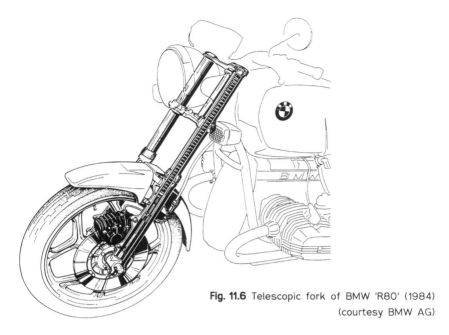

**Fig. 11.6** Telescopic fork of BMW 'R80' (1984) (courtesy BMW AG)

Modern telescopic forks provide considerable wheel travel which would otherwise be best achievable by swinging-fork suspensions. The brake caliper has to be positioned near the steering pivot axis to obtain the minimum moment of inertia about it; for this purpose, the caliper in **Fig. 11.6** is mounted above the wheel axis. In contrast with two-track vehicles, this measure is unobjectionable on solo motorcycles where the lateral forces on the wheels are virtually nil and therefore cannot cause increased 'released' play at the brake pads through elastic deformations and play in the wheel bearings, which would dangerously increase the brake-lever travel.

An efficient recent development of the telescopic fork is shown in **Fig. 11.7**. The two (upper) 'stanchions' are connected above the frame of the machine by a single fork bridge which is articulated to the frame by a ball joint. The (lower) sliding tubes that carry the wheel axle are connected above the tyre by another bridge which is guided by the ball-joint of a forward-pointing triangular link. The latter is connected to the frame by turning joints with a transverse axis. Springing and damping are effected by a separate strut acting on the triangular link. The small right-hand sketch indicates the kinematic functioning of the suspension. The two ball-joints replacing the 'kingpin' have the maximum possible spacing at any wheel position, thus securing the optimum stiffness against longitudinal forces - very important for good functioning of brake anti-lock systems. As in Fig. 11.1 b, the

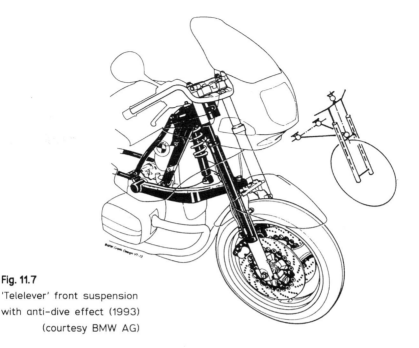

**Fig. 11.7**
'Telelever' front suspension
with anti-dive effect (1993)
(courtesy BMW AG)

'longitudinal pole' results in side view from the intersection of the plane of the triangular link and the line normal to the telescopic guidance through the upper ball-joint, so deliberate choice of the braking-force support angle – and thus anti-dive – is possible.

From the very early stages of motorcycle design there have been occasional attempts to make a 'hub-centre steering' system with a kingpin inside the wheel bearing; these arrangements too were intended to improve longitudinal stiffness. Quite apart from the technological expenditure and limited steering angle, the need for additional steering arms and rods leads to increased risk of steering backlash and friction. Whatever the layout, though, precise guidance and moderate steering self-centring are essential properties of a good motorcycle front suspension.

The most important rear-wheel suspension types for motorcycles are represented schematically in **Fig. 11.8**.

The 'plunger-type' suspension (a) corresponds to the telescopic front fork. Its 'longitudinal pole' L is at infinity and, as the brake is usually fixed to the plunger, represents the 'braking pole', too. With chain drive, the 'traction pole' $L_A$ is the intersecting point of the line normal to the guidance through the wheel centre and of the upper chain strand – see also Chapter 6,

Motorcycle Suspensions

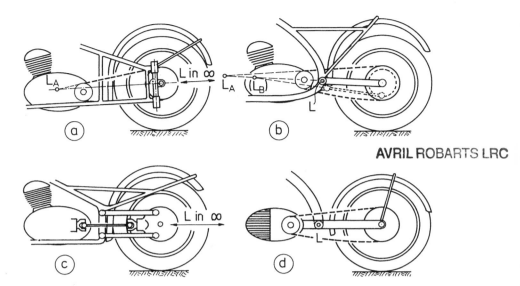

**Fig. 11.8** Rear suspensions for motorcycles (schematically)

Fig. 6.18. If drive is via a shaft and bevel gearing, the traction pole is of course at infinity, like the braking pole.

Because of its better springing properties, the trailing-arm layout (b) has become predominant, normally combined with strut-type springing elements. With chain drive, the intersecting point of the trailing arm and the upper chain strand is the 'traction pole' $L_A$. If the brake is supported directly by the trailing arm, its pivot joint L at the machine's frame is the 'braking pole', too. With a rotatably mounted brake support and a reaction rod connected with the frame (dotted lines), the braking pole $L_B$ follows from the trailing arm and the reaction rod.

A very rare layout is wheel guidance by a four-joint chain of links (c) which is at best suitable for shaft-and-bevel drive; if the links are parallel, the pole of the wheel carrier (here the final-drive housing) is at infinity.

If the engine is fixed to the wheel carrier (d) (as on some lightweight two-wheelers), the mode of the drive-train is of no importance for the traction pole, the pole L of the wheel carrier being the traction pole also.

Suspensions (b) to (d) in Fig. 11.8 show 'progressive' braking anti-lift effect and 'degressive' traction anti-squat effect, because the braking pole is displaced upward with wheel rebound and the traction pole downward with bump.

On the rear suspension of the Vespa scooter, **Fig. 11.9**, the air-cooled engine is carried on the trailing arm together with the gearbox and the

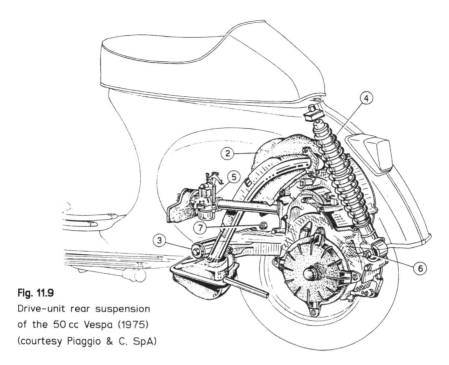

**Fig. 11.9**
Drive-unit rear suspension
of the 50 cc Vespa (1975)
(courtesy Piaggio & C. SpA)

exhaust system ('drive-unit wheel carrier'). As the driveline needs no movable or slidable transmission parts to compensate for wheel travel, a robust and sealed unit is achieved, the wheel being mounted (and detachable) as on a car. The axis of the float in the carburettor (5) intersects the axis of rotation (3) of the trailing arm, thus largely shielding the float from axial acceleration under wheel movement.

A plunger-type suspension that carries the final-drive bevel gearing of a shaft-driven wheel is shown in **Fig. 11.10**. The shaft's engine-side joint (not visible) is a flexible rubber coupling.

Shaft drive is applied also to the swinging-arm suspension in **Fig. 11.11**. The sole cardan joint (not visible) virtually coincides with the axis of rotation of the swinging arm. Ahead of the bevel gear, and not visible in the drawing, is a curved-tooth coupling to give compliance with minor angular and longitudinal deflections. In the middle of the shaft is an overload clutch consisting of meshing claws that are preloaded by a coil spring. Friction forces resulting from clutch action serve also to damp torsional vibrations. The suspension has a single arm for simple mounting and detachment of the wheel (see also Fig. 11.9). Inclining the strut allows considerable wheel travel in spite of moderate strut length, and its effective lever arm about the trailing-arm

Motorcycle Suspensions

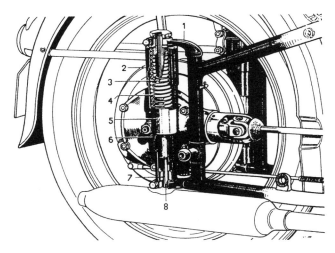

**Fig. 11.10** Plunger-type rear suspension of the BMW 'R 68' (1952) (courtesy BMW AG)

joint increases with wheel bump, thus generating a progressive spring characteristic with an essential share of the 'kinematic spring rate' – see also Chapter 5, Fig. 5.20. The swinging-arm joint marks the 'longitudinal pole', too, and leads to high braking- and traction-force support angles and thus about 100% anti-lift and anti-squat properties.

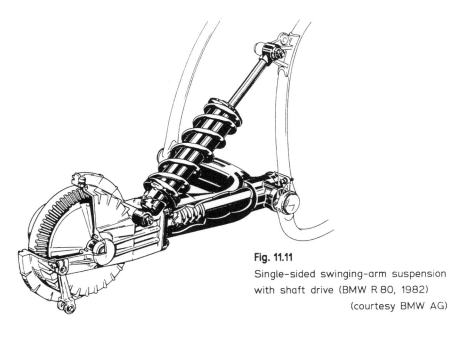

**Fig. 11.11**
Single-sided swinging-arm suspension with shaft drive (BMW R 80, 1982)
(courtesy BMW AG)

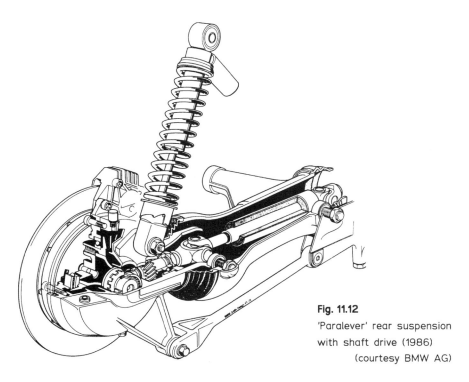

**Fig. 11.12**
'Paralever' rear suspension
with shaft drive (1986)
(courtesy BMW AG)

Sometimes swinging-arm suspensions embody several anchorages for the upper strut mounting(s) – or even slots, as used on post-war Veloce motorcycles – for simple altering of the angle of attack of the strut(s) to cope with variations in rider/passenger weight or type of riding.

A further development of the swinging-arm suspension towards a four-joint chain is illustrated in **Fig. 11.12**. Here the swinging arm is connected to the final-drive unit by a turning joint, and a reaction rod below it completes the kinematic chain. The pole of the wheel carrier in side view is the intersecting point of the swinging arm and the reaction rod, and is here positioned much farther from the wheel than the swinging-arm joint. This leads to a smaller support angle than for the layout in Fig. 11.11 which, however, still provides adequate anti-squat properties but reduces circumferential slip and tyre wear on bad road surfaces. The cardan shaft runs inside the swinging arm and is provided with a rubber coupling to protect the driveline from torque peaks.

# Chapter 12

# Independent Wheel Suspensions

## 12.1 General remarks

Independent wheel suspensions have only one degree of freedom; consequently, they have the same wheel motion relative to the vehicle body for both parallel and antimetric wheel travel. The achievement of advantageous wheel camber with respect to the road in cornering therefore leads to considerable camber change with parallel wheel travel or with variations in vehicle load. A roll centre above or below the road surface requires the track to change with wheel travel. Advantages of independent suspensions are their low weight, plenty of scope for achieving favourable elasto-kinematic effects, no coupling of the masses and, of course, no suspension parts that run right across the vehicle.

As each spring normally is displaced from the line of action of the wheel load, the mountings of the suspension links may well carry considerable preload forces.

For driven wheels the torque is commonly transmitted by transverse driveshafts; provided there are no hub reduction gears, the traction-force support angle is then equal to the positive wheel-travel angle on rear wheels and negative wheel-travel angle on front ones.

## 12.2 Front suspensions

Perhaps the oldest type of steerable independent suspension is the rectilinear sliding guidance or 'pillar type' suspension – a knuckle that slides up and down on the kingpin (see also Chapter 2, Fig. 2.8b). Front suspensions like that in **Fig. 12.1** were often used at the beginning of the 20$^{th}$ Century. In the example shown, the middle of the transverse leaf spring is supported by a vertical shackle (transverse compound spring, see Chapter 5, Fig. 5.14a). The caps below the spring ends enclose auxiliary coil springs. Steering is achieved via track-rods which are not yet fitted with the ball-joints so common today but have cardan joints with displaced axes of rotation for simpler construction. This suspension type was applied up to the 1950s on Lancia and (with globoid rollers for lower friction) American Motors cars, and

304                                                              Road Vehicle Suspensions

**Fig. 12.1** Pillar-type suspension, 'Wartburg-Wagen' (1898), BMW Museum, Munich, Germany (Photo by Werner Schwarzbach)

it has survived on the Morgan roadster. Steering errors are inevitable as the wheel travel increases (see Chapter 7, Fig. 7.32b).

A special front-suspension type that was quite often applied in former times is the Dubonnet design, invented in the 1920s by the eponymous amateur racing driver to transform rigid-axle suspensions into independent layouts which were coming into fashion at that time.

Normally, the wheel carrier is embodied in a leading longitudinal arm rotatably connected to a housing that swivels about the kingpin; the suspension is, therefore, positioned between the wheel and the kingpin and swivels completely about the latter when steered, in similar fashion to a motorcycle fork – see Chapter 6, Fig. 6.10.

When four-wheel brakes became standard, the Dubonnet design lost its simplicity, needing a rotatably mounted brake support in order to avoid jacking-up under braking. Again similarly to a motorcycle suspension, a Dubonnet system can be given considerable anti-dive properties without noticeably affecting the castor change with wheel travel.

**Fig. 12.2** shows in plan view the Opel 'Admiral' Dubonnet front suspension of 1938. The axle tube is fixed to the vehicle body and carries two cast housings that swivel about the kingpins; inside the housings are coil springs and telescopic dampers. A track-rod ahead of them connects the two

Independent Wheel Suspensions 305

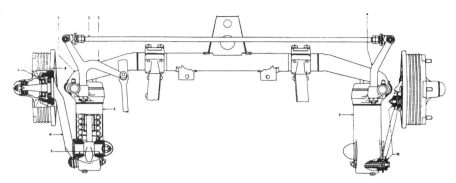

**Fig. 12.2** Opel 'Admiral' Dubonnet front suspension (1938) (courtesy Adam Opel AG)
Left: section through the longitudinal arm or wheel carrier
Right: section through the brake reaction rod

housings (forming a 'steering trapezium'), and a longitudinal drag-link acts on the left housing. The complete steering linkage is unaffected by wheel travel - see also Chapter 7, Fig. 7.32a. In this form, the Dubonnet system was at that time almost the standard front suspension of many of the passenger cars made by the General Motors Corporation in America and Europe (on Vauxhall models even with torsion-bar springing).

One of the last examples of a Dubonnet layout was the front suspension of the BMW '700'. **Fig. 12.3** shows an 'exploded view' of this mechanism with the longitudinal arm (i.e. the wheel carrier), the brake reaction rod and a vertical telescopic damper, the installed system being shown in the upper right-hand corner.

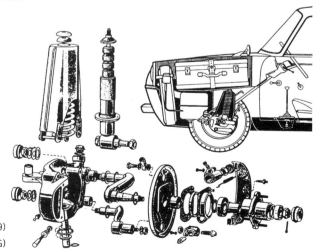

**Fig. 12.3**
BMW '700' Dubonnet front suspension (1959)
(courtesy BMW AG)

Early theoretical investigations into rigid-axle tramp and wheel-shimmy oscillations revealed the coupling of steering angles and moments of inertia or gyroscopic moments to be disadvantageous. This knowledge led to a proliferation of independent wheel suspensions, in particular those with minimum camber and track change with wheel travel - e.g. suspensions with longitudinal arms. Besides the Dubonnet system, double-trailing-arm suspension was often adopted, achieving record production numbers with Volkswagen.

In its first realization, the front suspension of the Volkswagen Type 1, popularly known as the 'Beetle', was of the planar type with parallel trailing arms - **Fig. 12.4**. Each wheel is sprung by two torsion bars (made as multi-leaf 'packs') which are connected directly to the trailing arms. The track-rods are of different lengths to avoid an additional slave arm and a medial rod. Since the trailing arms and the track-rods appear in side view with the same length and direction, the (theoretical) toe-in curve is rectilinear. As seen in Fig. 12.4, the trailing arms are connected to the 'coupler' of the suspension by turning joints, and the wheel carrier or 'knuckle' is articulated at the coupler via a 'physical' kingpin. Later, the wheel carrier was directly connected to the arms via ball-joints. In the Volkswagen Type 3, the axes of revolution of the trailing arms had opposing transverse inclinations, thus generating a camber change with wheel travel and a roll centre above the road. This version of course has a nearly 'spherical' layout.

Before reliable ball-joints became available, steerable independent wheel suspensions normally featured a 'physical' kingpin connecting the wheel

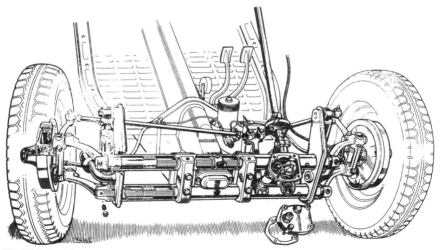

**Fig. 12.4** Double-trailing-arm suspension, Volkswagen Type 1 or 'Beetle'

(courtesy Volkswagen AG)

Independent Wheel Suspensions 307

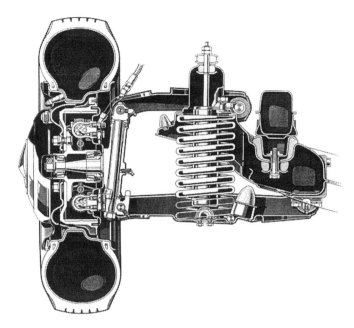

**Fig. 12.5** Double-wishbone suspension of Mercedes-Benz types 190, 219 and 220S (1954) with rubber-mounted subframe  (courtesy Daimler-Benz AG)

carrier (the 'knuckle') to the mechanism's coupler, as just visible in Fig. 12.4. **Fig. 12.5** shows a (planar) double-wishbone suspension with a section taken through the kingpin. The pivots of the wishbones can be seen above the kingpin and beside its lower end; the wishbones are not directly connected to the vehicle body but through a subframe attached to the body by rubber mountings, for better noise insulation and ride comfort.

In the past, transverse leaf springs often served for wishbones, **Fig. 12.6**, about $7/9$ of the leaf length corresponding to the effective kinematic wishbone length (see Chapter 5, Fig. 5.26). Visible at approximately the height of the wheel centres are the housings of lever-type dampers, mounted on the chassis and having their arms connected to the couplers of the suspension via short shackles. Lever-type dampers were almost standard until the 1950s and are still used today for door-closers.

Due to their simple and space-saving design and the wide separation of the joint positions on the vehicle body (and, consequently, the low reaction forces), strut and damper-strut suspensions are widely used today for front wheels; they hold their ground on all classes of passenger cars and even light trucks beside often elaborate multi-link systems. The earlier

**Fig. 12.6** Mercedes-Benz '170 V' front suspension with transverse leaf springs (1935) (courtesy Daimler-Benz AG)

handicap of strut suspensions – namely the transverse forces acting on the piston and the piston rod causing increased friction and wear – has been overcome by the invention of a spring layout that compensates for those transverse forces, at least in the case of vertical wheel loads, and by measures for minimizing friction; strut suspensions therefore are no longer inferior to other types in respect of the response properties of the springing.

Strut suspensions are mechanisms where a triangular link has been replaced by a turning-and-sliding link, i.e. the piston rod of the damper. The latter has therefore become a suspension link and so carries bending moments under the external loading. Already the offset of the wheel-load force vector from the strut upper mounting causes a moment at the wheel carrier that exerts a bending moment on the piston rod.

If the suspension links and the track-rod lie approximately in a common plane, which is true for many of the strut systems, the spatial attitude of the force $F_K$ at the strut mounting, resulting from the wheel-load force $F_Z$, can easily be estimated, **Fig. 12.7**a: the line of action of $F_K$ must run through the intersection point of the wheel-load force $F_Z$ in the plane under consideration.

# Independent Wheel Suspensions

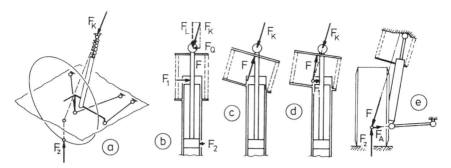

**Fig. 12.7** Compensation of transverse forces at the piston rod of a strut suspension

The force $F_K$ has a transverse component $F_Q$ that causes reaction forces $F_1$ and $F_2$ at the piston-rod guidance and at the piston itself, Fig. 12.7b. These forces result in friction and wear and impair the response of the springing.

On steerable strut suspensions, the upper and lower spring abutments are usually fixed to the piston rod and the damper tube respectively, and the spring swivels together with the strut under steering action. Total compensation of the transverse forces at the piston rod then requires that the centre line of the spring coincides with the line of action of the force $F_K$, Fig. 12.7c. This leads, of course, to a considerable inclination of the spring axis, and to lateral deflection of its ends with wheel bump and rebound. A helpful compromise is shown in Fig. 12.7d: the spring's axis is inclined by a significantly smaller angle and is displaced relative to the strut mounting in order to intersect with the external force $F_K$ in the working plane of the piston-rod guidance. This allows equilibrium to be established between the force $F_K$, the spring force F and the (considerably reduced) guidance force $F_1$. The piston force $F_2$ is zero, so the portion of the rod that has plunged into the damper is free of bending moment, and the piston rod slides in its guidance without elastic jamming.

On non-steerable strut suspensions, the upper end of the spring is usually supported at the vehicle body independently from the strut mounting; to avoid transverse forces at the piston rod under wheel load, the axis of the spring should run through the intersecting point of the wheel-load vector and the common plane of the suspension links, Fig. 12.7e.

In a damper-strut suspension the spring usually acts on the transverse link and loads the wheel carrier via its supporting ball-joint, **Fig. 12.8**. The supporting ball-joint should be as near to the wheel centre-plane as possible to minimize the moment generated by the wheel load, and hence the transverse forces at the piston rod. This optimal positioning follows almost

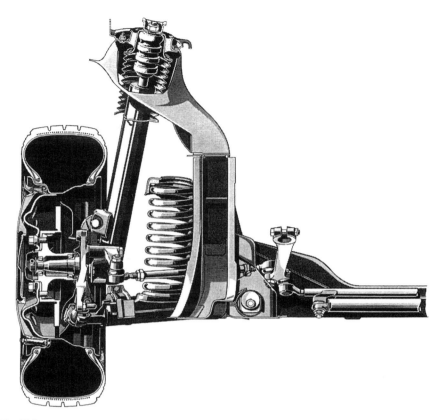

**Fig. 12.8** Cross-section of damper-strut front suspension, Mercedes-Benz Type 190 (1982) (courtesy Daimler-Benz AG)

automatically if the scrub radius is desired to be very small or even negative. The 'kingpin axis' is the line between the supporting ball-joint and the strut's upper mounting.

The idea of using a 'slider crank mechanism' – i.e. a 'crank mechanism with oscillating slider' – for wheel suspension, is quite old, having appeared first in a patent of Cottin & Desgouttes (1925) (**23**), and a year later a Fiat patent (**24**) proposed the use of a telescopic damper as the slider. However, this suspension type was not introduced on vehicles until 1948 (Ford 'Anglia'), and since the design involved a linkage consisting of transverse arms and the ends of the anti-roll bar according to a MacPherson patent (**22**) (see also Chapter 5, Fig. 5.13b), the name 'MacPherson suspension' has been adopted generically for strut suspensions even if they do not incorporate that particular feature.

Independent Wheel Suspensions                                    311

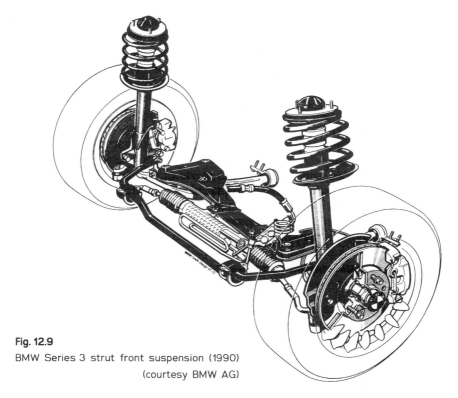

**Fig. 12.9**
BMW Series 3 strut front suspension (1990)
(courtesy BMW AG)

A typical strut front suspension is shown in **Fig. 12.9**, the coil spring being arranged according to Fig. 12.7d. The 'triangular' links are here made sickle-shaped, consisting of transverse and longitudinal arms. At its apex each link is connected to the subframe by a ball-joint, and longitudinal compliance is achieved by lateral deflection of the rubber joint at the rear end of the link's longitudinal arm. The angle of incidence between the transverse arm and the track-rod is chosen for optimum elasto-kinematic self-steering behaviour. The subframe is rigidly attached to the vehicle body because the force level on a strut suspension is low and there is no noticeable noise transfer via the suspension joints. The power-assisted rack-and-pinion steering is situated approximately at the same height as the triangular links and in front of the axle line, thus securing 'understeering' behaviour - see also Chapter 9, Fig. 9.9. The anti-roll bar is connected to the transverse arms of the suspension links via short shackles and serves only for springing purposes.

On front-drive cars it is often difficult to mount a steering linkage or a rack-and-pinion gear housing below the engine and gearbox. Hence the steering system may have to be in the upper region of the suspension and above the gearbox, **Fig. 12.10**. As the sliding guidance of the piston rod corresponds to a transverse link of infinite length, the trajectories of the outer track-rod joints with wheel travel show considerable radii of curvature, thus requiring lengthy track-rods. This problem can be solved by using a rack-and-pinion unit with medial take-off; the inner rod joints are connected to the rack via an arm that protrudes through a slot in the housing.

The axes of the springs are of course inclined relative to the piston rod, to reduce friction under wheel load.

On a driven front suspension with 'conventional' steering geometry – i.e. 'fixed' kingpin axis during a steering event – as in Fig. 12.10, the outer joint of the driveshaft is best positioned on the kingpin axis in order to avoid unwanted lengthening or shortening of the shaft during steering, in addition to the length changes that occur normally with bump or rebound wheel travel. And if the shaft joint is on the wheel axis, as on a suspension without hub reduction gear (or with planetary reduction gearing), this means of course that the castor offset at wheel centre $n_\tau$ must be zero. For any value $n_\tau \neq 0$ in the straight-ahead position, a steering angle $\Delta\delta$ will cause a shaft length variation $\Delta l = n_\tau \Delta\delta$.

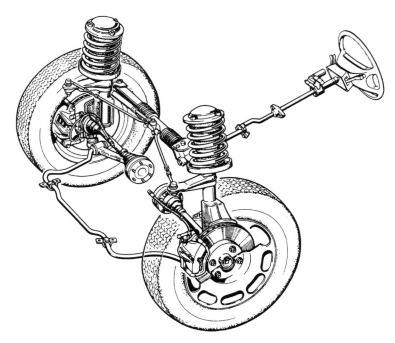

**Fig. 12.10** MacPherson strut suspension, Audi '100' (1976)   (courtesy Audi AG)

Independent Wheel Suspensions 313

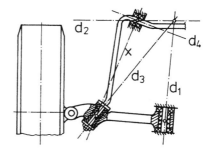

**Fig. 12.11**
Good arrangement of the rubber joints on a MacPherson anti-roll-bar linkage

The Fig. 12.10 suspension shows a special feature that was quite often applied in the past on both strut and double-wishbone suspensions: the arms of the anti-roll bar are attached to the transverse links and act as 'tension rods' (MacPherson principle, see Chapter 5, Fig. 5.13b). It is noteworthy that the axes of the rubber joints that connect the anti-roll bar to the vehicle body are not aligned, and **Fig. 12.11** explains the reason for this feature.

The axis $d_4$ of the front rubber joint is rectangular to the line x connecting the rubber joints at the vehicle body and at the transverse link. Any force acting between the latter and the anti-roll bar therefore loads the front rubber joint with a 'radial' force; this avoids axial force components that would lead to lateral displacement of the anti-roll bar and hence to a displacement of the transverse link, in plan view, which could cause unwanted steering angles depending on its angle of incidence with the track-rod, not drawn - see also Chapter 9, Fig. 9.10.

Even without carrying external forces, a MacPherson anti-roll bar will continuously be subject to lateral displacement because the suspension's transverse links guide its ends on circular paths which cause not only considerable bending moments but also - when occuring non-symmetrically, e.g. in cornering or on an uneven road - lateral movement of the bar. From this point of view, a purely 'kinematic' explanation of the arrangement is possible, too: any lateral travel of the anti-roll bar occurs along the line $d_4$, and consequently in a direction rectangular to the line x, and so cannot cause any displacement of the transverse link in plan view.

The axes $d_1$ and $d_3$ of the heavy-duty rubber joints on the transverse link intersect the effective axis of rotation $d_2$ of the anti-roll bar at a common point. This secures minimum 'coning angles' (see Chapter 5, Fig. 5.41) of the rubber joints and enables them to be provided with the desirable radial stiffness.

Because of the problems just mentioned, and because today the roll spring rate of a passenger-car front suspension is provided to a considerable

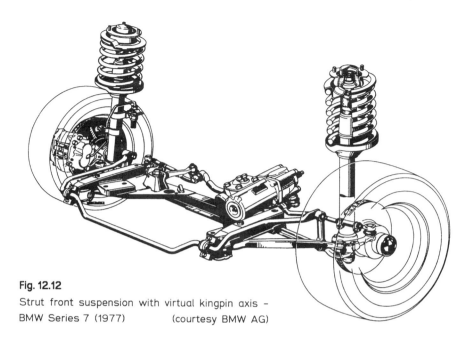

**Fig. 12.12**
Strut front suspension with virtual kingpin axis –
BMW Series 7 (1977)   (courtesy BMW AG)

degree by the anti-roll bar, the linkage of strut systems is nowadays realized more and more in a 'conventional' manner – i.e. by the strict separation of the suspension mechanism and the springing system.

Brake anti-lock systems require a very small or even negative scrub radius which causes clearance problems between the brake disc and the wheel-side joints of the suspension links. For this reason it may be helpful to create a 'virtual' kingpin axis, **Fig. 12.12**, by splitting triangular links into separate rod links – see also Chapter 8, Fig. 8.6. This measure is especially easy to provide on a strut suspension since it features only one triangular link on each side. Once a triangular link is split into a transverse and a diagonal link – e.g. a 'tension rod' – these two components can also be arranged at different heights for optimum use of space and for achieving wanted kinematic properties such as braking anti-dive effect.

With a pure steering event, and the spring assumed to be 'locked', this suspension represents a spherical steering mechanism, **Fig. 12.13**, with the strut mounting SL acting as the central point. The planes defined by the transverse link $QQ_0$ and the tension rod $ZZ_0$, and by the strut mounting or central point SL, intersect the road surface in the lines $Q'Q'_0$ and $Z'Z'_0$ which in turn intersect at the 'pole' P through which runs the virtual kingpin axis i. P and the tyre contact point A being known, the scrub radius $r_S$ and the castor offset n can be determined in the road plane.

Independent Wheel Suspensions                                                             315

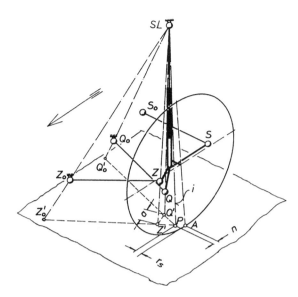

**Fig. 12.13**
Steering geometry of the
suspension of Fig. 12.12

Since the virtual kingpin axis i, in contrast with suspensions having 'fixed' kingpins, is displaced relative to the wheel carrier (the knuckle and the strut) when steered, several of the equations established in Chapter 8 for a fixed kingpin axis, and several of the familiar experimental values of steering geometry attunement, are no longer valid. The steering-geometry characteristics therefore have to be calculated according to Section 8.3.2.

Unlike that of conventional steerable suspensions, the scrub radius is variable with the steering angle, and its minimum value is best positioned near the straight-ahead position, at an 'inner' steering angle of a few degrees. Such geometry ensures that, in cornering, the outer scrub radius is always greater than the inner in order to avoid steering moments that try to steer the wheels to the inside of the corner under braking forces. This can be achieved by suitable arrangement of the suspension links and the joints (**14**).

With strut suspensions, a scrub radius around zero requires a substantial kingpin inclination angle and causes a considerably inclined path of the outer track-rod joint with steering movement. For this reason, the steering linkage of the suspension in Fig. 12.12 is designed as a 'spherical' four-joint chain, and the drop-arm and slave-arm axes are mutually inclined in cross-section — see also Chapter 8, Fig. 8.20e.

The wheel motion of this suspension, with spring travel and 'locked' steering is, as with all independent front suspensions (except the Dubonnet system), of a spatial character. To compensate transverse forces at the piston rod, the springs' axes are inclined to the strut axes, as already shown in Fig. 12.7.

With double-wishbone suspensions it is easy to achieve advantageous kinematic effects by suitable spatial arrangement of the links and joints. Splitting each triangular link into two separate rod links creates a spatial five-link suspension with a relatively free choice of kinematic attunement, **Fig. 12.14**. The kinematic layout is shown in **Fig. 12.15**.

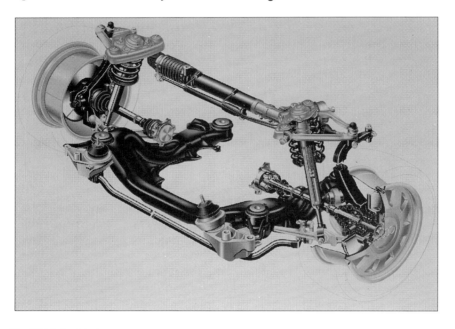

**Fig. 12.14** Audi 'A8' front suspension with virtual kingpin axis (1994) (courtesy Audi AG)

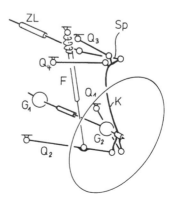

**Fig. 12.15**
The suspension mechanism of Fig. 12.14, shown schematically

Independent Wheel Suspensions 317

Damper and spring form a strut F that is not part of the suspension mechanism and is carried by the front lower transverse link $Q_2$. When the system is steered, assuming a 'locked' spring, the wheel motion is characterized by an instantaneous screw, and the virtual kingpin axis can be used merely for determining the kingpin inclination and the castor angle. Other steering characteristics such as the scrub radius, wheel-centre offset, traction-force radius, castor offset, castor offset at wheel centre and wheel-load lever arm follow from the generalized definitions given in Section 8.3.2. This is especially the case for the wheel-load lever arm: the vehicle's vertical lift with steering and a 'locked' strut depends on the chosen attitude of the strut and its mounting position at the transverse link $Q_2$; hence it is no longer defined by the attitude of the (virtual) kingpin axis alone!

As the upper linkage ($Q_3$ and $Q_4$) is positioned above the tyre tread, soft longitudinal compliance of the wheel is possible without excessive elastic wind-up of the wheel carrier K under braking force – see Chapter 9, Fig. 9.12. Unlike the situation with strut suspensions, a moderate kingpin inclination can be achieved and consequently a low wheel-centre offset $r_c$ (or a low traction-force radius $r_T$). High mounting of the power-assisted rack-and-pinion steering ZL and the track-rod Sp behind the axle favours elastic understeer, see Section 9.2.

The lower links $Q_1$ and $Q_2$ are mounted on a subframe that carries the engine too and is flexibly mounted to the vehicle body.

The inner joints of the driveshafts are of the tripod type (see Chapter 3, Fig. 3.15) to reduce the transmission of driveline oscillations to the suspension and the steering.

Unlike a fixed-axis kingpin, a virtual axis as in Figs 12.12 and 12.14 moves with the steering relative to the wheel carrier and will therefore usually not be suitable for geometrically locating the outer driveshaft joint. The optimum position of that joint is the one point on the wheel axis whose path during steering shows a centre of curvature in (or near to) the inner shaft joint; it will normally be found on the inner side of the virtual kingpin, and somewhere within a 'tube' that embraces the upper and lower outer ball-joints of the suspension links.

## 12.3 Rear suspensions

After its major success for front wheels, independent wheel suspension has largely succeeded on rear wheels too - at least in passenger cars. In spite of this, the rigid axle is still popular at the rear, because on non-driven wheels it can be made so light that its principal disadvantages (as for instance the coupling of masses) become unimportant in comparison with its advantages, especially constant track and camber. However, one serious disadvantage remains, namely the up-and-down travel of the axle beam below the vehicle floor - something that imposes considerable restrictions on the exhaust system and luggage compartment. On rigid-axle suspensions, moreover, soft longitudinal compliance for riding comfort can be achieved only within narrow limits; to mount such a suspension to a subframe would of course be much more expensive than an independent system.

Because of the poor durability of driveshaft joints in former times, the preference for independent suspension fell on types requiring the smallest number of such joints, namely swing-axle layouts.

A 'pure' swing-axle with its axis of revolution exactly in the straight-ahead direction requires either a precise and space-wasting connection of the swing-arms to the vehicle body or a rotational connection of the axle tubes and the final-drive housing. Each driveshaft joint must, of course, lie on the axis of rotation of the swing-arm. The main advantage of the second solution, **Fig. 12.16**, is the dust-proof sealed construction; the reduced basic length of the turning joints leads to high reaction forces under longitudinal forces at the road wheels.

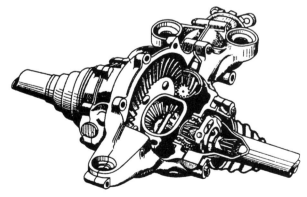

**Fig. 12.16**
Final drive of the Mercedes-Benz '170' swing-axle (1931) (courtesy Daimler-Benz AG)

Independent Wheel Suspensions 319

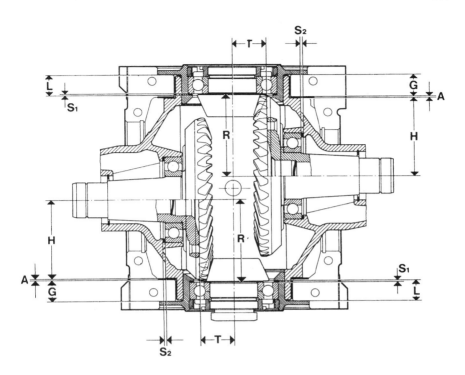

**Fig. 12.17** Bevel-gear drive and swing-arm suspension of the 'Pinzgauer' off-road car
(1972) (courtesy Steyr-Daimler-Puch AG)

The special single-joint version of the swing-axle, without any driveshaft joint but with bevel gears oscillating about their pinions with wheel travel (see Chapter 6, Fig. 6.16), requires a coaxial arrangement of the two swing-arm joints in the gear housing and in the common axis of the pinions. In **Fig. 12.17**, the wheel axes are displaced in the straight-ahead direction; this causes different wheelbases on the two sides but allows the bevel gears to be of equal size. As one pinion is in front of the axle and one behind it, the traction-force support angles of the two wheels are of equal amount but opposite sign, and the resulting support angle is zero. The traction torque acts on the suspension and the springs like the roll moment of a rigid axle - see Chapter 6, Fig. 6.7. Being of robust construction, this system allows a central-tube frame and completely sealed driveline.

If a driven swing-arm suspension features single driveshaft joints, the axis of rotation of each swing-arm can be arranged skewly in plan view, in order to react longitudinal forces more readily. The suspension will then, of course, show toe-in change with wheel travel in addition to camber change.

**Fig. 12.18** Semi-trailing swing-axle suspension, 'Goggomobil' (1954) (courtesy BMW AG)
d = effective axis of rotation of the wheel carrier

A very ingenious yet simple solution is illustrated in **Fig. 12.18**: the semi-trailing swing-arms are formed by longitudinal arms that contain the wheel bearings and support the struts, and by the half-shafts, their joints serving as the inner suspension joints. Since normal constant-velocity joints are not suitable for carrying axial loads (as will occur under wheel loading, braking/traction and lateral forces at the wheels), only the cardan joints come into consideration. The forces resulting from external loading at the wheels stress the cardan joints' needle bearings far less than do those caused by the traction torque.

The traction-force support angle has to be determined according to the rule of a 'vehicle-fixed' torque support – see Chapter 6, Fig. 6.15.

With a skew arrangement of the swing-arm's axis of rotation (which no longer runs through the inner driveshaft joint, and which perhaps does not intersect the wheel axis either) the generalized semi-trailing-arm suspension provides greater possibilities for geometrical suspension attunement. However, it requires an extensible driveshaft with two joints. **Fig. 12.19** shows a semi-trailing-link rear suspension with space-saving 'barrel springs' and separate telescopic dampers. The links are mounted to the subframe via stiff rubber joints, while the subframe and the final-drive housing are bolted together as a unit. The latter's elasto-kinematic properties are determined by the spatial arrangement of the directions of the principal spring rates of the rubber mountings between the subframe and the vehicle body. With a resultant 'elastic centre' (see Section 5.3) behind the wheel axes, the suspension unit will tend towards understeer when subjected to

Independent Wheel Suspensions

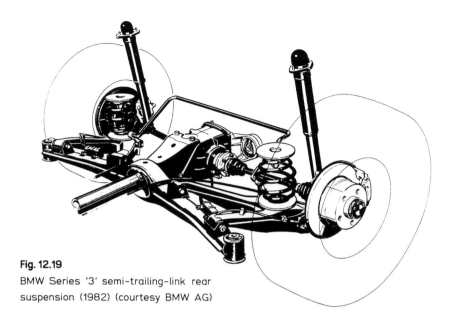

**Fig. 12.19**
BMW Series '3' semi-trailing-link rear suspension (1982) (courtesy BMW AG)

lateral forces. The rubber mounting at the rear of the final-drive housing is positioned to the left to counter the torque of the longitudinal driveshaft and to secure equal vertical deflections of the subframe's front mountings under traction torque, thus avoiding lateral inclination of the subframe and consequent camber change at the wheels. Contrary to the situation with rigid-axle suspensions, this measure does not affect the wheel-load distribution under traction or braking forces (see also Chapter 6, Fig. 6.7b).

As a planar suspension mechanism, the semi-trailing-link layout has only limited geometrical attunement possibilities. By replacing the turning joint between each of the links and the vehicle body or subframe by a 'turning and sliding joint', and generating axial shift depending on wheel travel, the suspension gets spatial characteristics that are based on an instantaneous-screw motion – see also Chapter 2, Fig. 2.8b. The screw-axis position is here fixed relative to the vehicle body, **Fig. 12.20**, and the axial shift is controlled by a short rod link connecting each semi-trailing link to the vehicle body or subframe. The distance of the rod link from the screw axis, its angle of incidence therewith and its length determine the instantaneous pitch of the screw and the pitch's change with wheel travel. The rubber joints connecting the semi-trailing arm to the subframe have high radial spring rates and very low axial rates.

With a semi-trailing-link suspension, all such characteristics as the roll centre, the camber and toe-in change with wheel travel and the support

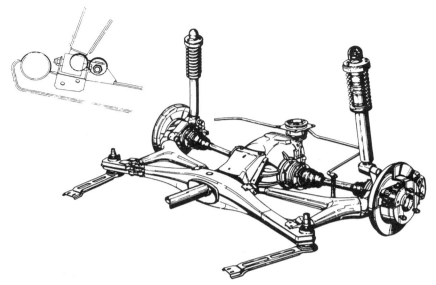

**Fig. 12.20** Semi-trailing-link rear suspension with instantaneous-screw geometry, BMW '528' (1981)  (courtesy BMW AG)

angles are defined by the position of the axis of revolution of the link, and so are interdependent; however, it is clear to see in Fig. 12.20 that the additional rod links mainly influence the track change with wheel travel and, thus, the position of the roll centre. The latter's movement with wheel travel is considerably greater than for normal semi-trailing-link suspensions, so the 'jacking-up' effect in cornering is reduced - see also Chapter 7, Fig. 7.20. The attitude of the screw axis can be varied to give a relatively free choice of suspension characteristics - within the constraints imposed by the available space in the vehicle only.

A spatial wheel suspension can also be produced by the rotatable mounting of a wheel carrier on a semi-trailing link, and control of its attitude by an additional rod link. In **Fig. 12.21**, the wheel carrier is connected to the trailing link by a transverse turning joint and to the vehicle body by a longitudinal rod link - here the lever of the anti-roll bar. In side view, the suspension acts in similar fashion to a Watt linkage, **Fig. 12.22** (see also Chapter 7, Fig. 7.14), and shows a positive wheel-travel angle $\varepsilon$. With drive via transverse shafts, the traction-force support angle $\varepsilon_A$ is equal to the wheel-travel angle $\varepsilon$. If the brake were fixed to the wheel carrier (3), the braking-force support angle $\varepsilon_B$ would be very low or even negative; for this reason, a rotatably mounted brake support (5) is incorporated, controlled by a rod link (6) connected to the trailing link (2). In Fig. 12.22, the 'poles'

Independent Wheel Suspensions                                          323

**Fig. 12.21** Mercedes-Benz '450 SE' rear suspension (1973) (courtesy Daimler-Benz AG)

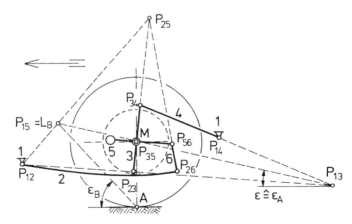

**Fig. 12.22** Diagram of the longitudinal kinematics of the suspension of Fig. 12.21 (simplified by a 'planar' model)

are determined by means of the 'polar ray method': if the relative pole $P_{ab}$ of two bodies a and b is known, and if the relative pole $P_{ac}$ of one of the bodies (e. g. the body a) and a third body c is known too, the relative pole $P_{bc}$ lies on the line through $P_{ab}$ and $P_{ac}$. Unlike the suspensions of Figs 12.16 to 12.20, though, that shown in Figs 12.21 and 12.22 represents a four-joint mechanism chain where the wheel carrier is no longer directly connected to the vehicle body (in fact it is a trapezoidal-link suspension).

Spherical mechanisms represent a transition stage between suspensions with direct and with indirect connection of the wheel carrier to the vehicle body, since the 'central joint' at the wheel carrier is directly mounted to the vehicle body (even if it is no longer realized by a 'physical' joint, but by a virtual point). These mechanisms may be attuned for quasi-spatial motion in all coordinate planes but show a revolution about an instantaneous axis instead of an instantaneous screw motion; hence they require compromises between the various suspension characteristics, as already explained in Section 3.3.

Several variants of spherical suspensions with a 'physical' central joint Z are depicted in **Fig. 12.23**. On the spherical 'double-wishbone' suspension of the Chevrolet 'Corvette Sting Ray' from 1963 (a) and the semi-trailing-link variant of the Ford 'Zodiac' from the 1960s (with one transverse link and a vertical shackle) (b), the constant-length cardan shaft represents a transverse link – a typical Anglo-American design feature that appeared first on Jaguar and Lotus suspensions. The instantaneous axis m is the intersecting line of the planes defined by the two links and the central joint Z. In Fig. 12.23b that axis does not coincide with the line through the joints at the trailing link and it moves with wheel travel. This 'spherical' character affects mainly the toe-in curve and the roll-centre geometry with wheel travel, both of which differ essentially from those of a normal semi-trailing-link suspension. Fig. 12.23c shows the rear suspension of the 1954 Mercedes-Benz racing car, a 'single-joint' swing-axle with longitudinal control

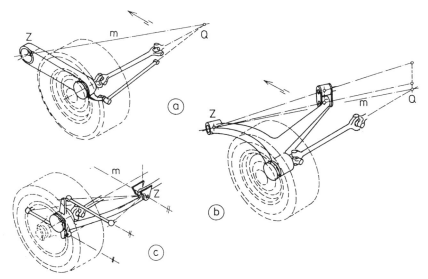

**Fig. 12.23** Schematic drawings of spherical rear suspensions with a 'physical' central joint Z

Independent Wheel Suspensions 325

by antimetric rod links. The suspension represents in effect a 'spherical' Watt linkage; the rectilinear guidance property of this linkage is used here to provide minimum toe-in change with wheel travel.

A quite recent spherical double-wishbone suspension with 'physical' central joint and providing 'conscious' elasto-kinematic attunement is shown in **Fig. 12.24** and its kinematic diagram in the left-hand drawing.

To achieve approximately constant toe-in with wheel travel, the instantaneous axis m must in cross-section always remain parallel to the wheel stub-axle (see Section 7.5); this would here require transverse links with a length half the distance of the transverse pole Q from the wheel. Camber change with wheel travel depends on the pole distance too (see Section 10.2). If a moderate camber change is wanted, this requires a large pole distance and, consequently, relatively long links that would cross the vehicle's medial plane (dashed lines). For this reason, the real links are arranged skewly in plan view, their inner joints being on the connecting lines of those of the theoretical links and the central joint Z. The real

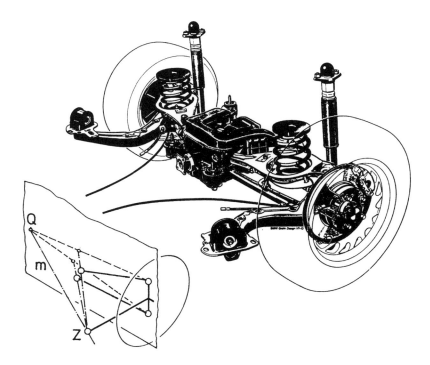

**Fig. 12.24** Spherical double-wishbone suspension with 'physical' central joint, BMW Series '3' (1990) (courtesy BMW AG)

(skew) links and the theoretical transverse links are equivalent in kinematic terms - see also Chapter 3, Fig. 3.8.

Because the central rubber joint Z has a low spring rate for longitudinal compliance of the suspension, the skew links would cause toe-out with braking and toe-in with traction force. To avoid this, the principal axes and spring rates of the central joint are dimensioned for simultaneous rearward and inward (or forward and outward) deflection under longitudinal forces, to compensate for the skew motion of the outer joints of the transverse links. And since the lines of action of the links intersect the wheel centre-plane behind the working point of the lateral force at the tyre, the central joint complies in the same sense as the force, making for understeering elastic behaviour.

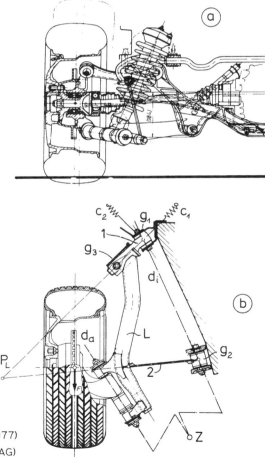

**Fig. 12.25**
Porsche '928' rear suspension (1977)
(courtesy Dr.-Ing. h. c. F. Porsche AG)

# Independent Wheel Suspensions

Also of the spherical type is a suspension consisting of an upper rod link and a lower trapezoidal link, with axes of rotation $d_a$ and $d_i$ intersecting at a point Z, **Fig. 12.25**. Elasto-kinematics are essentially controlled by the trapezoidal link L, which here represents a special 'four-joint chain': its inner joint at the front (which provides longitudinal compliance) comprises a short rod 1 connected to the chassis by a rubber joint $g_1$ and to the trapezoidal link by a rubber joint $g_3$. The rod 1 can swivel about the joint $g_3$ against a rubber seating in the link L. In relation to the rubber joint $g_1$, this arrangement has the effect of two skewed principal spring rates $c_1$ and $c_2$. The rear arm 2 of the trapezoidal link L is in fact a flexible leaf and completes the four-joint chain mechanism of L. To minimize axial load and shift and, consequently, additional stressing of the leaf 2, its inner rubber joint $g_2$ is slightly inclined relative to the inner axis of revolution $d_i$. The rod 1 and the leaf 2 define a 'pole' $P_L$ of the trapezoidal link's body L with respect to the vehicle body. Since $P_L$ is outside the vehicle and behind the tyre contact point, the suspension will on the one hand elastically react by toe-in with braking and toe-out with traction force, to compensate for other elastic deflections of other components and to provide an advantageous starting point for power-change events in cornering, and on the other hand will ensure elastic understeer in cornering. For toe-in adjustment, incidentally, the joint $g_3$ can be moved via a slot in the link L.

With three transverse links and one longitudinal arm, the suspension in **Fig. 12.26** looks similar to the spherical double-wishbone layout of Fig. 12.24

**Fig. 12.26**
Ford 'Focus' rear suspension (1998)
 (courtesy Ford Motor Company)

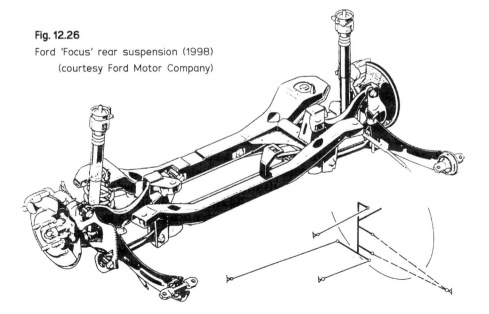

but actually belongs in the spatial category. Suitable positioning of the joints of the transverse links and the latters' angular inclination in plan view enables the achievement of the desired kinematic properties and elasto-kinematic effects under braking, traction and cornering forces – see Chapter 9, Figs 9.9 to 9.11. Because the three transverse links cancel three degrees of freedom, the longitudinal arm has – unlike the rigid one in Fig. 12.24 – to be replaced by a flexible vertical leaf or 'sword' which corresponds to a triangular link (dashed lines in the inset drawing), see Chapter 9, Fig. 9.14 b.

In side view, the wheel carrier swivels approximately about the front joint of the longitudinal arm. For this reason, the vertical displacement of the front lower transverse link's outer joint with wheel travel is smaller than that of the corresponding joint of the rear link. To avoid toe-out with wheel travel, the transverse components of the circular paths of both joints must, of course, be equal, and this requires a front transverse link considerably shorter than the rear one.

Although the suspension, as already mentioned, is not a spherical but a spatial mechanism (and therefore offers a relatively free choice of kinematic and elasto-kinematic attunement), its kinematic properties in side view (wheel-travel angle, support angle) are similar to those of the spherical suspension shown in Fig. 12.24.

Apart from semi-trailing-link suspensions and their variants, and rare spherical systems, rear-wheel suspensions are often derivatives of the double-wishbone principle, while strut or damper-strut layouts are of less importance than for front suspensions. Since nearly all rear suspensions incorporate vertical dampers or struts, strut suspensions (as such) do not offer advantages in occupied space around the luggage compartment, and their only real advantage is that their transverse links are arranged in a single plane low down in the vehicle.

A 'dead' planar damper-strut suspension is shown in **Fig. 12.27** in cross-section. The damper tube is attached to the wheel carrier which is guided by a lower trapezoidal link with inner and outer turning joints. The axle is sprung by a transverse leaf spring which runs from one wheel to the other and is supported twice at the body structure near the inner joints of the trapezoidal links, while its hooked ends embrace rubber buffers attached to the links. A leaf spring mounted in this way serves also as an anti-roll device – see Chapter 5, Fig. 5.29.

Rear suspensions based on the double-wishbone principle played a relatively minor role in sports and luxury cars from the 1930s onward, normally with 'planar' geometry, until in 1982 a *spatial* suspension appeared that was deliberately designed for elasto-kinematic properties. The series

Independent Wheel Suspensions 329

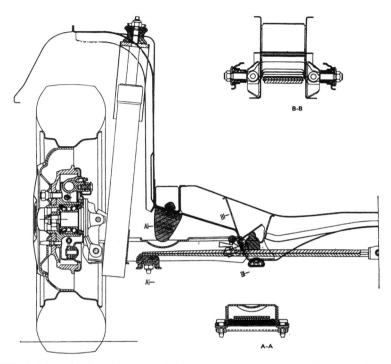

**Fig. 12.27** Autobianchi 'A 112' damper-strut rear suspension with twice-supported leaf spring (1969) (courtesy FIAT Auto SpA)

production of this five-link system proved the abilities of sophisticated suspension systems in everyday use and encouraged the development of almost a flood of multi-link designs even for low-price vehicles. Its latest version is shown in **Fig. 12.28**. Two lower and two upper transverse links, mutually inclined in plan view, carry the longitudinal and lateral forces, and a 'track link' Sp at roughly axle height determines the steering or toe-in angles; it may have been positioned deliberately in the 'neutral' axis of the elastic camber change – see also Chapter 9, Fig. 9.9.

If Sp is envisaged as the 'track-rod' of a steerable suspension, the four transverse links would generate a 'virtual' kingpin axis (i) during a steering event. In plan view, the lower links intersect outside their outer joints, while the upper ones cross each other inboard of their outer joints. The 'virtual' kingpin (i) runs approximately through these intersections or crossing points – see the upper drawing in Fig. 12.28. So a deliberate choice of the 'kingpin inclination angle' is possible – so important for the proper attunement of the elastic steering angles under traction and braking forces, as already

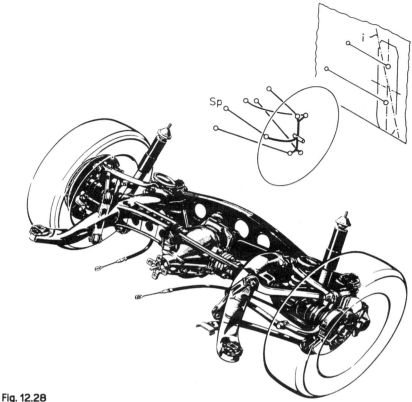

**Fig. 12.28**

Mercedes-Benz '500 SEL' five-link rear suspension (1991) (courtesy Daimler-Benz AG)

explained in relation to Figs 9.10 and 9.11 in Chapter 9. This design demonstrates once again the wide variety of kinematic harmonization possible with multi-link suspensions: with respect to the 'steering geometry', the upper (crossing) links correspond to a notably shorter triangular link; but, while such a triangular link would lead to progressive camber change and reduced roll-centre-height change with wheel travel, the full transverse length of the upper links is effective, thus providing optimum roll-centre and camber geometry.

The five-link principle allows the elasto-kinematic attunement to be optimized for all important cases of external loading. However, as the braking torque is reacted by links in two planes one above the other, there is a limit to the suspension's longitudinal compliance in terms of the wind-up effect – see Chapter 9, Fig. 9.11. On the other hand, an elastically mounted subframe is usually provided for better noise insulation and easier assembly,

# Independent Wheel Suspensions

so its rubber mountings take over a significant share of the longitudinal compliance.

As already shown in Chapter 9, elastic wind-up can be avoided or at least reduced on double-wishbone suspensions by reacting the braking torque by one special suspension link alone – e.g. a trapezoidal link or a stiff longitudinal link – see Chapter 9, Figs 9.13 and 9.14. A variant of the latter method is applied to the five-link suspension illustrated in **Fig. 12.29**.

The kinematic concept has already been shown in Chapter 2, Fig. 2.13b. Braking torque is reacted by the longitudinal link and (via the upper transverse link) by a vertical 'intermediate coupler' that connects the upper link with the longitudinal link. In contrast to spherical suspensions (see Fig. 12.24), the wheel carrier is not fixed to the longitudinal link, and the choice of the joint positions of the intermediate coupler allows the nearly deliberate achievement of a 'longitudinal pole' (and, consequently, support angle) in side view, independently of the position of the longitudinal link's front joint.

Because the longitudinal link, unlike the transverse links, is mounted not on the subframe but directly on the vehicle body, the wheel carrier does not share the elastic angular movement of the subframe in its rubber mountings under traction torque, and the power-change reaction of the axle can be optimized as already described by Fig. 9.17 in Chapter 9.

It would, of course, be simpler to use a trapezoidal link, which could easily be mounted on a subframe and so would reduce the number of connection points to the body structure (and consequently the risk of noise transfer). If the axes of the 'turning joints' of the trapezoidal link are

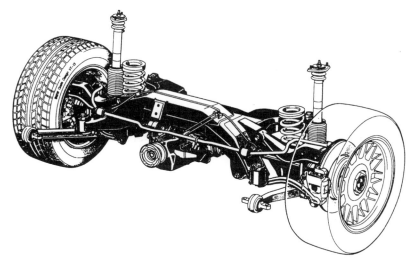

**Fig. 12.29** BMW '850i' rear suspension (1989) (courtesy BMW AG)

parallel or intersect in a common plane, the suspension is of the 'spherical' type (see Fig. 12.25); if they operate askew in space, though, this leads to a 'spatial' system.

In the suspension of **Fig. 12.30** both trapezoidal links are mounted on a subframe formed by a front transverse sheet-metal bridge with widely spaced soft rubber mountings, and by the final-drive housing. Since the trapezoidal links are very long, soft rubber joints for good longitudinal compliance would cause excessive toe-in change with traction or braking. For this reason, a special measure is applied for elasto-kinematic attunement: the inner axes (1) of the trapezoidal links are guided by a rectangular front frame (2) which can swivel about a transverse axis through its upper rubber joints at the subframe, thus allowing the inner axes (1) to move longitudinally, while rubber bearings with high axial compliance support them on a rear transverse beam (3) at the final-drive housing. This layout clearly results in the preferred parallel longitudinal motion of both wheels, even under one-sided longitudinal forces.

The upper rod link is substituted here by the constant-length cardan shaft already mentioned as a typical Anglo-American design feature.

At its rear end the final-drive housing is supported not by the usual rubber mounting but by two diagonal rods whose point of intersection represents the 'elastic centre' - see also Chapter 9, Fig. 9.7b. As this point can obviously be positioned very low, the elastic tilting of the subframe under cornering forces (and the consequent loss of camber) can be minimized.

Since the axes of the inner and outer joints of the trapezoidal links do not intersect, the suspension is of the spatial type. This spatial character

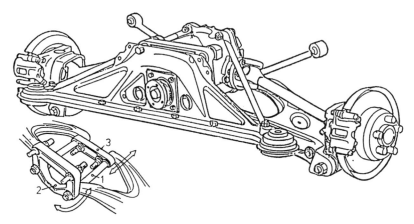

**Fig. 12.30** Jaguar XJ 6 rear suspension (1988)   (courtesy Jaguar Cars Ltd)

Independent Wheel Suspensions

enables, for example, a nearly deliberate choice of the bump-steer function: in side view, the outer joint axes of the trapezoidal links are inclined to the road surface by a greater angle than the inner ones. While the inclination of the inner joints generates a positive support angle, the gradient of toe-in change against wheel travel is determined by the difference between the angles of the inner and outer joint axes and the length of the trapezoidal links.

More freedom of kinematic and elasto-kinematic attunement – especially of elastic steering angles generated by the angles of incidence of suspension links – naturally requires more than one transverse rod link. However, with two such links and one trapezoidal link the suspension would be kinematically 'over-constrained'.

To avoid this, the outer 'turning joint' of the trapezoidal link in **Fig. 12.31** is replaced by a ball-joint at the rear end and a vertical rod link (an 'intermediate coupler') at the front – see also Chapter 9, Fig. 9.13c. This arrangement ensures a stiff support of the braking torque by the trapezoidal link but at the same time allows the wheel carrier to swivel about the rear ball-joint under the influence of the steering-angle control of the two

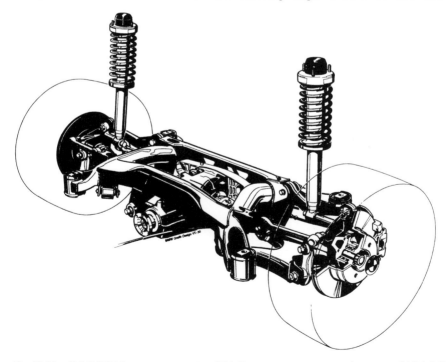

**Fig. 12.31** BMW '750i' rear suspension (1994) (courtesy BMW AG)

transverse links. Elastic steering angles under longitudinal forces can be generated by the angle of incidence of the transverse rod links in plan view; the links being mounted in the stiff upper region of the subframe (see also Fig. 9.15b), this requires only small angles of incidence, with the advantage of a reduced sensitivity to production tolerances.

The rear transverse beam of the subframe shows triangular apertures to save weight, as already illustrated in Chapter 10, Fig. 10.20c.

The 'longitudinal pole', and with it the support angles, are determined mainly by the angle of incidence between the inner and outer turning-joint axes of the trapezoidal link, while the upper transverse links determine the gradient of toe-in change against wheel travel.

Longitudinal compliance of the wheel is defined by the spring rate of the front inner trapezoidal link's rubber joint; the rear inner joint is relatively stiff and under longitudinal forces it acts as the pivot point of the trapezoidal link in plan view. This suspension allows very large longitudinal wheel travel without excessive elastic wind-up under braking torque; together with the compliance of the subframe mountings, the possible longitudinal deflection of each wheel is as much as 15 mm.

# Chapter 13

# Rigid-Axle Suspensions

## 13.1 General remarks

A rigid axle shows no track change with wheel bump and rebound; the wheel camber remains constant with parallel wheel travel and, relative to the road surface, in cornering too if the radial tyre deflection due to the wheel load transfer is neglected. A higher roll centre can be adopted than for an independent suspension, since no lateral movement of the tyre contact point occurs with parallel wheel travel and, consequently, no lateral forces are excited. The 'elastic centre' of the suspension is in the medial plane of the vehicle; for this reason, the elasto-kinematics of a rigid-axle suspension are similar to those of an independent suspension with a subframe - see also Chapter 9, Fig. 9.7. Since the axle suspension has two degrees of freedom, the need for space in the wheel-arch is greater than with independent suspensions which show only one mode of wheel motion under all conditions.

Reaction forces at the suspension links can be avoided if the spring forces act on the axle beam in the cross-sectional plane through the tyre contact points.

If a 'live' rigid axle incorporates the final-drive housing, the percussion point $T_s$ of each wheel is situated in the region between the middle of the axle and the other wheel, **Fig. 13.1**, and a vertical impact force $F_1$ on one

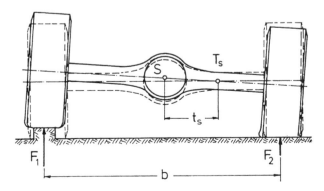

**Fig. 13.1** Coupling of masses on a live rigid axle

wheel causes a reaction force $F_2$ on the other; the axle begins to 'tramp'. The distance between the centre of gravity and the percussion point follows from the radius of inertia $i = \sqrt{\Theta/m}$ according to equation (5.23), Chapter 5, as $t_s = 2i^2/b$.

The reaction force $F_2$ can be avoided if 'decoupling of the masses' can be achieved: $t_s = b/2$ requires $i = b/2$, this being possible if the mass of the axle is small in relation to its moment of inertia, as on 'dead' axles or de Dion axles, see Chapter 6, Fig. 6.11b. Moreover, a de Dion axle does not show the wheel-load transfer that occurs on normal live axles, due to the torque of the longitudinal driveshaft, and that reduces the transferable power unless a differential lock is used - see Fig. 6.7. Since with a de Dion axle the traction torque is transferred via transverse driveshafts, the traction-force support angle is equal to the wheel-travel angle in the case of a rear axle.

A kinematically 'exact' rigid-axle suspension features four rod links or can at least be derived from a four-link guidance - see Chapter 2, Fig. 2.6g. In addition, though, there are 'over-constrained' suspensions (see Section 13.3) and, rarely, 'under-constrained' ones.

**Fig. 13.2** shows on the left an under-constrained rigid-axle suspension; one longitudinal link each side serves for longitudinal control, while lateral control may be achieved by either a Panhard rod or the leaf springs. The braking or traction torque is supported by the springs, and the axle 'winds up' in side view - see also Chapter 5, Fig. 5.28. The wind-up angle is inversely proportional to the square of the leaf length, see equation (5.48); therefore the preference is for long leaf springs which, moreover, need a smaller number of leaves and so help to reduce friction.

The support angle follows from the vertical wheel travel, assuming strictly vertical forces at the wheels; as the symmetrical leaf springs lead to a parallel movement of the axle in side view, the axle is displaced, still parallel,

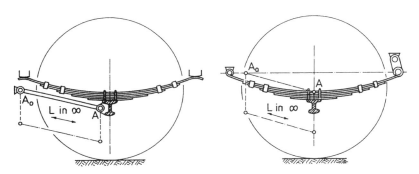

**Fig. 13.2** Under-constrained rigid-axle suspensions

# Rigid-Axle Suspensions

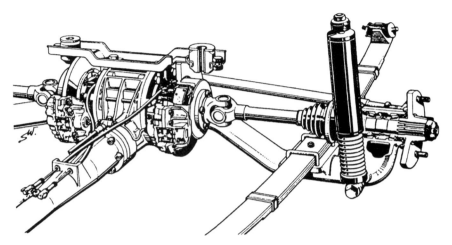

**Fig. 13.3** Glas '2600' de Dion rear axle (1965) (courtesy BMW AG)

along a circular path with the radius $AA_0$ of the rods, and this corresponds to a parallelogram guidance (dashed lines) with the 'longitudinal pole' L at infinity.

The 'classic' rigid-axle suspension by leaf springs alone, Fig. 13.2 (right), is kinematically equivalent to that of the left-hand drawing if the rod links are replaced notionally by the radii of curvature of the spring halves – see Chapter 5, Fig. 5.26.

Under-constrained rigid-axle suspensions are occasionally used on heavy trucks to reduce the reaction forces, as already mentioned in Chapter 2.

**Fig. 13.3** shows an under-constrained axle suspension with longitudinal control by leaf springs and lateral control by a Panhard rod. The final-drive unit is mounted to the chassis, as are the brakes (so-called 'inboard brakes'). Here, the traction torque as well as the braking torque is transmitted to the wheels via the transverse driveshafts, so determination of the traction and braking-force support angles has to take account of the case of a vehicle-fixed torque support. This design is a 'classic' de Dion axle in which longitudinal forces load the axle beam via the wheel bearings only. As the leaf springs are situated near the wheel centres, the wind-up torques resulting from longitudinal forces are minute, so wind-up angles are negligible.

## 13.2 Kinematically exact systems

The simplest method of longitudinal control of a rigid axle, and of torque support about the transverse axis, is the 'torque-tube' which is rigidly fixed to the axle beam and supported on the vehicle body by a 'thrust' ball-joint, **Fig. 13.4**. The longitudinal driveshaft runs well sealed within the torque-tube. Concentric with the thrust ball (by which the transverse axis for parallel wheel motion is defined) is the sole driveshaft joint. The leaf springs are attached to the chassis by front and rear shackles, providing lateral control of the axle. This robust suspension type is still encountered today, preferably with precise lateral guidance by links (for lateral control by Panhard rod, see also Chapter 7, Fig. 7.33).

In the de Dion axle layout shown in **Fig. 13.5**, the torque-tube is replaced by a frame of 'drawbar' shape. The thrust-ball mounting can be seen on the chassis crossmember (lower right of photograph). To avoid the steering angles caused by a Panhard rod, rectilinear guidance by a 'scissors' mechanism was chosen – see also Chapter 2, Fig. 2.14d: two triangular links are connected by a ball-joint, and their turning joints are on the chassis and the axle beam.

**Fig. 13.4** Mercedes-Benz '8/38' chassis (1926) with torque-tube rear axle
(courtesy Daimler-Benz AG)

# Rigid-Axle Suspensions

**Fig. 13.5**
Mercedes-Benz '770'
de Dion rear axle (1937)
(courtesy Daimler-Benz AG)

If the front end of the torque-tube is mounted with horizontal compliance and rigidly supported only in the vertical direction, the mounting is equivalent to a 'ball-and-surface' joint, as illustrated in Chapter 2, Fig. 2.2e. The rear-axle suspension of **Fig. 13.6** therefore needs two additional longitudinal rod links for fore-and-aft control, see also Fig. 2.14b. In side view, the 'longitudinal pole' will be the intersecting point of the longitudinal links and the vertical line through the 'ball-and-surface' joint. Forward positioning of the coil springs lowers the effective spring rate for parallel wheel travel relative to the rate in roll, and an anti-roll bar is fitted as well.

An example of a rigid-axle suspension by two rod links and one triangular link is the de Dion system shown in **Fig. 13.7**. The triangular link is arranged

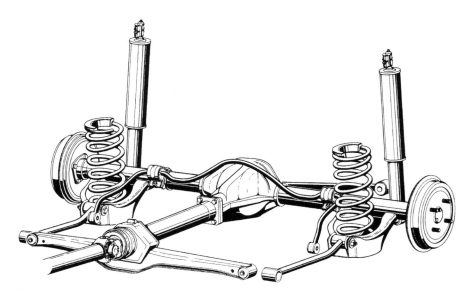

**Fig. 13.6** Opel 'Ascona' rear suspension (1970)  (courtesy Adam Opel AG)

in the opposite direction to the longitudinal links, producing in side view a mechanism similar to a Watt linkage. This measure, which is of use only on 'dead' and de Dion axles, leads to 'progressive' anti-lift properties under braking forces (unlike the Fig. 13.3 layout, the brakes are outboard here), and (on De-Dion axles) allows a positive traction-force support angle. The

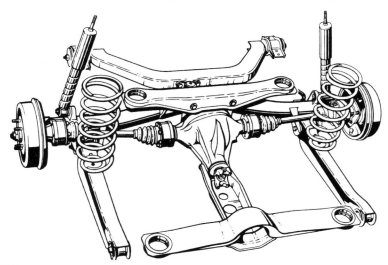

**Fig. 13.7** Opel 'Admiral' de Dion rear axle (1969)  (courtesy Adam Opel AG)

Rigid-Axle Suspensions

almost parallel longitudinal rod links essentially determine the direction of the instantaneous axis of antimetric wheel travel, see Chapter 7, Fig. 7.15b; since, on the other hand, the 'longitudinal pole' of parallel wheel travel is determined by the centre lines of the rods also, the wheel-travel angle (and hence, on a de Dion axle, the traction-force support angle) must be about zero to avoid kinematic oversteer in cornering.

Free choice of bump-steer properties and of the support angles of a de Dion axle can be achieved only by a suspension with all the links arranged askew - e.g. a four-link system as shown in Figs 7.15 and 7.28, Chapter 7 (and carried out on a BMW prototype of the 1970s).

In Fig. 13.5, as already mentioned, a scissors mechanism consisting of two triangular links replaces a Panhard rod. It follows that any rod link *can* be replaced by such a mechanism. If the two longitudinal links of the suspension in Fig. 13.7 are replaced by scissors, the result is a rigid-axle suspension consisting of five triangular links or equivalents thereof. In **Fig. 13.8**, each side's scissors mechanism consists of a rigid arm attached rotatably to the chassis and connected to a torsion-bar spring, and a short shackle rotatably connected to the axle beam. For alteration of the vehicle ride height, the angular position of the front ends of the torsion bars can be reset via locking bolts.

This suspension represents an example of extreme 'kinematic' spring attunement (see also Section 5.4) with a springing characteristic that shows a very low spring rate in the 'normal position' of the vehicle and considerable progression with wheel bump and rebound. If the angle between the shackle and the arm of the torsion bar approaches 180° (right side, in the illustration), the effective springing force rises towards infinity while the

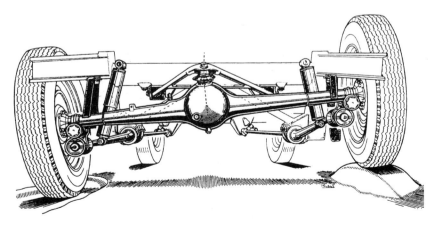

**Fig. 13.8** BMW '501' rear axle with torsion-bar springing (1952)    (courtesy BMW AG)

**Fig. 13.9** Four-link rigid-axle suspension, Ford 'Taunus' (1970) (courtesy Ford Motor Co.)

torsion-bar torque remains limited; the 'kinematic spring rate' finally reaches 100% of the total spring rate.

A kinematically influenced spring characteristic requires considerable directional changes for the lines of action of the forces (here clearly visible from the very skewed attitude of the shackle of the right wheel) and causes variable horizontal reaction forces. For this reason, and as already mentioned in Section 5.4, such a measure is no longer recommendable on modern wheel suspensions which are deliberately designed to have elasto-kinematic properties.

The basic mechanism of an 'exactly guided' rigid-axle suspension is the four-link system, an example being given in **Fig. 13.9**. As the intersecting point of the centre-lines of the upper rod links B is near the axle, the roll-centre height will be nearly constant with wheel travel. The lower links D are nearly longitudinal and define the direction of the instantaneous axis of antimetric wheel travel.

Replacing a triangular link by two separate rod links has two advantages. First it avoids extreme angular deflection of the apex joint of a triangular link and divides the angle of distortion of the axle-side and the vehicle-side joints of the rod link into 'coning' angles of approximately half the value of that angle. Second it avoids additional bending stress at the apex of a triangular link, maybe caused by compliance of the vehicle frame - e.g. in trucks.

# 13.3 Over-constrained systems

Rigid-axle suspensions with more than four rod links are 'over-constrained' and show an overall degree of freedom less than 2. Reasons for their application were given in Section 2.3.3.

The five-link suspension shown in **Fig. 13.10** consists of four longitudinal rod links and one transverse (Panhard) rod. As the upper longitudinal rods are essentially shorter than the lower, this results in an approximately rectilinear path of the tyre contact point in side view, assuming a 'locked' brake or final drive (see also the rectilinear guidance – though shown there in cross-section of the vehicle – in Fig. 10.2, Chapter 10). Consequently, the support angles are virtually constant with wheel travel.

The short upper links save space in the passenger and luggage compartments. Since they are arranged nearer to the centre plane of the vehicle than the lower links, mutual constraint of their mountings under vehicle roll can largely be avoided – see also Chapter 9, Fig. 9.18b.

**Fig. 13.11** shows a 'live' front axle controlled by two longitudinal arms and a Panhard rod. The instantaneous axis for parallel wheel travel runs through the chassis-side joints of the arms, thus providing considerable traction and braking-force support angles. Running from one wheel carrier to the other, the track-rod is behind the axle, while a transverse drag-rod (not drawn) acts on the steering arm in front of the right wheel and is arranged parallel to the Panhard rod to avoid steering errors – see also Chapter 7, Fig. 7.35b.

If the longitudinal arms were rigidly attached to the axle beam, the latter would receive high torque loadings in roll, meaning that roll angles would be very small. For this reason, each arm is connected to the axle

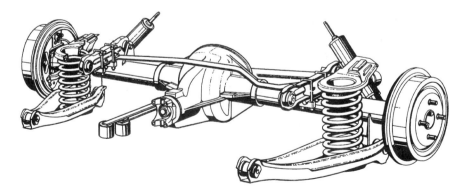

**Fig. 13.10** Five-link rear-axle suspension, Opel 'Rekord Caravan' (1966)

(courtesy Adam Opel AG)

**Fig. 13.11** Live front axle of the all-terrain Mercedes-Benz type '240GD/300GD' (1979) (courtesy Daimler-Benz AG)

beam by two rubber joints, one behind the other. These rubber-joint pairs are able to transmit braking and traction moments from the axle to the vehicle body (accepting, of course, some elastic wind-up of the axle) and allow some relative deflection of the arms, while acting, too, like stabilizer (anti-roll) springs depending on their spring rates. The vehicle's rear axle is designed on the same principle but, of course, with trailing longitudinal arms.

If the axle were not driven, the longitudinal arms could be rigidly fixed to the axle beam and the latter made 'torsional', e.g. by an 'open' profile section – see also Chapter 2, Fig. 2.18b and Chapter 7, Figs 7.25 and 7.36. Viewed in this light, the suspension suggests a link between rigid-axle suspensions and compound suspensions, which latter will be treated in the following (and last) chapter.

# Chapter 14

# Compound Suspensions

A compound suspension is the generalized form of the kinematic guidance of the two wheels of an 'axle'; such a mechanism needs two degrees of freedom in total, as explained in Chapter 2. On independent suspensions this is achieved by having two separate suspension mechanisms, each with one degree of freedom, while on rigid-axle suspensions a single wheel carrier mounts two wheels and is guided with two degrees of freedom; whereas the two wheels of an ordinary rigid-axle system cannot move independently of each other, compound suspensions permit relative movement of the two wheels.

Compound suspensions are used to achieve a compromise between the properties of independent and of rigid-axle suspensions – e. g. small changes of track, camber and toe-in with parallel wheel travel, but favourable wheel camber and a relatively high roll centre, and perhaps noticeable bump-steer, when the vehicle rolls.

The compound suspension shown in **Fig. 14.1** can be regarded as deriving from both an independent and a rigid-axle system – first from a swing axle and second from a rigid axle which has in effect been 'sawn through' in the middle and then coupled by a turning joint. The latter's height is controlled by a swinging vertical 'triangular link' and its lateral attitude by a short transverse rod link, so to speak a residual 'Panhard rod', with relatively soft rubber mountings to protect the vehicle body from lateral accelerations caused by track changes on uneven roads. Two longitudinal links react longitudinal forces.

**Fig. 14.2** shows the low position of the turning joint that connects the two axle tubes; an essential aim of this design was to make the roll-centre lower than for normal swing axles. The final-drive housing is in one with the left axle tube, and the half-shaft of the right wheel is connected to the final drive by a cardan joint and an extensible sliding coupling with balls for reduced friction. Behind the gear housing, a 'compound spring' connects the two axle halves, see also Chapter 5, Fig. 5.14b.

Assuming the engine to be 'locked', and consequently the final-drive pinion too, the crown-wheel of the final drive swivels about the pinion with wheel travel; depending on the gear reduction rate, a 'longitudinal pole' and a considerable traction-force support angle appear at the left wheel – see

**Fig. 14.1** Mercedes-Benz '220' compound rear suspension (1959)

(courtesy Daimler-Benz AG)

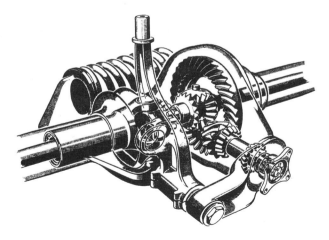

**Fig. 14.2** Final drive and connection of the axle tubes for the suspension shown in Fig. 14.1

(courtesy Daimler-Benz AG)

Compound Suspensions 347

Chapter 6, Fig. 6.16. The same is valid for the right wheel, as it copies this motion via the cardan joint. Unlike a normal swing axle (= independent suspension), this system clearly shows strong anti-squat properties. The longitudinal driveshaft generates a roll moment on the left axle tube which carries the final drive, and consequently on the vehicle's springing too, as on rigid axles (Chapter 6, Fig. 6.7).

The instantaneous axis of antimetric wheel travel and the resulting steering angle are determined by the longitudinal links in a similar manner to rigid axles – see also Fig. 7.28 of Chapter 7 – while a resulting change of the wheels' toe-in is impossible as long as the medial turning-joint axis is parallel to the road surface.

Compound suspensions can be derived from independent systems by connecting links on one side to the mechanism at the other side – see also Chapter 2, Fig. 2.18a. On a sports car's rear suspension, **Fig. 14.3**, trailing arms bear wheel carriers rotatably connected to them via turning joints with longitudinal axes. Each upper transverse rod link is connected to the relevant opposite suspension. With parallel wheel travel the whole system swivels about the transverse axis of the trailing arm joints at the chassis. However, with antimetric wheel travel the medial points of the transverse rods remain approximately 'fixed' in space due to the condition of antimetric motion, and the 'transverse pole' of each wheel results in cross-section as the intersecting point of the appropriate rod link and the line that runs parallel to the axis of the chassis-side turning joint through the wheel-carrier/trailing-arm turning joint.

In relation to vehicle production volumes, compound suspensions are found much more frequently than might at first be expected. They are favoured for rear-axle suspensions, especially of front drive cars, based on

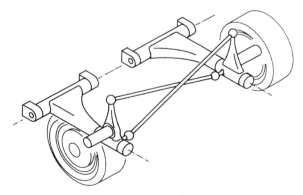

**Fig. 14.3** Fairthorpe 'TX 1' rear suspension (1965)
(schematically, according to *Motor*, October 30, 1965, page 43)

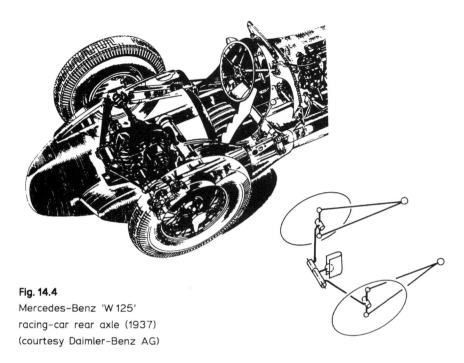

**Fig. 14.4**
Mercedes-Benz 'W 125'
racing-car rear axle (1937)
(courtesy Daimler-Benz AG)

the principle of a rotatable or twistable transverse connection of the two wheel carriers, as already mentioned several times, notably in Chapter 2, Fig. 2.18b, and Chapter 7, Figs 7.25 and 7.36.

The first application of this principle seems to have been in the racing-car rear suspension of **Fig. 14.4** which was obviously derived from a de Dion rigid axle. The rotatable connection of the axle tubes is here sited behind the wheels, and - in contrast with the layout of Fig. 7.25 - the instantaneous axis of antimetric wheel travel achieves an attitude similar to that of two-joint swing axles, showing a 'transverse pole' between each wheel and the medial plane of the vehicle. This leads to a rather high roll centre and, in cornering, to both wheels leaning inward, as with a motorcycle. In contrast with independent suspensions, this behaviour is not influenced by any 'jacking-up' effect in cornering. The longitudinal flexible leaves or 'swords' are rigidly attached to the wheel carriers and correspond, from the kinematic viewpoint, with triangular links as shown in the small schematic drawing. Lateral control is achieved by a 'crosshead' mechanism that corresponds to a 'ball-and-surface' joint.

Similar cornering behaviour is shown by the suspension in **Fig. 14.5** which has the axle tubes connected by a 'turning-and-sliding' joint instead of a

Compound Suspensions 349

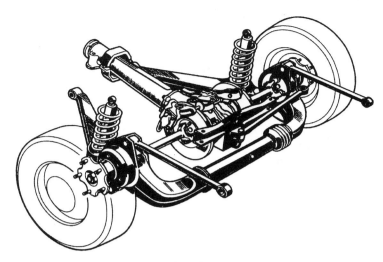

**Fig. 14.5** Rover '2000' rear suspension (1963)
(from *Ein Jahrhundert Automobiltechnik - Personenwagen*, VDI, Germany 1986)

simple turning joint. This requires a transverse rod for the lateral control of each of the wheel carriers - see also Chapter 7, Fig. 7.17a; the rods are here realized as constant-length cardan shafts, already mentioned as a typical Anglo-American design feature. Longitudinal control of the wheel carriers is effected by opposed rod links similar to a Watt linkage (see Chapter 2, Fig. 2.18b, as well).

Much simpler design is possible if the wheels are not driven. The suspension of **Fig. 14.6**, which is closely derived from a rigid axle, represents the zenith of an evolution that had started in 1959. The longitudinal arms (flexible leaves) are rigidly connected to the axle beam, which is an 'open' U-profile, and a Panhard rod behind the axle provides lateral control (in the first version the Panhard rod was arranged obliquely to avoid lateral displacement of the axle with parallel wheel travel - see also Chapter 7, Fig. 7.33c). With parallel wheel travel, the suspension swivels about the axis of the front joints, and with antimetric wheel travel the axle beam is twisted about its shear axis, see Fig. 7.36.

The opposite end of the 'compound scale' is represented by the system shown in **Fig. 14.7**, which falls near to an independent suspension. The twistable transverse beam has been sited between the front joints of the longitudinal arms and has a T-section. The latter's shear-centre is the intersecting point of the flange and the web and clearly is displaced from the joint axis by a very small distance. There is no Panhard rod, and the longitudinal arms are made resistant to bending and torsion to be able to

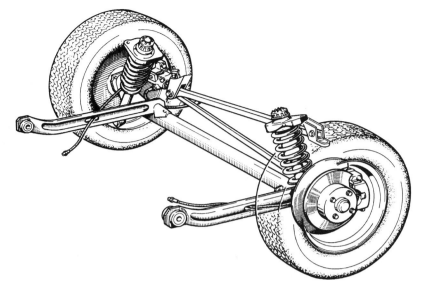

**Fig. 14.6** Audi '100' rear suspension (1976) (courtesy Audi AG)

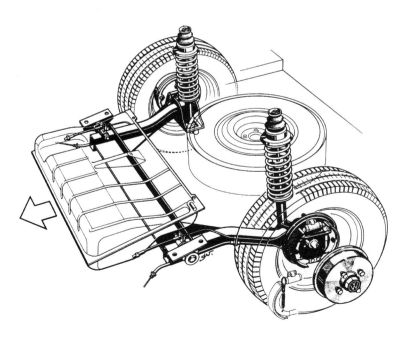

**Fig. 14.7** Volkswagen 'Scirocco' rear suspension (1974) (courtesy Volkswagen AG)

Compound Suspensions

react camber moments and lateral forces. In terms of its kinematic function, this design is roughly equivalent to an independent trailing-arm suspension; therefore the roll centre is at the road surface (or closely above it), and camber change with antimetric wheel travel is minute, causing the wheels to lean with the vehicle body in cornering. As there is virtually no up-and-down movement of the transverse beam, the suspension occupies little space in the vehicle.

The rear suspension of **Fig. 14.8** has very obvious 'compound' properties. The U-profile transverse beam is located between the wheel axes and the front joints but nearer to the latter. With antimetric wheel travel or vehicle roll the suspension behaves like a semi-trailing-link system, and with parallel wheel travel like a pure trailing-link layout (see also Chapter 7, Fig. 7.36). On account of the shape in plan view the suspension has also become known as the 'H-frame' layout, and in this form it has gained considerable popularity worldwide for light front-drive vehicles.

The rigid connection of the bending- and torsion-resistant longitudinal arms to the twistable open-section transverse beam requires skilled design and advanced production methods to avoid stresses caused by the warpage of the beam under torsion – see also Chapter 10, Fig. 10.14a. (As is widely known, minimum warpage results from T and V profiles.)

Since an open transverse beam cannot be cranked to let a driveshaft pass through, suspensions such as those in Figs 14.7 and 14.8 are hardly

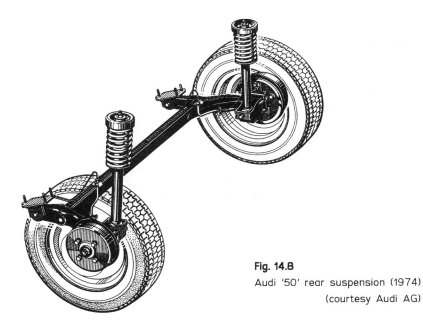

**Fig. 14.8**
Audi '50' rear suspension (1974)
(courtesy Audi AG)

viable for driven wheels. There are narrow limits, too, on vehicle size: heavier ones normally require greater wheel travel for comfort, and the external forces increase proportionally to the vehicle weight, while the track, and with it the beam length, increase much more slowly. The use of high-duty metallic materials, on the other hand, is discouraged by their more complicated welding conditions and their greater notch sensitivity. However, the H-frame suspension type appears well suited for construction from fibre-reinforced plastic. If the stress at the connection of the arms and the beam cannot be mastered, rubber mountings could be used, as in Fig. 13.11, Chapter 13.

On the suspensions of Figs 14.7 and 14.8, lateral forces are transferred to the vehicle body by the front joints of the longitudinal arms alone, and, consequently, ahead of the wheels. This causes an 'oversteering' moment at the suspension in plan view. To counteract this, the front rubber mountings can be designed with skewed principal axes 1 and 2 for an elastic centre behind the axle, similar to the mountings of a subframe, **Fig. 14.9** (see also Chapter 9, Figs 9.6b and 9.7); in contrast to the subframe situation, of course, the mountings are here additionally loaded torsionally under wheel travel.

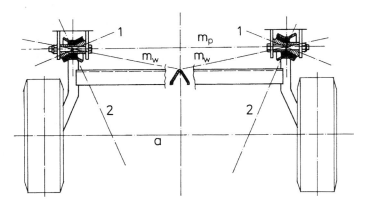

**Fig. 14.9** Schematic of anti-steer rubber mountings on an 'H-frame' suspension (according to *Automobiltechnische Zeitschrift* (1982) **84**, page 65)

# Final Remarks

A book can do no more than survey a particular field; however, the successful design of wheel suspensions requires a deep personal preoccupation with the subject and a suitable working environment. Nevertheless, an essential aim will be achieved if the book is able to draw the reader's attention to the various questions and problems that arise during the development of a suspension system and to show up – or at least to indicate – methods of solving them.

Hopefully, some of the principles underlying all the methods presented in this book will have been spotlighted.

The different types of independent, rigid-axle and compound suspension can be analysed and judged by standard calculation methods and criteria.

Wheel suspensions represent a fascinating section of spatial kinematics, showing truly non-linear forms of motion, in contrast with the usual – though also spatial – mechanisms found in, for instance, the textile and food-processing industries.

Nearly all wheel suspensions are 'statically determinate' mechanisms. For this reason, the concept of 'virtual work' applied consistently in this book proves to be an unrivalled simple, clear and always successful method of analysing forces and characteristics, even on complex systems. Moreover, for the same reason, the influence of the compliance of suspension joints on the diagram of forces in a system, and on its characteristics, is negligible. Hence, kinematic and elasto-kinematic analysis are best done separately, though in close interaction.

While the definitions of the suspension characteristics that are in part more than 100 years old seem inexact at first sight, they prove to be very accurate when scrutinized closely. The virtual-work concept allows compatible and understandable definitions of all the familiar suspension and steering characteristics, even for spatial mechanisms.

Analysis of the influence on suspension of driveshafts and hub reduction gears derived from the application of the virtual-work principle requires only minimal additional programming efforts on all the known driveline variations, and reveals very clear conformities.

# References

(ATZ = *Automobiltechnische Zeitschrift*, Franckh-Kosmos-Verlag, Stuttgart, Germany)

(1) **Bastow, D.** (1987) *Car Suspension and Handling*, 2nd edition, Pentech Press, London, UK.
(2) **Behles, F.** (1962) Möglichkeiten und Grenzen zur Verbesserung der Federweichheit von Kraftfahrzeugen, Dissertation, Technische Universität München, Germany.
(3) **Behles, F.** (1964) Die Beherrschung des Brems- und Anfahrnickens, *ATZ*, **66**, 225.
(4) **Beyer, R.** (1963) *Technische Raumkinematik*, Springer-Verlag, Berlin, Germany.
(5) **Bittel, K.** (1951) Anlenkung der Radaufhängung im Stoßmittelpunkt, *ATZ*, **53**, 117.
(6) **Braess, H.-H.** (1967) Beitrag zur Stabilität des Lenkverhaltens von Kraftfahrzeugen, *ATZ*, **69**, 81.
(7) **Braess, H.-H. and Ruf, G.** (1976) Influence of tire properties and rear axle compliance on power-off effect in cornering, 6th Int. Conference on *Experimental Safety Vehicles*, Washington, USA.
(8) **Burckhardt, M. and Glasner von Ostenwall, E.-C.** (1974) Beitrag zur Beurteilung des Beschleunigungs- und Bremsverhaltens eines Kraftfahrzeugs, *ATZ*, **76**, 103.
(9) **Forkel, D.** (1961) Ein Beitrag zur Auslegung von Fahrzeuglenkungen, Deutsche Kraftfahrtforschung und Straßenverkehrstechnik (booklet 145), VDI-Verlag, Düsseldorf, Germany.
(10) **Göbel, E.-F.** (1974) *Rubber Springs Design*, Butterworth Publishers, London, UK.
(11) **Gross, S.** (1966) *Calculation and Design of Metal Springs*, Chapman & Hall, London, UK.
(12) **Helms, H.** (1974) Grenzen der Verbesserungsfähigkeit von Schwingungskomfort und Fahrsicherheit an Kraftfahrzeugen, Dissertation, Technische Universität Braunschweig, Germany.
(13) **Kolbe, J.** (1937) Der Kurvenlegerwagen, *ATZ*, **40**, 146.
(14) **Matschinsky, W., Dietrich, C. and Winkler, E.** (1977) Die Doppelgelenk-Federbeinachse der neuen BMW-Sechszylinderwagen der Baureihe 7, *ATZ*, **79**, 357.
(15) **Matschinsky, W.** (1992) Bestimmung mechanischer Kenngrößen von Radaufhängungen, Dissertation, Universität Hannover, Germany.

(16) **Mitschke, M.** *Dynamik der Kraftfahrzeuge*,
(1995) *Vol. A: Antrieb und Bremsung*, 3$^{rd}$ edition,
(1997) *Vol. B: Schwingungen*, 3$^{rd}$ edition,
(1990) *Vol. C: Fahrverhalten*, 2$^{nd}$ edition,
Springer-Verlag, Berlin, Germany.
(17) **Schmelz, F., Seherr-Thoss, H.-Chr. Graf von, and Aucktor, W.** (1992) *Universal Joints and Driveshafts. Analysis, Design, Applications*, Springer-Verlag, Berlin, Germany.
(18) **Weber, R.** (1981) Beitrag zum Übertragungsverhalten zwischen Schlupf und Reifenführungskräften, *Automobil-Industrie*, Vogel-Verlag, Würzburg, Germany, **26**, 449.
(19) **Winkelmann, O.-J.** (1961) Anforderungen an das Fahrverhalten von Kraftfahrzeugen, *ATZ*, **63**, 121.
(20) German patent 1925347.
(21) French patent 826275.
(22) US patent 2660449.
(23) German patent 455779.
(24) German patent 460548.

# Index

A-arm *see* Triangular link
A-bracket axle *see* Thrust-ball axle
Accelerating centre 114
Ackermann 195, 196, 224-226, 229-232, 234, 237, 238
Active suspension 105, 106, 128, 170
All-wheel drive 111, 139
All-wheel steering 188, 191, 224
Anti-dive 1, 40, 70, 113, 126, 127, 139, 149, 233, 294, 295, 298, 304, 314
Anti-lift 126, 128, 129, 137, 139, 266, 299, 301, 340
Anti-roll bar 1, 56, 72, 91, 104, 106, 148, 151, 170, 171, 266, 311, 313, 314, 322, 328, 339, 344
Anti-squat 1, 70, 113, 129, 130, 139, 266, 299, 301, 302, 347
Aquaplaning 106, 135, 200
Articulated vehicle 195, 239
Attitude angle 188

Barrel spring 92, 320
Bobillier's method 21, 22, 160, 270, 271
Braking anti-lock system 110, 116, 203, 217, 297, 314
Braking centre 114
Braking-power control device 111, 128
Bredt formulae 284

Cantilever spring 85
Cardan joint, Cardan shaft 31, 42-45, 50, 122, 198, 260, 300, 302, 303, 320, 324, 332, 345, 347, 349
Castor angle 117, 118, 199, 202, 204, 207, 208, 222, 224, 258, 293, 317
Castor offset 117, 118, 199-201, 205, 206, 208-211, 215, 226, 233, 236-238, 258, 293, 294, 314, 317
Castor offset at wheel centre 143, 199, 205, 206, 209, 312, 317
Castor trail *see* Castor offset
Chain drive 124, 298, 299
Compound spring 70-72, 140, 142, 151, 171, 345
Conicity 58
Coning angle 7, 96, 97, 249, 283, 313, 342
Constant-velocity joint *see* Homokinetic joint
Copying vibration 106
Coriolis force 192
Coupling of masses 69, 244, 303, 306, 318, 335, 336

Critical speed 187
Culmann's method 24

Damper-strut suspension 307, 309, 310, 328, 329
de Dion axle 118, 182, 336-341, 348
Double-trailing-arm suspension 306
Double-wishbone suspension 12, 13, 29, 58, 157, 159, 170, 178, 179, 182, 203, 245, 251, 258, 260, 270-272, 307, 313, 316, 324, 325, 328
Drive-unit wheel carrier 115, 124, 244, 300
Dubonnet suspension 118, 148, 181, 220, 270, 304-306, 315

Earles fork 294, 295
Elastic centre 68, 75-77, 249, 250, 320, 332, 335, 352
Elastic wind-up 16, 87, 88, 246, 257-260, 317, 330, 331, 334, 336, 337, 344

Fifth-wheel steering 195, 239
Fixed polode 22, 23

Gas-pressure damper
   *see* Single-tube damper
Gimbal error 43, 198
Girder fork 295, 296
Gough diagram 55, 56
Gyroscope 152, 191, 192
Gyroscopic moment 152, 191-194, 306

Hartmann's method 21, 22
H-frame suspension 184, 185, 351, 352
Homokinetic joint 43-47, 50, 122, 198, 213, 214
Hub reduction gear 2, 41, 47, 49, 50, 119-121, 123, 133, 214, 218-220, 303, 312, 353
Hypoid gear 123

Inboard brake 116, 337
Instantaneous screw 12, 31, 32, 36, 38, 196, 203, 204, 211, 274, 276, 317, 321, 322, 324
Intermediate coupler 11, 14, 274, 331, 333

Jacking-up effect 58, 106, 158-160, 162, 163, 168-170, 172, 175, 176, 272, 294, 304, 322, 348

Kinematic spring rate 63, 80, 135, 136, 301, 342
Kingpin inclination 199, 202, 204, 207, 208, 210, 216, 217, 220, 222, 224, 226, 238, 258, 317, 329
Kingpin offset *see* Scrub radius
Kingpin offset at wheel centre *see* Wheel-centre offset

Level control system 99, 103-106, 128
Lockable differential 235, 237, 336
Longitudinal compliance 82, 184, 221, 243, 245, 254, 258, 259, 294, 311, 317, 318, 326, 327, 330-332, 334

MacPherson suspension 72, 91, 310, 312, 313
Moving polode 22, 23

Open profile 18, 151, 173, 282-285, 289, 290, 344, 349, 351
Oversteer 137, 176, 180, 187, 188, 194, 254, 255, 265, 341, 352

Panhard rod 15, 16, 162, 182, 183, 266, 336-338, 341, 343, 345, 349
Parabolic spring 85
Percussion point 69, 190, 191, 244, 335, 336
Pillar-type suspension 303, 304
Ply steer 58
Pneumatic trail 54-56, 201, 210, 211, 232, 233, 236, 237, 247, 253, 254
Polar-ray method 323
Power change (power-off effect) 127, 188, 189, 256, 264, 265, 327, 331
Power-divider transfer gear 111

Rear-wheel steering 188
Restoring torque due to weight 211-213, 231, 234, 237
Rolling oscillation 238, 239, 293
Rotating masses 112, 137, 138
Rotational contour 290-292

Scissors mechanism 15, 338, 341
Scrub radius 199, 200, 203, 205, 210, 215-220, 238, 257, 258, 314, 315, 317
Scrub torque 236, 237
Semi-trailer truck 110, 111, 239, 241, 242
Semi-trailing-link suspension 12, 13, 130-132, 180, 320-322, 324, 328, 351
Shear centre 174, 184, 290, 349
Shimmy 271
Shore hardness 95, 98, 262
Single-tube damper 101-103

Skyhook 106
Slip angle 54-58, 143, 150, 151, 153, 156, 167, 168, 176, 186-190, 193, 194, 201, 225, 232, 234
Spacing spring 85
Spherical suspension (or mechanism) 13, 30-32, 35, 37, 42, 154, 180, 222, 259, 260, 273, 306, 314, 315, 324, 325, 327, 328, 331, 332
Stabilizer bar *see* Anti-roll bar
Steering damper 239
Steering input 188-190, 193
Steering kickback 244, 245
Steering trapezium 222, 226, 228, 305
Strut suspension 9, 14, 29, 170, 238, 258, 263, 307-314, 317, 328
Subframe 75, 82, 249, 250, 259, 261-265, 307, 317, 318, 320, 321, 330-332, 334, 335, 352
Swing-axle suspension 6, 12, 78, 122, 123, 155, 169, 170, 175, 318-320, 324, 345, 347
Sword 260, 328, 348

Telescopic fork 107, 294, 297
Thrust-ball axle 15, 141, 142, 162, 182, 338
Toe-in (toe-out) 50, 51, 57, 143, 150-153, 156, 158, 167, 178-182, 185, 226, 234, 248, 252-254, 256, 257, 259, 261, 262, 265, 278, 306, 319, 321, 324-329, 332-334, 345, 347
Traction-force radius 213-220, 232, 236, 237, 317
Traction-slip control system 129
Trailing-arm suspension 12, 80, 81, 107, 118-120, 124, 131, 133, 134, 136, 137, 142, 170, 175, 351
Transmission angle 226-230, 238
Trapezoidal link 8, 11, 13, 14, 30, 31, 248, 249, 259, 260, 263, 327, 328, 331-334
Triangular link 8, 9, 11, 13-15, 34, 38, 45, 162, 163, 178, 182, 203, 223, 251-254, 256, 258, 276, 279, 311, 314, 316, 328, 330, 338, 339, 341, 342, 345, 348
Tripod joint 43, 317
Twin-tube damper 101-103
Tyre non-uniformity 58, 238
Tyre slip angle *see* Slip angle

Understeer 137, 148, 176, 180, 186-190, 194, 218, 248, 254, 256, 257, 261, 265, 311, 317, 320

Index

Unsprung masses 59, 66, 67, 78, 112, 118, 119, 137, 138, 146, 147, 149, 152, 238, 243, 245

Virtual kingpin 143, 196, 203, 210-212, 224, 231, 314-317, 329
Virtual work 38, 39, 41, 47, 118, 147, 153, 201, 204-206, 353

Watt linkage 15, 160, 161, 174, 322, 325, 340, 349
Wheel-centre offset 199, 200, 205, 207, 213, 215, 216, 219, 236-238, 255, 257, 258, 317
Wheelfight 238, 239
Wheel-load lever arm 201, 204, 206-209, 211, 212, 215, 232, 234, 237, 238, 317